T0264499

SYNTHESIS

USING

VILSMEIER

REAGENTS

Charles M. Marson
Department of Chemistry
University of Sheffield
United Kingdom

Paul R. Giles
Department of Organic Chemistry
Catholic University of Louvain
Belgium

CRC Press
Taylor & Francis Group
Boca Raton London New York

CRC Press is an imprint of the
Taylor & Francis Group, an **informa** business

First published 1994 by CRC Press
Taylor & Francis Group
6000 Broken Sound Parkway NW, Suite 300
Boca Raton, FL 33487-2742

Reissued 2018 by CRC Press

Library of Congress Cataloging-in-Publication Data

Marson, C. M. (Charles M.)
    Synthesis using Vilsmeier reagents / C. M. Marson, P. R. Giles.
        p. cm.
    Includes bibliographical references and index.
    ISBN 0-8493-7869-9
    1. Vilsmeier reagents. 2. Organic compounds—Synthesis. I. Giles, P. R.
(Paul R). II. Title.
QD77.M44 1994
547.2—dc20
                                                                      94-5797

A Library of Congress record exists under LC control number: 94005797

Publisher's Note
The publisher has gone to great lengths to ensure the quality of this reprint but points out that some imperfections in the original copies may be apparent.

Disclaimer
The publisher has made every effort to trace copyright holders and welcomes correspondence from those they have been unable to contact.

ISBN 13: 978-1-315-89797-4 (hbk)
ISBN 13: 978-1-351-07707-1 (ebk)

Visit the Taylor & Francis Web site at http://www.taylorandfrancis.com and the
CRC Press Web site at http://www.crcpress.com

Preface

Vilsmeier reagents are probably best known in the context of the Vilsmeier-Haack-Arnold formylation of aromatic and heteroaromatic rings. Remarkably, over 50 different functional group transformations have been effected using Vilsmeier reagents. In addition, over 50 different heterocyclic ring systems have been prepared by means of Vilsmeier reagents. It is hoped that this book will serve as a reminder of the versatility of Vilsmeier reagents, a versatility that extends well beyond aromatic formylation.

Chapter 1 provides the background to Vilsmeier reagents: their formation, structures, and general patterns of reactivity. Chapter 2 discusses the variety of functional groups that can be assembled using Vilsmeier reagents; by inspection of the contents pages, it should be possible to locate quickly information on how a particular functional group can be formed using Vilsmeier reagents. Chapter 3 and subsequent chapters deal with the synthesis of ring systems; again, an inspection of the contents pages should allow in most cases a rapid location of information on how a given ring sytem has been synthesized. In this way, the book can be used as a work of reference by practicing organic chemists.

On careful study, a considerable measure of understanding of patterns of reactivity in Vilsmeier chemistry is possible. It is hoped that the rationalizations offered will stimulate further thought, and possibly provide the basis for new reactions, undoubtedly lying in wait for the synthetic organic chemist.

Charles M. Marson   Sheffield
Paul R. Giles   September 1993

NEW DIRECTIONS in ORGANIC and BIOLOGICAL CHEMISTRY

Series Editor : C.W. Rees, FRS
Imperial College of Science, Technology and Medicine
London, UK

New and Forthcoming Titles

Chirality and the Biological Activity of Drugs
Roger J. Crossley

Enzyme-Assisted Organic Synthesis
Manfred Schneider and Stefano Servi

C-Glycoside Synthesis
Maarten Postema

Organozinc Reagents in Organic Synthesis
Ender Erdik

Activated Metals in Organic Synthesis
Pedro Cintas

Capillary Electrophoresis: Theory and Practice
Patrick Camilleri

Cyclization Reactions
C. Thebtaranonth and Y. Thebtaranonth

Mannich Bases: Chemistry and Uses
Maurilio Tramontini and Luigi Angiolini

Vicarious Nucleophilic Substitution and Related Processes in Organic Synthesis
Mieczyslaw Makosza

Radical Cations and Anions
M. Chanon, S. Fukuzumi, and F. Chanon

Chlorosulfonic Acid: A Versatile Reagent
R. J. Cremlyn and J. P. Bassin

Aromatic Fluorination
James H. Clark and Tony W. Bastock

Selectivity in Lewis Acid Promoted Reactions
M. Santelli and J.-M. Pons

Dianion Chemistry
Charles M. Thompson

Asymmetric Methodology in Organic Synthesis
David J. Ager and Michael B. East

Chemistry of Pyridoxal Dependent Enzymes
David Gani

The Anomeric Effect
Eusebio Juaristi

Chiral Sulfur Reagents
M. Mikołajczyk, J. Drabowicz, and P. Kiełbasiński

Contents

chapter one

Formation, Structure, and General Reactions of Vilsmeier Reagents

1.1 Introduction

Soon after the beginning of the 20th century Dimroth and Zoeppritz[1] showed that formanilide, PhNHCHO, behaves as a formylating agent when in the presence of phosphorus oxychloride ($POCl_3$). However, only resorcinol afforded a high yield of a formylated product, N,N-dialkylanilines failing to react. It was Vilsmeier who showed that generally high yields of formylated products could be obtained by the use of a 1:1 complex of N,N-dimethylformamide (DMF)-$POCl_3$ or methylformanilide (MFA)-$POCl_3$.[2,3] Subsequently, Witzinger[4,5] compared the Vilsmeier-Haack formylation reaction with other electrophilic aromatic substitutions, such as Friedel-Crafts acylations.

An example of the classical formylation is the reaction of 2-methylindole **1** with DMF-$POCl_3$.[6] This at once illustrates that the formylated product (here **4**) is not the primary reaction product (scheme 1.1). In the main, an iminium salt is the first isolable product. In the case of 2-methylindole, both the intermediate iminium salts **2** and the enamine **3** are particularly stable, and may be isolated.[6]

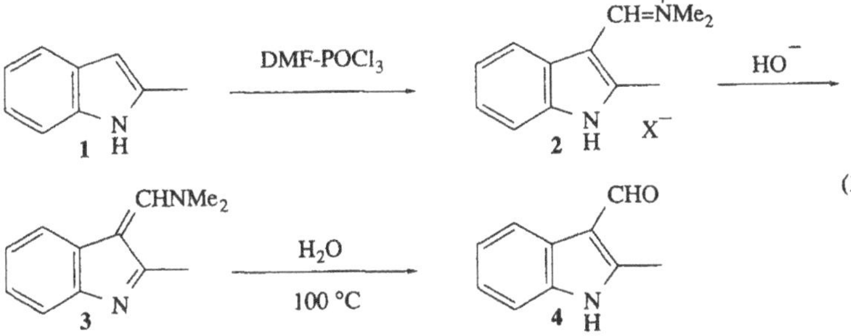

(1.1)

Before 1950, the scope of the Vilsmeier-Haack reaction had been extended little beyond the formylation of reactive aromatic and hetero-aromatic compounds. Over the next two decades, Arnold showed that numerous aliphatic compounds underwent a variety of substitution reactions, evidently the result of complex and multi-step processes.[7-13]

Halomethyleneiminium salts,[14] *e.g.* N,N-dimethylchloromethylene-ammonium chloride, $[Me_2N=CHCl]^+$ Cl^-,[15] are perhaps best known as reagents for, or intermediates generated in, the Vilsmeier-Haack reaction,[16,17] one of the most common methods of formylating activated aromatic rings. A Vilsmeier reagent is produced when a disubstituted amide, typically DMF,[18] is treated with an acid halide, typically phosphorus oxychloride, though to a lesser extent, phosgene.

The potential for carbon-carbon bond-forming reactions[19-23] of halo-methyleneiminium salts in organic synthesis is by no means confined to Vilsmeier-Haack formylation of activated aromatic nuclei. The synthetic value of halomethyleneiminium salts is exemplified by their reactions with compounds containing a C=O linkage, *e.g.* the conversion of acetone **5** into the β-chlorovinylaldehyde **6** (scheme 1.2). A number of varied and otherwise inaccessible compounds have been so prepared.

$$\underset{\textbf{5}}{\overset{O}{\underset{\|}{\bigwedge}}} \quad \xrightarrow{\text{DMF-POCl}_3} \quad \underset{\textbf{6}}{\overset{Cl}{\bigwedge}}\diagup\text{CHO} \quad \textbf{(1.2)}$$

Vilsmeier-Haack-Arnold reactions are not even encompassed solely by the term 'acylation', although this does recognize that amides other than formamides can be used to effect substitution of R^1H to give R^1COR^2. One of the chief aims of this book is to show the enormous synthetic potential[14-18] of Vilsmeier reagents which, as will become apparent in the following pages, have provided access to over 60 functional groups, and over 50 ring systems, and extends far beyond the formylation of an activated aromatic nucleus. The reagents allow carbon-carbon bonds to be formed in many other contexts. Recent work has shown that the course of such reactions can often be controlled by substrate, temperature or conditions under which the reactions are worked up. Although Vilsmeier reagents are known to be capable of somewhat unexpected transformations, considerable rationalization of the products is now possible. Some of the topics that will be discussed in this book include:

(i) Vilsmeier-Haack-Arnold formylation of activated aromatic rings[16,17]
(ii) electrophilic formylating agents in general[24,25]
(iii) the chemistry of β-chlorovinylaldehydes[26] (common products of the reaction of Vilsmeier reagents with ketones)
(iv) cyclizations under Vilsmeier conditions.[27]

The importance of β-chlorovinylaldehydes arises from their considerable synthetic versatility and generality of preparation. The reaction of Vilsmeier reagents with ketones containing methyl or methylene groups adjacent to the carbonyl group, reported in the late 1950s by Arnold and co-workers[28,29] affords substituted β-chloroacrylaldehydes such as **6** (scheme 1.2 and section 2.11.3).

It is instructive to compare derivatives of formic acid with each other, in terms of thermodynamic stability and formylating power. Formyl chloride **7** has a half-life of 1 hour at -60 °C and decomposes into hydrogen chloride and carbon monoxide.[30] Formyl chloride can be regarded as the formylating agent in the Gattermann-Koch synthesis of aldehydes. On the other hand, chloromethylene dibenzoate **8** is a stable analog of formyl chloride, but only alkylates aromatic compounds such as mesitylene and anisole in the presence of AlCl$_3$.[31] Complexation to a Lewis acid induces dissociation of **8a** to give the fairly powerful electrophile **8b**.

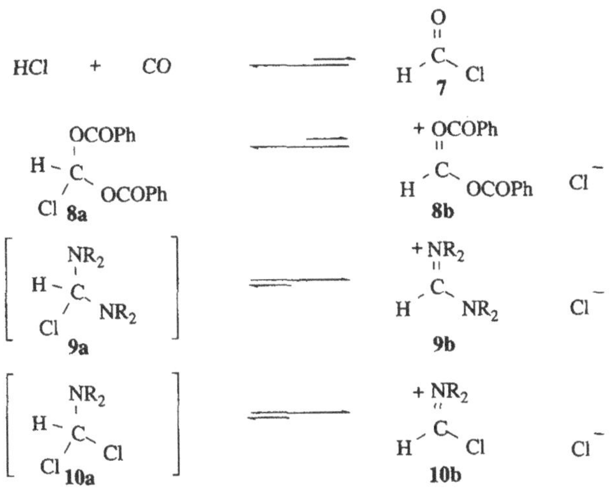

$$(1.3)$$

Formal replacement of the OCOPh groups in **8a** by either by NR$_2$ or Cl gives the compounds **9** and **10**; these are only known in their dissociated forms **9b** and **10b** respectively. The formamidinium salts **9b** are weak electrophiles that react only with strong bases such as carbanions (by addition and subsequent elimination of a secondary amine to give aminomethylene compounds). However, chloromethyleneiminium salts of the form **10b** are highly electrophilic (and are decomposed by moisture and protic solvents). Salts of the type **10b**, the so-called Vilsmeier-Haack-Arnold complexes, are found to form the most reactive class of ionized analogs of formyl chloride, and their reactivity unsurprisingly confers upon them a unique place in organic synthesis.

1.2 Formation and Structure of Halomethylene-iminium salts (Vilsmeier Reagents)

1.2.1 Physical Properties of Vilsmeier Reagents

It is well-known that inorganic acid halides (*e.g.* $SOCl_2$, $COCl_2$ and $POCl_3$) react with DMF to form active complexes, referred to as Vilsmeier-Haack reagents[32-35] which have found extensive use as formylating, halogenating and dehydroxylating agents.[36] DMF has been shown to react with $SOCl_2$ at room temperature giving a 1:1 complex $[Me_2N=CHOS(O)Cl]^+$ Cl^- which loses SO_2 reversibly to give the crystalline complex $[Me_2N=CHCl]^+$ Cl^-; both salts have been characterized,[32] and the latter has been frequently used in organic synthesis.

Iminium chlorides are insoluble in non-polar solvents such as petroleum ether, benzene and diethyl ether, but soluble in polar ones including chloroform and dichloromethane, facts that support an ionic structure,[33] as do the IR spectra, a band at ~1670 cm^{-1} being attributed to the $C=N^+$ bond stretch,[33-35,37,38] and which is almost invariant with the counteranion. Imino-methyleneiminium bromides show bands in the range 1560-1640 cm^{-1}.[38] IR spectra of *N,N*-disubstituted chloromethyleneiminium chlorides imply strong hydrogen bonding between the chloride anions and the hydrogen atoms of the "NH_2" group.[37] NMR spectra also support the ionic structure of bromomethyleneiminium bromides.[12,37] The non-ionic nature of the analogous fluoro compounds is displayed in their physical properties and reactions; thus, (difluoromethyl)dimethylamine **11** is a liquid (b.p. 47-51°C),[39-41] whereas *N,N*-dimethylchloromethyleneiminium chloride is a solid (m.p. 140-145°C).[10,42] Moreover, ionization of the fluoro compound **11** is achieved by boron trifluoride (scheme 1.4).[39-41]

$$Me_2N-\underset{\underset{F}{|}}{\overset{\overset{H}{|}}{C}}-F \quad + \quad BF_3 \quad \longrightarrow \quad Me_2\overset{+}{N}=\underset{F}{\overset{H}{C}} \quad BF_4^- \qquad (1.4)$$

$$\mathbf{11} \qquad\qquad\qquad\qquad \mathbf{12}$$

The physical properties of many halomethyleneiminium halides have been described[10,38,42-44] including many bromomethyleneiminium bromides[38] and perbromides.[38] The halogen atom bonded to carbon can be the same as or different from the counter ion. The interesting *N,N*-dimethyliodomethylene-iminium iodide has also been reported, m.p. 110°C.[37] The existence of *N,N*-unsubstituted iminium salts in the ionic form has been established by conductivity measurements.[45] Zollinger has suggested that iminium salts may exist partly as free ions and partly as ion pairs.[33,34]

1.2.2 Outline of Formation of Vilsmeier Salts

The action of PCl_5,[42,44,46] $COCl_2$,[37,42,44] $SOCl_2$,[33,42] and $(COCl)_2$[33] on R^1CONR_2 can all give $R^1C(Cl)=NR_2^+$ Cl^-, by respective loss of $POCl_3$, CO_2, SO_2, and a mixture of CO and CO_2 from intermediate adducts. There are many other halide-transferring reagents including $COBr_2$[47] and PBr_5;[47] a comprehensive review is available.[48] Carbonyl chloride isocyanate reacts with tertiary amides, but the chloromethyleneiminium isocyanate is not isolated; addition of $SbCl_5$ affords instead the amino-substituted 1-oxa-3-azalactatrienium hexachloroantimonates.[49] Only a few methods of formation, chiefly relevant to synthetic transformations described elsewhere in this book, will be described.

The generally accepted representation[50-52] of the Vilsmeier reagent derived from DMF and $POCl_3$ or $SOCl_2$ corresponds to the respective structures **13a/13b** and **14a/14b** (scheme 1.5). However, the β-phosphoryliminium chloride **15** has been suggested[53] as being more reactive than the β-chloroiminium phosphate **13a**, an equilibrium mixture of those salts being generated. Raman spectra have been held[54] to establish the structure **15**, formed in the reaction of DMF with $POCl_3$. The alternative structure **14b** (formed from DMF and $COCl_2$) was not Raman active. However, this claim for structures other than **13** has been considered to be erroneous.[27]

$$(1.5)$$

The mechanism of formation and the structures of the Vilsmeier reagents (derived from DMF and $POCl_3$, $SOCl_2$ or $COCl_2$) have been studied by ^{1}H NMR[54-57] and ^{31}P NMR[54,55] spectrometry. Rate constants and activation parameters of formylation by the complex $HCONMe_2 \cdot COCl_2$ in $CHCl_3$ of furan, thiophene, selenophene and tellurophene have been determined.[58]

For the generation of halomethyleneiminium salts, DMF is the amide most commonly used; the acid chloride employed is usually $POCl_3$, although phosgene and thionyl chloride also find use. As with the Vilsmeier-Haack-Arnold reaction, a number of dialkyl or aryldialkyl amides has also been used, including *N*-phenyl-*N*-methylformamide. Advantages of DMF[59] over *N*-methylformamide include the cost and weight of formylating agent required. Solvents commonly employed[27] are DMF, a chloroalkane, or chloroalkene (*e.g.* C H_2Cl_2, $CHCl_3$ or $ClCH_2CH_2Cl$), or $POCl_3$. Temperatures used are normally in the range of 0-100°C, 70-80°C being a satisfactory general working temperature.[27]

Bromoformylation and iodoformylation procedures are usually similar to those for chloroformylation. DMF does not react with carbonyl bromide; the desired bromomethyleneiminium bromide **16** and the analogous iodo compound **18** are usually prepared from the chloromethyleneiminium salt in chloroform with gaseous HBr or HI. Salt **18** exhibits the same properties as the adducts of DMF with either POBr$_3$ or PBr$_3$,[60,61] the structure **17** having been proposed for the latter adduct.[60]

Klages and co-workers[62] studied the addition of HBr and HCl to nitriles, to adducts of nitriles and Lewis acids, and to imidoyl halides. These workers disproved Hantzsch's postulate of a 1:1 nitrile halide adduct; Hantzsch had also erroneously described the "adducts" as nitrilium salts. Klages' formulation of these compounds as halomethyleneiminium salts was confirmed by Allenstein.[41,63]

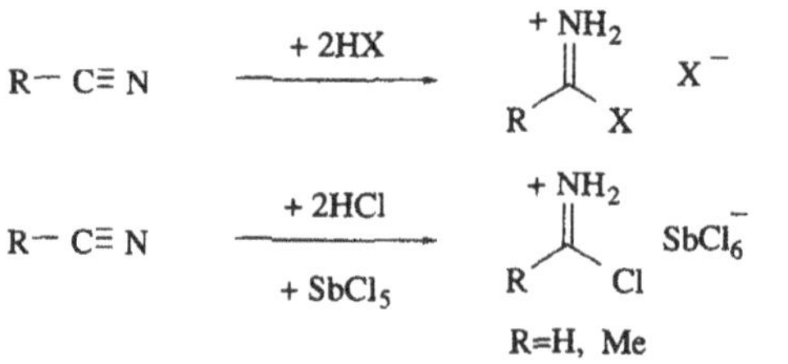

$$(1.6)$$

R=H, Me

The reaction of chloromethyleneiminium chlorides with hydrogen halides is referred to in sections 2.11.2 and 2.11.4. When HBr,[61] HI,[37] or HF[39-41] are passed through a solution of **14b** in chloroform, halogen exchange occurs.[37] The two salts **16** and **18** were first prepared in this way. The related fluoromethyleneiminium salt **12** has to be prepared indirectly (scheme 1.7).

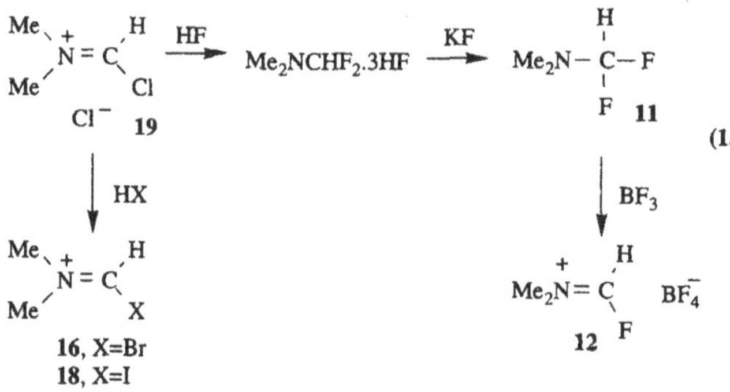

$$(1.7)$$

1.2.3 DMF-POCl$_3$ and Related Adducts

Early studies by Dimroth and Zoeppritz,[1] and Vilsmeier and Haack[2,3] established the ability of complexes of *N,N*-disubstituted formamides with POCl$_3$ to formylate aromatic and heteroaromatic compounds. However, these reactions were known and used for a long time before the constitution of such reagents found a sound basis in experimental fact.

The adduct of *N*-methylformanilide and POCl$_3$ was prepared in a pure state by Vilsmeier and Haack.[2] Later, Bredereck and co-workers obtained an elementally pure and crystalline sample of the adduct of *N,N*-dimethylformamide and POCl$_3$.[64]

$$
\begin{array}{ccc}
\underset{\underset{\underset{\textstyle \mathbf{15}}{\textstyle Cl^-}}{\textstyle Me}}{\overset{\textstyle Me_{\diagdown}+}{\diagup}} \mathrm{N}\!=\!\!\!<\!\!\!\begin{array}{l}\textstyle H \\ \textstyle OP(O)Cl_2\end{array}
&
\underset{\underset{\underset{\textstyle \mathbf{19}}{\textstyle Cl^-}}{\textstyle Me}}{\overset{\textstyle Me_{\diagdown}+}{\diagup}} \mathrm{N}\!=\!\!\!<\!\!\!\begin{array}{l}\textstyle H \\ \textstyle Cl\end{array}
&
\underset{\underset{\textstyle \mathbf{20}}{\textstyle PO_2Cl_2^-}}{\overset{\textstyle Ph_{\diagdown}+}{Me\diagup}} \mathrm{N}\!=\!\!\!<\!\!\!\begin{array}{l}\textstyle H \\ \textstyle Cl\end{array}
\end{array}
\qquad (1.8)
$$

The nature of the group or halogen atom bonded to the iminium carbon atom, and the nature of the counter anion proved to be difficult to establish. For the *N*-methylformanilide-POCl$_3$ adduct, a chloromethyleneiminium dichlorophosphate **20** had been proposed by Lorenz and Witzinger.[5] However, Smith,[65] Silverstein,[66] Jutz,[67] and Ziegenbein and Francke[68] had proposed a dichlorophosphatoiminium chloride structure **15** for the DMF-POCl$_3$ adduct. Arnold and Holy[37] pointed out that iminium salts such as **16**, **18**, and **19** also contain bands at 1040 cm^{-1} and 1160 cm^{-1} so that the assignments as P-C-O bond stretches by Bredereck[71] and Bosshard and Zollinger[34] were by no means unequivocal.

Other evidence as to the constitution of amide-POCl$_3$ complexes was available. The DMF-POCl$_3$ adduct was found to be much *more* reactive than chloromethyleneiminium chlorides such as **19**. These qualitative results obtained from synthetic studies were confirmed by the kinetic measurements of the formylation of thiophene derivatives.[69,70] The difference in reactivity between **19** and DMF-POCl$_3$ could be attributed to a different solubilities of the iminium salts, or to differing degrees of association of the ion pairs. Further evidence inclining to the existence of a C-O-P linkage, and hence the assignment of structure **15** to the DMF-POCl$_3$ adduct came from the work of Ziegenbein and Franke[68] who established the constitution of the product **21**, formed from phenylacetylene and DMF-POCl$_3$ (scheme 1.9).

$$
\mathrm{Ph}\!-\!\mathrm{C}\!\equiv\!\mathrm{CH} \xrightarrow{\;\;\text{DMF-POCl}_3\;\;}
\underset{\underset{\textstyle \mathbf{21}}{\textstyle \underset{Ph}{\diagup}}}{\overset{Cl_{\diagdown}}{}}\mathrm{C}\!=\!\overset{\textstyle H}{\underset{\textstyle H}{\mathrm{C}}}\!-\!\overset{\textstyle OP(O)Cl_2}{\underset{\textstyle NMe_2}{\mathrm{C}}}
\qquad (1.9)
$$

The structure **15** was also consistent with the IR data known, even though such IR data did not allow unambiguous assignments.

The generally accepted formulation of the DMF-POCl$_3$ adduct as **13a/13b** may be an oversimplification. Arnold and Holy[37] showed that both chloride and dichlorophosphate anions exist in the DMF-POCl$_3$ adduct. The most satisfactory interpretation of these results is in terms of an equilibrium between **15** and **13a/13b** (scheme 1.10).

$$\text{Cl}_2(\text{O})\text{PO} \overset{\text{H}}{\underset{\textbf{15}}{\diagdown}}\!\!=\!\!\overset{+}{\underset{\text{Me}}{\text{N}}}\!\!\overset{\text{Me}}{\diagup} \quad \text{Cl}^- \quad \longrightarrow \quad \text{Cl}\overset{\text{H}}{\diagdown}\!\!=\!\!\overset{+}{\underset{\text{Me}}{\text{N}}}\!\!\overset{\text{Me}}{\diagup} \quad \text{PO}_2\text{Cl}_2^- \quad (1.10)$$

$$\textbf{13a/13b}$$

This interpretation is also consistent with NMR studies made by Martin.[50,55,71] However, ^{31}P NMR spectra show that in solution the equilibrium is almost completely in favour of the *N,N*-dichloromethylene-iminium dichlorophosphate **13a/13b**.[55] More recently, conclusions based on other spectroscopic data[50] and on the studies of vinylogous amides[72] have also been in favour solely of structure **13a/13b**.

In the ^{1}H NMR spectra of the DMF-POCl$_3$ adduct in chloroform, the iminium hydrogen signal has been variously reported as δ 10.17[50] and δ 9.75.[65]

The kinetic studies on the DMF-POCl$_3$ adduct by Martin[55,71] showed the reaction to be first order, in respect to both [DMF] and [POCl$_3$]. The reaction velocity increases in the following series of solvents: CHCl$_3$ < ClCH$_2$CH$_2$Cl < CH$_2$Cl$_2$. Activation of energies in chloroform and dichloro-methane have been estimated at ~8 and ~5 kcal mol^{-1}, respectively;[55,71] however, a more recent study of the kinetics in 1,2-dichloroethane calculated the energy of activation to be 15.8 kcal mol^{-1}, and the entropy of activation to be -20.7 kcal mol^{-1}.[69] Martin's studies also showed that the phosphorus moiety is not bound to the adduct irreversibly and can be transferred to other molecules of DMF.[55,56,71]

The exchange of halogen and oxygen between DMF and the DMF-POCl$_3$ adduct has been demonstrated by using a deuterated iminium species (scheme 1.11).[72] The kinetics of this reaction have been studied by Martin.[56]

$$\underset{(\text{CH}_3)_2\text{N}}{\overset{\text{O}}{\|}}\!\!\underset{\text{H}}{\diagdown} \; + \; \underset{\text{Cl}}{\overset{\text{H}}{\diagdown}}\!\!=\!\!\overset{+}{\underset{\text{CD}_3}{\text{N}}}\!\!\overset{\text{CD}_3}{\diagup} \quad \longrightarrow \quad \underset{(\text{CD}_3)_2\text{N}}{\overset{\text{O}}{\|}}\!\!\underset{\text{D}}{\diagdown} \; + \; \underset{\text{Cl}}{\overset{\text{H}}{\diagdown}}\!\!=\!\!\overset{+}{\underset{\text{CH}_3}{\text{N}}}\!\!\overset{\text{CH}_3}{\diagup} \quad (1.11)$$

1.2.4 Higher Amide-POCl₃ Adducts

The dialkylamides of aliphatic and aromatic monocarboxylic acids have been shown to form adducts with POCl$_3$.[64] There have been few or no reports on the isolation of POCl$_3$ complexes with higher acid amides, or with urea. However, their existence can be inferred from the numerous derivatives that can be found, for example with primary amines[64] and amides.[73-76]

The proximity of the two carboxamide groups in the bis(dialkylamides) of dicarboxylic acids determines whether a complex is formed, and if so its constitution.[74] No complex is formed with the bis(dimethylamide) of oxalic acid, but a 1:1 adduct is formed from the bis(dimethylamide) of malonic acid and POCl$_3$.[74] The bis(dimethylamide) of succinic acid forms a 1:2 adduct with POCl$_3$ in low yield, as do the bis(dimethylamides) of adipic acid and terephthalic acids.

Adducts with POCl$_3$ have been formed with five-, six- and seven-membered lactams, and also with the *N*-alkylated and *N*-arylated derivatives (α-pyrrolidinones, α-piperidinones, ε-caprolactams and 1-methylquino-linone[74]).

1.2.5 Thioamide-POCl$_3$ and Related Adducts

The adduct formed from *N,N*-dimethylthioformamide and POCl$_3$ is considered to be more reactive than the DMF-POCl$_3$ adduct, and has been used to formylate heterocycles.[77]

DMF and PSCl$_3$ react to give Me$_2$NC(=S)CHO; a mechanism that initially gives Me$_2$NCHS and POCl$_3$ has been proposed.[78] The POCl$_3$ generated then reacts with Me$_2$NCHS to give [Me$_2$N=CHC(=S)NMe$_2$]$^+$ PO$_2$Cl$_2^-$, which on hydrolysis affords Me$_2$NC(=S)CHO.[78]

1.2.6 Acid Amide-Pyrophosphoryl Chloride Adducts

The adduct of DMF and pyrophosphoryl chloride has been recently reported by Heaney.[79] The DMF-P$_2$O$_3$Cl$_4$ adduct is more electrophilic than that derived from DMF-POCl$_3$, and also more hindered, allowing some regiochemical control. Thus, anisole was formylated to give predominantly *p*-methoxybenzaldehyde in 70% yield, as compared to 34% with DMF-POCl$_3$.

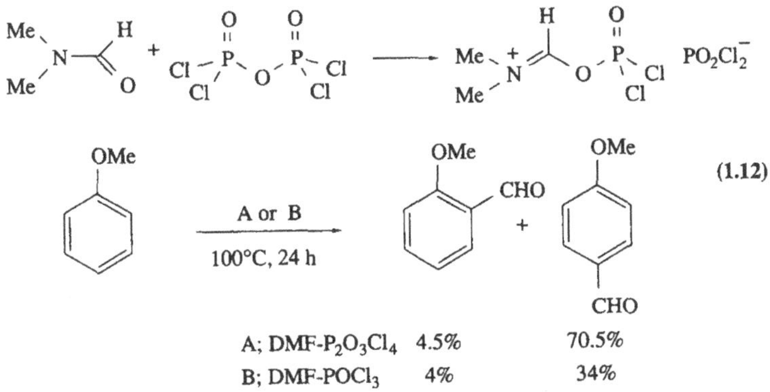

(1.12)

| | A; DMF-P$_2$O$_3$Cl$_4$ | 4.5% | 70.5% |
| | B; DMF-POCl$_3$ | 4% | 34% |

The poor formylation of anisole has been interpreted as supporting the view that a series of equilibria exist in the classical reaction mixture with POCl$_3$. Only when the PO$_2$Cl$_2^-$ anion is present is the reagent sufficiently electrophilic for reaction to occur with moderately activated systems such as anisole. For electron-rich systems *e.g.* pyrrole, little advantage in terms of the regiochemistry or yields of the reaction were noted using P$_2$O$_3$Cl$_4$ in the place of POCl$_3$.

1.2.7 Acid Amide-Carbonyl Halide Adducts

Phosgene, despite its toxicity, reacts with acid amides to give some of the purest samples of chloromethyleneiminium chlorides[42-44,68,69,80,81] obtained by any method. The reaction proceeds through a polar adduct, which at higher temperatures loses CO_2 with the formation of the iminium chloride **22** (scheme 1.13).

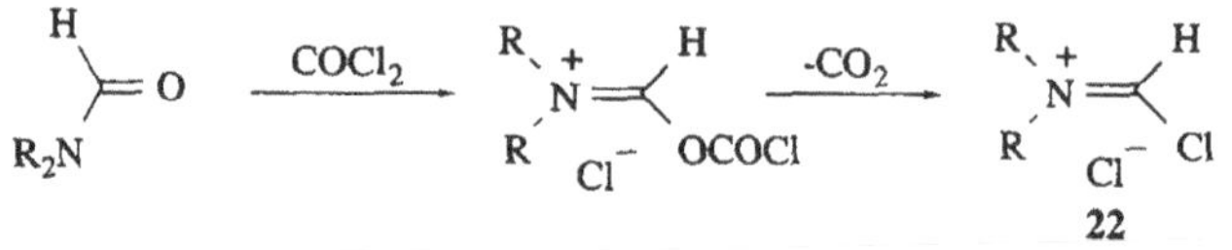

(1.13)

Carbonyl bromide reacts with acid amides to give bromomethylene-iminium salts that are usually very labile.[38,82] The bromomethyleneiminium perbromides can usually be converted into the corresponding bromide by treatment with cyclohexene (scheme 1.14).[38]

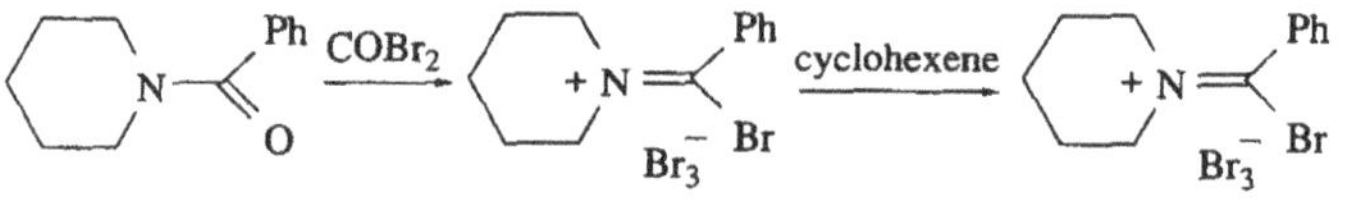

(1.14)

Carbonyl fluoride does react with acid amides, but to give the α,α-difluoroamines[83,84] which exist as such, rather than in the form of an iminium fluoride. On the other hand, N-substituted amides undergo N-acylation (scheme 1.15).[84]

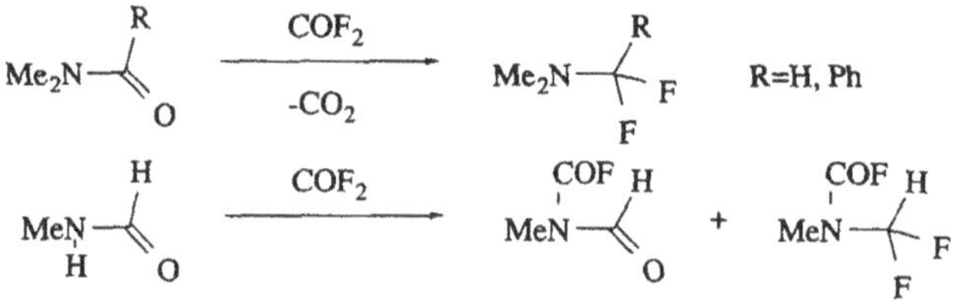

(1.15)

1.2.8 Acid amide-Phosphorus Trihalide Adducts

Bredereck prepared triformamidomethane **23** by heating formamide with PCl_3.[85]

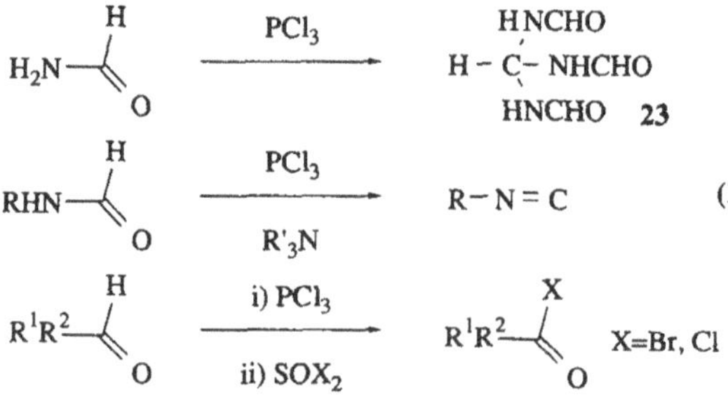

(1.16)

On the other hand, *N*-monosubstituted formamides are dehydrated to nitriles by PBr$_3$, PCl$_3$ or BCl$_3$ in the presence of a tertiary amine (scheme 1.16).[86,87] *N,N*-Disubstituted formamides react differently again, the adducts with PBr$_3$ or PCl$_3$ being converted into carbamoyl halides by thionyl bromide or chloride.[88]

Smith[65] prepared a crystalline adduct in which two molecules of DMF had combined with one of PCl$_3$. He formulated the adduct as **24** which, however, is not supported by its high solubility in CCl$_4$ and its low conductivity. The IR spectrum exhibits a band at 1660 cm^{-1}; the ^{1}H NMR spectrum shows that the two molecules of DMF are bound in the same manner. It is known that other Lewis acids such as BF$_3$, BCl$_3$ and SbCl$_5$ can be coordinated to DMF, as in structure **25**, which further work may establish as being the best representation (scheme 1.17).

$$\text{(1.17)}$$

The reaction of acid amides and primary or secondary amines with PCl$_3$ provides a general synthesis of amidines (section 2.27). It is still not clear whether such reactions involve acid amide-PCl$_3$ adducts, or whether they proceed *via* phosphoric acid trihalides or dialkyl amides which may be formed as intermediates from PCl$_3$ and amines.

1.2.9 Acid Amide-Thionyl Chloride Adducts

Thionyl chloride reacts with DMF to give an adduct **26**.[37,43,68,69,89] The adduct **26** can lose SO$_2$ to give chloromethyleneiminium chlorides **19**. This loss of SO$_2$ is reversible, as was shown by Kikugawa.[90]

$$\text{(1.18)}$$

The adduct **26** was prepared in crystalline form by Ferre and Palomo[91] and was examined by infra-red spectroscopy. The adduct **26** can be re-formed from the iminium chloride **19** in liquid SO$_2$. The ^{1}H NMR spectrum[65,71] of **26** displays a singlet at δ 7.8 (iminium hydrogen atom) and two singlets at δ 2.46 and δ 2.25, assigned as the two methyl groups that are nonequivalent, owing to restricted rotation around the iminium double bond.

1.2.10 Acid Amide-Sulfuryl Chloride Adducts

The reaction of formamide with SO_2Cl_2 affords triformamidomethane (*cf.* PCl_3, section 1.2.8).[85] *N*-Monoalkylated formamides react with SO_2Cl_2 more rapidly than they do with $SOCl_2$.[92] It is believed that the final products, the isocyanates **28**, are formed by elimination of HCl form intermediary carbamoyl chlorides **27**,[92-94] since the latter were isolated when R is methyl or ethyl.[92]

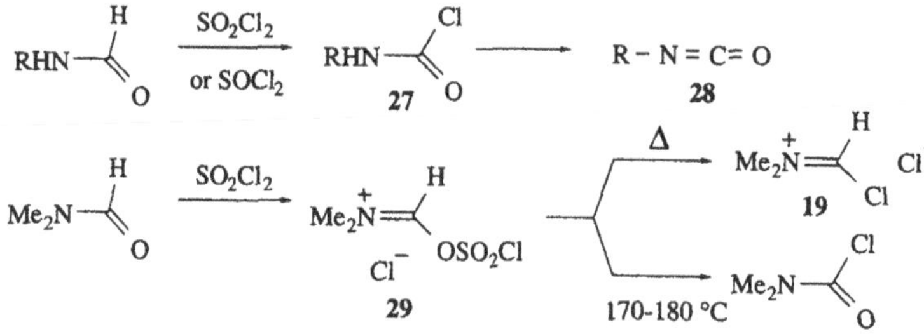

Kuehle first prepared an adduct from DMF-SO_2Cl_2, m.p. 40-41 °C.[92] The adduct was assigned the structure **29** based on its IR spectrum, and the later work by Kojtscheff[95] confirmed this assignment. At 170-180 °C adduct **29** decomposes into dimethylcarbamoyl chloride.[92] An alternative thermal degradation to give the chloromethyleneiminium chloride **19** has been proposed by Kojtscheff[95] and explains the formation of arenesulfonyl chlorides from moderately activated aromatic compounds such as anisole, and acid amide-SO_2Cl_2.

1.2.11 Acid Amide-Carboxylic Acid Chloride Adducts

DMF and acid chlorides react to give ionized products of constitution **30**. Crystalline adducts **30** have been prepared (R=Me, Ph, X=Br).[64,96] Thioformamide forms an analogous crystalline adduct $[H_2N=CHSCOPh]^+$ Cl^-, with benzoyl chloride.[97]

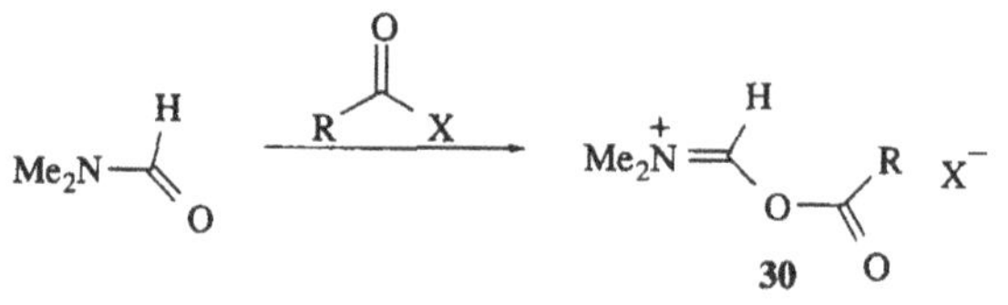

The adduct **31** is initially formed from DMF and ethyl chloroformate, but decomposition occurs giving DMF and ethyl chloride *via* the intermediate **33**.[98] The intermediate **32**, analogous to **33**, can be isolated by the action of fluoroborate on the adduct **31**.[99]

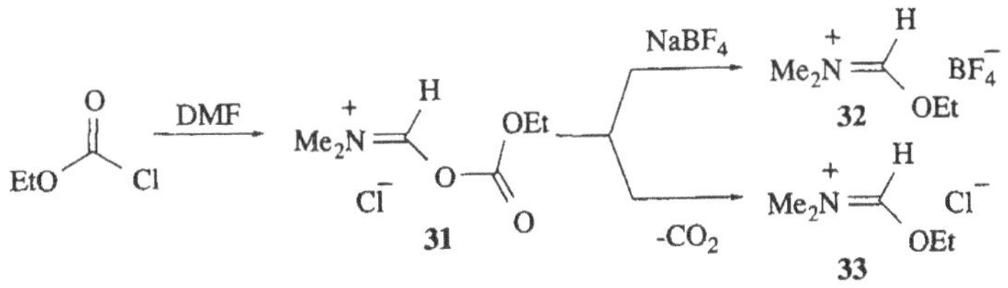

$$(1.21)$$

DMF and some other amides have also been shown to afford the adducts with aryl chloroformates[100-103] and trichloroacetyl chloride.[104] Thioamides may be acylated using ethyl chloroformate. A general procedure for the preparation of adducts **29** from the sodium or ammonium salt of the carboxylic acid RCO_2H, and $[Me_2N=CHCl]^+$ Cl^- has been described.[105]

1.2.12 Acid Amide-Acid Anhydride Adducts

Acylation of amides with carboxylic acid anhydrides yields nitriles **35** and imides **36**; this can be explained on the assumption that the O-acylated species such as **34** are the primary intermediates.[106] Kroehnke's work[107-109] on vinylogous acid amides showed that they undergo O-acylation by acetic anhydride. A mixture of DMF-Ac$_2$O allows dimethylaminomethylene groups to be introduced into active methylene compounds (see section 2.6).

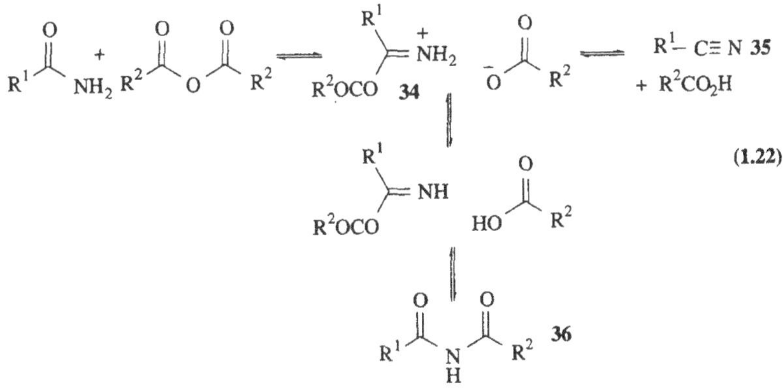

$$(1.22)$$

The complex of DMF and $(CF_3SO_2)_2O$ formylates relatively weakly activated aromatic compounds more efficiently and under milder conditions than conventional Vilsmeier reagents.[110]

1.3 General Reactions of Halomethyleneiminium Salts

1.3.1 Introduction

The reactions of halomethyleneiminium salts with a wide variety of nucleophiles have been studied, particularly reactions involving O-, and S-

nucleophiles. Several of these reactions are described in the appropriate sections concerning synthesis of the particular functional groups. A brief survey is given here.

Chloromethyleneiminium salts react violently with water to re-form the amide;[42] reaction with salts of primary alcohols affords isolable *N,N*-dimethylalkoxymethyleneiminium salts *e.g.* $[Me_2N=CH(OEt)]^+ SbCl_6^-$, from $[Me_2N=CHCl]^+ SbCl_6^-$.[37,64] Tertiary alcohols also react.[8] The conversion of alcohols into alkyl halides by a variety of iminium salts is efficient and has found wide use in synthesis.[42,43,112] It proceeds with complete inversion of configuration. The synthesis of alkyl chlorides form alcohols and $[Me_2N=CHCl]^+ Cl^-$ is especially noteworthy. The hydroxyl group of a variety of heterocyclic compounds can be converted into chloro groups. (A more lengthy discussion is given in section 2.2.4.)

Carboxylic acids and sulfonic acids are converted into the corresponding acid chlorides (section 2.14); the reactions are of very wide scope. Epoxides and ethers are cleaved, for example by DMF-COCl$_2$, to give α,ω-dichloro compounds (section 2.2.1). The action of hydrogen sulfide (H$_2$S) on chloromethyleneiminium salts affords the corresponding thioamides (section 2.20.2). Reaction with thiols followed by hydrolysis affords thioesters, but if thiolysis of the *N,N*-dialkylmercaptoalkylmethyleneiminium chlorides is effected with H$_2$S, the corresponding dithioesters are obtained (section 2.19.3).

Halomethyleneiminium salts react with aromatic amines to give *N,N,N*-trisubstituted amidines (section 2.27). Primary amides undergo dehydration, giving nitriles (section 2.23), often in yields of greater than 80%, in a few minutes at room temperature; the reaction is of wide scope. Some oximes are also converted into amidinium salts (section 2.29.2); thioureas react analogously.

N-Monosubstituted formamides react with COCl$_2$, SOCl$_2$, or PCl$_5$ to give chloromethyleneiminium salts that in the presence of a tertiary amine undergo α-elimination to give isocyanides (section 2.24); the reaction, discovered by Ugi, is very general. Arylthiocyanates are the final products form the reaction of *N*-formylsulfenamides with phosgene and triethylamine.[113]

Treatment of a halomethyleneiminium salt, that has at least one α-hydrogen atom, with base effects deprotonation to give the α-chloroenamine (section 2.21.1).

Reaction of halomethyleneiminium salts with alkoxides affords the highly reactive amide acetals (section 2.20.3). Reaction of the salts with lithium dialkylamides afford ynamines (section 2.22).

Exchange of the counteranion of halomethyleneiminium salts can often be effected with Lewis acids. Counteranions of $AlCl_4^-$,[33-35,114] $SbCl_6^-$,[37,115,116] and BF_4^-,[37] have been introduced using the respective Lewis acid. Bromide counteranions have been exchanged for hexafluoroantimonate and tetrafluoroborate.[38]

Chlorination of halomethyleneiminium chlorides affords the α,α-dichloroamido compounds, which may be hydrolyzed to the corresponding α,α-dihaloamides.[42,117] In contrast, bromination gives side reactions, chiefly the monobromo derivatives contaminated with chlorine.[117]

The aryl hydrazones of α-ketoamide chlorides are obtained by coupling halomethyleneiminium chlorides with diazonium salts.[70]

The thermal decomposition of halomethyleneiminium halides was extensively studied by von Braun and other workers. The decomposition into nitriles occurs stepwise, *via* the imide halides. Certain bromomethylene-iminium bromides decompose more cleanly than the corresponding chloro-methyleneiminium chlorides.

1.3.2 Electrophilicity of Halomethyleneiminium salts and Selection of the Vilsmeier Reagent

The numerous papers concerning the structure and reactivity of Vilsmeier-Haack-Arnold complexes[1,3,33,34,49,50,55,64,69-72] can be summarized by scheme 1.23.

$$(1.23)$$

The formamide **37** and the acid halide **38** react in a second-order process of association of a base and an acid to give a transition state, here represented as **39**. This can then proceed to intermediates **40**, **41**, and **42**, by a series of equilibria. The rate of the reaction is influenced by the basicity of **37**, the Lewis acidity of **38**, and the polarity of the solvent. Which of the intermediates **40**, **41**, and **42** is favored depends on the nucleophilicity and leaving group tendency of $^-$OX or halide anions. Fluoride being a poor leaving group means that intermediates **41** are formed, *e.g.* difluoromethyl-dimethylamine.[10,84] The covalent C-F bond can be cleaved by addition of BF_3, the equilibrium being immediately displaced to give the tetrafluoro-borate analog of **42**. Since $POCl_3$ forms salts of the type **42** with DMF or *N*-methylformanilide,[37,55,56,71] but with these amides $SOCl_2$ forms salts of the type **40**,[97,98] it has been concluded that the basicity of the counteranions are as follows: $F^- \gg SO_2Cl^- > Cl^- > PO_2Cl_2^- > BF_4^-$.

The shift in equilibrium positions away from intermediates **40**, **41**, and **42** is exemplified in the rapid exchange of chloride ion in $[Me_2N=CHCl]^+$ Cl^-

by bromide, iodide or fluoride[10,37,61] and in the reactions of DMF with phosgene and oxalyl chloride, giving intermediates **43** and **44**, respectively, which rapidly decompose to give the same iminium salt **19** (scheme 1.24).

An equilibrium[55,56,72] that is particularly important when DMF is the solvent is its attack by chloromethyleneiminium salts such as **19**. This equilibrium weakens the electrophilic potential of **19**, but it can be suppressed by the presence of HCl in the reaction mixture, or by using a proton-donating solvent, which leads to protonation on the oxygen of DMF as a competing reaction (scheme 1.25).

In preparative work, the requirement is an iminium cation that is sufficiently electrophilic to react, but sufficiently stable so as not to dissociate into starting materials or precursors. Substituents that increase the basicity of the formamide **37** also stabilize the cations **40** and **42**, thus weakening their electrophilic properties. Conversely, iminium salts with electron-attracting groups on nitrogen tend to undergo dissociation. An example is the preparation of 2-naphthylsulfonyl chloride from the corresponding acid; the respective percentage yields using adducts of $POCl_3$ or $SOCl_2$ with Ph(Me)NCHO and Ph_2NCHO are 98, 24, and 0. The weakly Lewis acidic thionyl chloride reacts essentially completely with the relatively basic DMF, giving **40**, but with the more weakly nucleophilic N-methyl-formanilide formation of the iminium salt is incomplete. With the much less basic Ph_2NCHO, little, if any, of the iminium salt is formed. The salt **40** has been established as the effective species.[33]

Phosgene appears to act as a stronger Lewis acid than thionyl chloride, probably on account of the fragmentation step in which **19** is formed by the irreversible loss of CO_2. However, phosgene is insufficiently reactive to convert N,N-dimethyl(2-cyano-3-dimethylamino)acrylamide into the corresponding vinamidinium salt, whereas the stronger Lewis acid, phosphorus oxychloride does so.[118]

The formylating ability of several iminium salts has been compared by reacting complexes of $POCl_3$ and various formamides with 3,4-methylenedioxythioanisole.[119] *N*-Methylformanilide gave better yields of 6-methylmercapto-3,4-methylenedioxybenzaldehyde, but the *p*-nitro derivative gave much lower yields than did *N*-methylformanilide itself. These results imply that iminium salt formation is incomplete for the *p*-nitro derivative. The important conclusion can be drawn that the Vilsmeier reagent with the electrophilic properties that are expected to be the strongest, may well not be the most satisfactory reagent.

Among Vilsmeier reagents, the same chloromethyleneiminium ion does not necessarily produce the same reactivity. Thus, $[Me_2N=CHCl]^+$ $PO_2Cl_2^-$ is a more powerful formylating agent than $[Me_2N=CHCl]^+$ Cl^- chiefly because the latter is only slightly soluble in the solvents employed, typically chlorinated hydrocarbons. In practice, $DMF\text{-}POCl_3$ is used with no additional solvent, and the reaction mixture is almost always homogeneous. Another reason for differing reactivity of the same chloromethyleneiminium ion may be the ability to form tight ion pairs, depending upon the solvent. It is otherwise difficult to explain the claim that *N,N*-dimethylthioformamide-$POCl_3$ is a more powerful formylating agent than $DMF\text{-}POCl_3$, on the assumption that the same cation $[Me_2N=CHCl]^+$, is involved.[84]

Several features may influence which Vilsmeier reagent is selected for preparative work. $DMF\text{-}COCl_2$ and $DMF\text{-}(COCl)_2$ mixtures both afford $[Me_2N=CHCl]^+$ Cl^-, but it usually more convenient to handle oxalyl chloride, being a liquid that can be added to a solution of DMF in chloroform. The alternative is the reaction of a stoichiometric quantity of phosgene with DMF.

Provided that a large excess of DMF is avoided, $DMF\text{-}POCl_3$ (*i.e.* $[Me_2N=CHCl]^+$ $PO_2Cl_2^-$) is a more powerfully electrophilic reagent than complexes that act as $[Me_2N=CHCl]^+$ Cl^-. However, the latter usually gives a more straightforward and cleaner work-up; additionally, only two equivalents of acid must be neutralized compared with six for one mole of the $DMF\text{-}POCl_3$ complex.

For certain reactions, DMF can be advantageously replaced by *N*-formylpyrrolidine or *N*-formylpiperidine, and $POCl_3$ by $POBr_3$. *N*-Methylformanilide-phosphorus oxyhalide complexes are more electrophilic reagents than the corresponding DMF complexes, but the former are more sterically hindered. This steric hindrance has, however, been used to control the degree of formylation in the products.[120,121] *N*-Methylformanilide complexes are, however, frequently limited to use below 80°C, in order to prevent extensive decomposition.

1.3.3 Substitution of Aromatic and Heteroaromatic C-H bonds

A number of examples of aromatic formylation are given in section 2.8.4; heteroaromatic formylations are further discussed on section 2.8.5. Such reactions involving chloromethyleneiminium salts have been shown to be second-order,[75,76,122] the rate-determining step being the formation of the Wheland intermediate **45**, as depicted for *N,N*-dimethylaniline (scheme 1.26). Rapid deprotonation affords the iminium salts of the type **46**; these are the products from the reaction, and not the covalently bound chlorine compounds that erroneously appear in some early publications. Hydrolysis of the salts **46**, usually with dilute alkali, affords the aldehyde. Reaction with H_2S affords thioaldehydes. Reduction with $NaBH_4$ gives the tertiary amine.

Adducts **14** and **19**, being moderately strong electrophiles of intermediate steric bulk, afford monosubstituted products in general, as does *N*-methyl-formanilide-$POCl_3$. The deactivating effect on the (hetero)aromatic ring on the cationic charge also prevents further electrophilic attack. However, under forcing conditions *N,N*-dimethylaniline and some derivatives can be diformylated using **14**,[123,124] which also converts azulene into azulene-1,3-dicarboxaldehyde (43%).[125,126] Intramolecular versions of the Vilsmeier-Haack-Arnold reaction are also known.[127,128]

There are valuable reviews on the Vilsmeier-Haack-Arnold formylation of hydrocarbons.[16,17,51] Benzene, hydrindene, naphthalene, some methyl naphthalenes, phenanthrene, dibenz[*a,h*]anthracene, and chrysene have been found not to react with **14** or **19**, although acenaphthene is attacked at the 5-position. The preferred site of attack can be reliably calculated using simple HMO methods.[129]

The π-excessive five-membered heterocycles such as pyrroles, furans, thiophenes and selenophenes are usually substituted at the 2- or 5-position by **14**, **19**, or *N*-methylformanilide-$POCl_3$. Indoles and benzo[*b*]thiophenes that are not highly substituted afford the 2- and 3-formyl derivatives respectively.[16,17,51] 2,5-Dimethylpyrrole and its *N*-phenyl and *N*-benzyl derivatives react with **19** to give the corresponding 3-carboxaldehydes.[130] Even α-unsubstituted pyrroles can react at the β-position. Strongly electron-withdrawing groups on nitrogen diminish the π-density of the pyrrole nucleus, but can lead to selective formyl derivatives (scheme 1.27).[131]

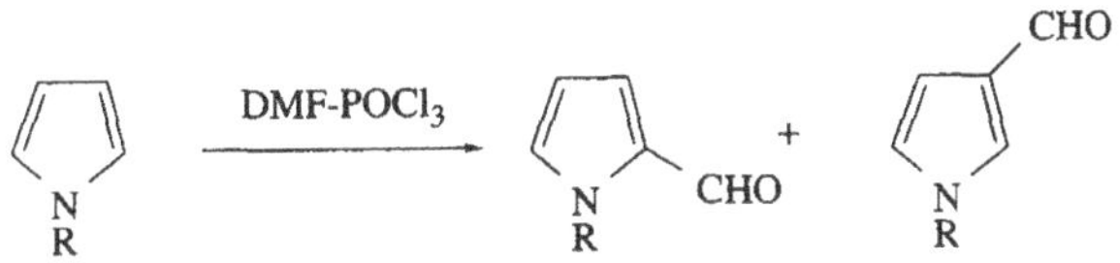

$$R=COMe, 61\% \quad R=COPh, 74\%$$
$$R=COOEt, 54\% \quad R={}^tBu, 69\% \ (1:14)$$

(1.27)

The same principles generally apply to fused heterocyclic systems. Thus, the parent compounds afford the aldehydes **47**[132] and **48**[133] (the latter by work-up with aqueous NaSH). A variety of the substituted thiathiophthenes gave the corresponding formyl derivatives **49**,[134-136] but a dimethyl derivative behaved differently, and as an active methylene compound, affording the enamine **50**.[134]

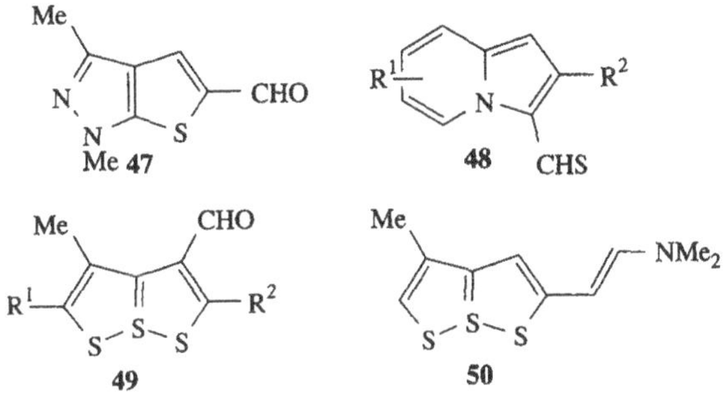

(1.28)

1.3.4 Reactions of Alkenic Double Bonds

Double bonds usually undergo iminoalkylation as the first step. If the resulting iminium cation can be deprotonated, it may do so, in which case further iminoalkylation may occur, to give polyiminium species, and hence polyformylated, or even ring-closed products. If deprotonation does not occur, hydrolysis is likely to give an α,β-unsaturated aldehyde. Consequently, the major (but not the only) reactions of alkenic double bonds are the formation of new ring systems, of which section 1.3.4 is illustrative. Detailed examples will be found under the section dealing with the appropriate functional group formed.

Styrene and its Derivatives

Witzinger showed that alkenic double bonds are susceptible to electrophilic substitution[4,5] and in general **14, 19**, or *N*-methylformanilide-POCl$_3$ are used. A slow step leads to a carbocation **51** (comparable with the Wheland intermediate formed in the formylation of a (hetero)aromatic ring), and a rapid and irreversible elimination of HCl affords the conjugated iminium species such as **52**.

(1.29)

A variety of groups can be tolerated on the aromatic ring, a *p*-methoxy group not inducing ring-formylation to any extent; the double bond may also be substituted at the α- or β-positions, *e.g.* by methyl.[137] Sometimes the iminium salt **52** crystallizes from the reaction mixture. Alkaline hydrolysis affords the unsaturated aldehyde. Similarly, reaction of 1-arylpenta-2,4-dienes with **19** affords the dienals **53**.[138]

53a R=H, 92%
53b R=Me, 91%

54

55

(1.30)

Systems with several double bonds between the aryl and aldehyde groups have been similarly prepared.[139] Vinyl groups attached to the pyrrole rings, for example, those in iron (III) haemin, have been shown to undergo formylation; attack at the methine positions was not observed.[140] Where the aromatic ring is strongly activated, as in 3,4-dimethoxystyrenes, the iminium species **52** may undergo ring-closure to give an enamine, and in some cases, alkaline hydrolysis leads to the formation of indanones.[141] Stilbene is unreactive towards **19**; however, 4-dimethylaminostilbene is converted into (2)-4-dimethylamino-α-phenylcinnamaldehyde in modest yields.[142] Under forcing conditions, the vinylogues of stilbene **54** (n=1, 2, 3) are formed. In certain cases, ring-closure of the intermediate iminium cations corresponding to **54** (n=1) occurs, with formation of a substituted cyclopentadiene ring **55**, bearing formyl groups.[138] The formation of pyridines from styrenes is described in section 4.4.2.3. Indene reacts with **14** or **19** at room temperature to give, after alkaline work-up, indene-2-carboxaldehyde; however, at 80°C further iminoalkylation occurs to give a cation that can be isolated as its perchlorate salt.[11]

Double Bonds of Alkenes and Cycloalkenes

Alkenes are rapidly attacked by chloromethyleneiminium salts, usually *beta* to the more substituted carbon atom of the double bond, for steric reasons, and because of greater inductive stabilization of the resulting carbocation. Polysubstitution is the rule, because sequential deprotonation and iminoalkylation is usually possible, and because the first iminoalkylation usually has a higher activation energy than subsequent ones, highly activated dienes being the intermediates. However, since there is often a limited number of protons that can be lost, and since polyiminoalkylated species have increasing steric hindrance, the reactions can be synthetically useful, and often lead to single products in reasonable yields.

A straightforward example is the reaction of camphene **56** with DMF-POCl$_3$; deprotonation of the iminium salt **57** is precluded by Bredt's rule, so that the salt can be isolated (as the perchlorate in 69% yield) or hydrolyzed to aldehyde **58**.

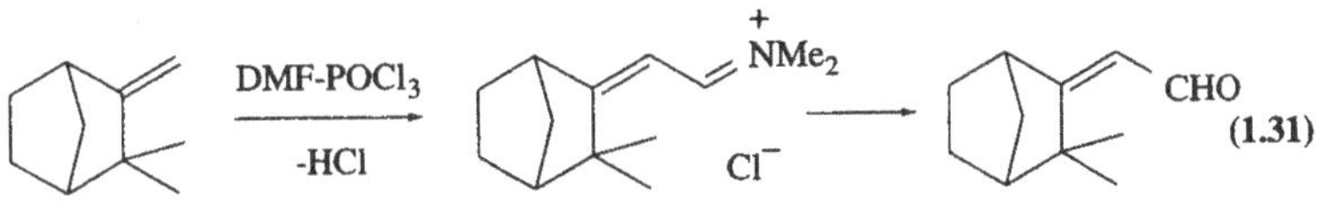

Isonorbonyl chloride (25%) is also formed by the addition of HCl (generated in the reaction) to camphene **56**.[143] Another example of mono-formylation is the reaction of 17-methylene-5α-androstan-3β-ol acetate with DMF-POCl$_3$, although deprotonation is in principle possible in that case.[144]

Even if only one site of the initially formed iminium species can be deprotonated, polyalkylation and even ring closure, can result. Thus, 2-methylenebornane **59** reacts with **14** to give the salt **60** (which can be isolated as the perchlorate in 45% yield).

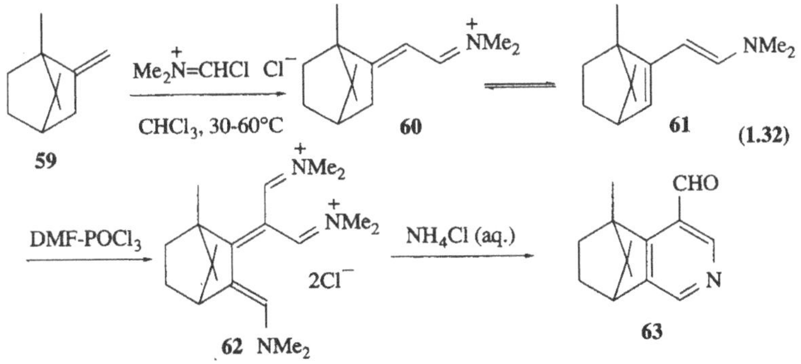

With the more reactive [Me$_2$N=CHCl]$^+$ PO$_2$Cl$_2$$^-$, the barrier to deprotonation of **60** to give **61** (containing a strained endocyclic double bond) is surmounted, and the dienamine **61** undergoes iminoalkylation at both the β- and γ-positions to give the dication **62**, which undergoes ring closure with NH$_4$Cl to give the fused pyridine **63** (93%).[143]

Methylenecyclohexane reacts with **14** to afford in low yield the trimethinium perchlorate **64** (work-up with perchlorate anion). It is presumed that the more basic β-position of the intermediate dienamine is first alkylated, and that the δ-position is not iminoalkylated because of diminished electrophilicity and/or steric hindrance.[143]

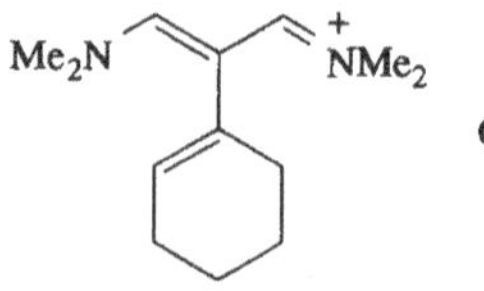

Me2N
NMe2
ClO4−

(1.33)

64

Simple aliphatic alkenes can give complex products, or more often, mixtures partly owing to the number of possible deprotonations and iminoalkylations that can occur, and partly because of the highly reactive carbocations generated. Thus, 2-methylpropene affords the polyiminium species **65** (isolable as the triperchlorate in 73% yield) that can be isolated or ring closed to 2,7-naphthyridine-4-carboxaldehyde **66**.[145,146]

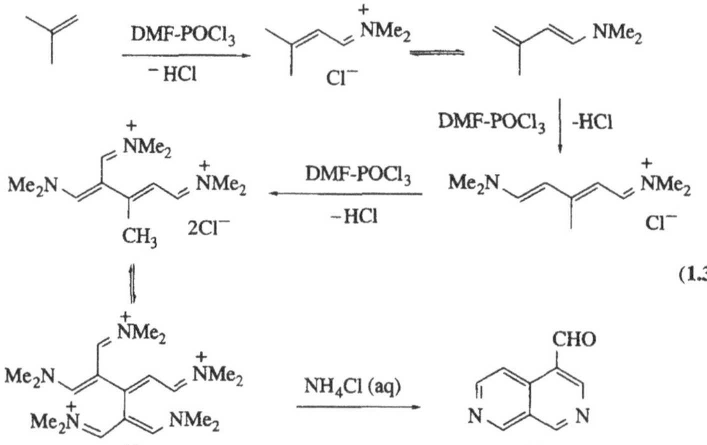

DMF-POCl3
- HCl
NMe2
Cl−
NMe2
DMF-POCl3
-HCl
NMe2
Me2N
NMe2
DMF-POCl3
-HCl
Me2N
NMe2
Cl−
CH3
2Cl−
NMe2
Me2N
NMe2
Me2N
NMe2
65
3Cl−
NH4Cl (aq)
CHO
N
N
66

(1.34)

The formylation of a number of alkenes to give α,β-unsaturated aldehydes is discussed in section 2.8.2. Monoformylation of 1,3,5-trienes is possible, even where deprotonation and further iminoalkylation can in principle occur. The fact that cyclopentadiene undergoes consecutive, albeit controllable, iminoalkylation implies that the conformation of the diene or triene is an important consideration. The differences between linear conjugation (as in polyiminium cations derived by iminoalkylation of polyenes) and cross-conjugation, as in the iminium cations formed by the iminoalkylation of cyclopentadiene (section 2.8.2) and 2-cyclohexen-1-one derivatives (section 2.11.3.4) also appear to be significant.

The reaction of Vilsmeier reagents with fulvenes and steroidal dienes and trienes leads generally, but not always, to conjugated aldehydes; a more detailed discussion is provided in section 2.8.2. The reactions follow the mechanistic patterns outlined in the present section.

1.3.5 Reactions of Ketones

The usual product of monoformylation of ketones is the salt **74**[28,147] (scheme 1.36); further formylation can arise only if deprotonation involving the R^1 group of the cation is possible. Polyformylation is sometimes observed (*e.g.* as for cyclopentanone (section 2.29.4). The most common products isolated (after hydrolytic work-up) are β-halovinylaldehydes *e.g.* **74**, and the preparation of these is extensively documented in section 2.11. This process, in which an aldehyde group is introduced and the oxygen atom is replaced by halogen, is sometimes referred to as 'haloformylation.' An early (vinylogous) example is the conversion of anthrone **67** into 10-chloro-9-anthracenecarboxaldehyde **69** (scheme 1.35).[148]

$$(1.35)$$

The intense red colour that develops in such reactions is due to iminium cations such as **68**. Since its discovery, the reaction of ketones with Vilsmeier reagents has found extensive application in synthesis. Many of the heterocyclic ring systems described in chapter 4 require β-chlorovinyl-aldehydes **68**, and some use the iminium salt precursors **67**. The reagents and ketones are usually readily available and inexpensive, and the β-halo-vinylaldehydes are extremely versatile. They are vinylogues of acid halides, and it is therefore unsurprising that the halogen atom is readily displaced by nucleophiles.

In the chloroformylation of enolizable ketones, it is usual to add the ketone gradually to the Vilsmeier reagent, with cooling. The Vilsmeier reagent is always used in excess (usually 2.5-5 equivalents per mole of ketone). A period of induction is frequently observed prior to an exothermic reaction. Solvents such as 1,2-dichloroethane or trichloroethene are essential to control the otherwise violent reaction, when preparations are carried out on a large scale. After the initial reaction has subsided, the mixture is heated for a further period, then quenched with ice, and the mixture neutralized, usually with cold aqueous sodium acetate or sodium carbonate. Preparations and properties of β-halovinylaldehydes are further documented in section 2.11.

Unlike the action of Vilsmeier reagents on α-hydroxyketones,[149] the reaction of Vilsmeier reagents with simple enolizable ketones has been extensively explored. No unambiguous mechanistic course for the reaction of ketones with Vilsmeier reagents has been developed. Arnold[28] suggested that the ketone enolizes prior to reaction with the Vilsmeier reagent; this is consistent with the fact that only sufficiently nucleophilic alkenes are formylated by DMF-POCl₃. Scheme 1.36 outlines the currently accepted mechanism.[14]

$$(1.36)$$

Electrophilic attack by the Vilsmeier reagent on the weakly basic carbonyl oxygen atom of ketone **70a** slowly forms salt **71** and HCl. Further substitution of salt **71** by the Vilsmeier reagent to give the dication **73** is improbable. The key role in this reaction is thought to be the liberation of HCl during the conversion of ketone **70a** into salt **71**. Aryloxymethylene-iminium chlorides, related to the proposed intermediate **71** are formed from aryl chloroformates and DMF[150,151] and are known to be considerably electrophilic. HCl catalyzes the equilibrium between tautomers **70a** and **70b**; the latter undergoes rapid substitution by the Vilsmeier reagent giving the β-*N,N*-dimethylvinylketones **72** which in certain cases are isolable provided that the reaction is performed at low temperatures. Additionally, salt **71** may formylate the enol **70b** giving ketone **72**. With increasing concentrations of HCl, autocatalytic acceleration of the reaction is observed. Reaction of ketone **72** with the Vilsmeier reagent gives the labile bisiminium chloride **73** which readily undergoes displacement to give the iminium precursor **74** of the β-chloroacrylaldehyde **75**. The perchlorate analogs of **74** have been isolated in very high yield.[147]

Chlorotrienes[152-155] (obtained during the chloroformylation of some steroidal dienones) can be converted by the Vilsmeier reagent into the same chloroformyltrienes which are derived form the steroidal dienones themselves. However, it appears unlikely that the chloroalkenes are the primary intermediates in the process of chloroformylation, and Arnold has shown that α-chlorostyrene does not react with Vilsmeier reagents.[28]

Typical patterns of regioselectivity for ketones are illustrated in scheme 1.37. For acyclic ketones, monosubstitution generally favors the regioisomer **76**, a course predominantly governed by the relative thermodynamic stabilities of the two possible enol intermediates. α,α-Disubstitution blocks formylation at the α-site so that only the aldehyde **77** can be formed, and that is usually the sole product. Karlsson and Frejd[156] have shown that the methyl group of 3-methylcyclohexanone has only a moderate effect on the regioselectivity (**78:79**=10:90). A cyclic ketone that contains a heteroatom in the ring can form the corresponding β-halovinylaldehyde such as **80** (X=O, S).[157]

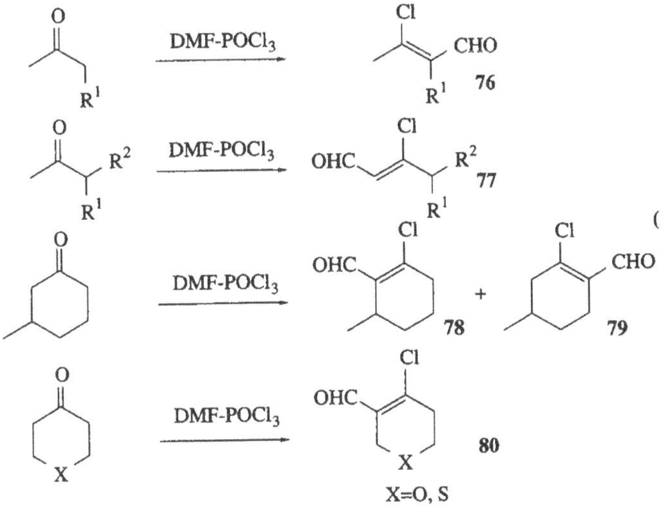

(1.37)

1.3.6 Reactions of Amidic Carbonyl Compounds

Similar mechanistic features apply to the reaction of amides with Vilsmeier reagents as apply to ketones, although a greater diversity of reactions is exhibited. The greater basicity of the carbonyl oxygen atom in amides as compared with ketones suggests that the chloromethyleneiminium cation will initially attack the carbonyl oxygen atom in the former cases (which may also apply to ketones). Since two early reviews[14,29] the area has grown substantially,[158] particularly in the reaction of lactams with Vilsmeier reagents. In this section are emphasized reactions in which the amide group undergoes transformations into other functionalities. However, Bischler-Napieralski type cyclizations, extensively reviewed elsewhere, are not considered here.

Scheme 1.38 outlines some general pathways involved when amides or lactams react with Vilsmeier reagents.[14] Chlorinated products are derived by initial *O*-acylation, followed by nucleophilic attack by chloride ion to give an enamine **84** which rapidly reacts to give the stable intermediates **86**; hydrolysis can afford either an amide **87** or the enaminoaldehyde **88**. For simple amides chlorinated products are not observed, so that the iminium species **81** may be deprotonated to give **83** followed by subsequent acylation.

(1.38)

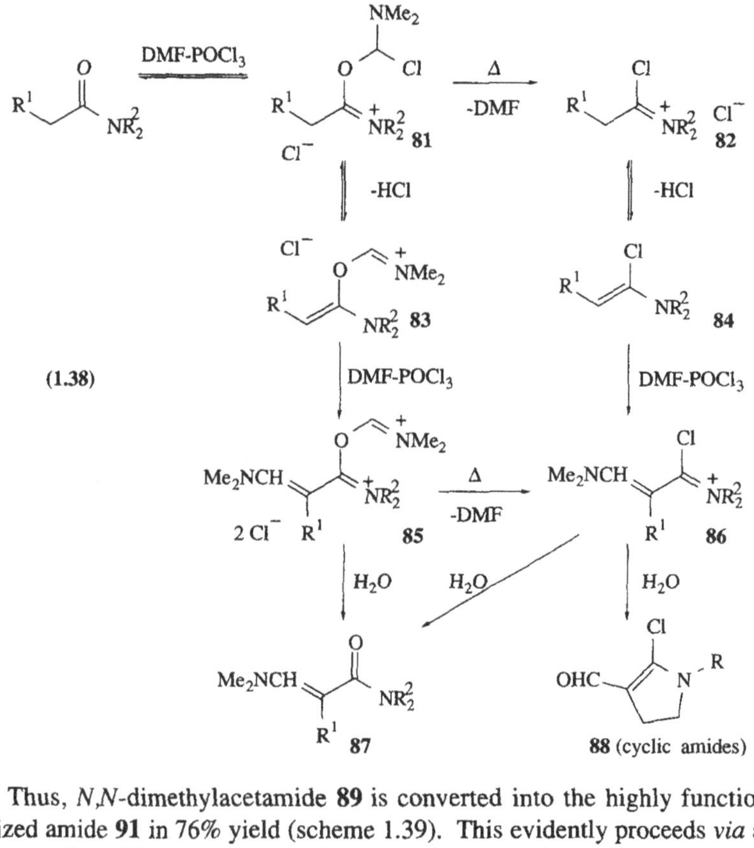

Thus, *N,N*-dimethylacetamide **89** is converted into the highly function-alized amide **91** in 76% yield (scheme 1.39). This evidently proceeds *via* an intermediate **90** that undergoes further iminoalkylation. The dehydrating properties of Vilsmeier reagents may lead to nitriles, presumably *via* the iminium species **82**.

$$(1.39)$$

Synthesis of the trimethinium salts **92** using Vilsmeier reagents has been extended from carboxylic acids to acetamides and thioamides (scheme 1.40).[159]

$$(1.40)$$

There is no clear-cut distinction between the pathways followed by acyclic *versus* cyclic amides (lactams), although the formation of β-chloro-vinylaldehyde moieties, as in **88** is chiefly and perhaps exclusively confined to lactam substrates. However, there are many lactams that give products other than those containing a β-chlorovinylaldehyde group.

As an illustration of the diversity of functional groups obtainable, compounds **93-103** will be considered. Whereas 10-acetylphenothiazine reacts with DMF-POCl₃ to give the corresponding trimethinium salt **86**[160] a series of 1-acetyl-Δ³-pyrrolidin-2-ones is converted by DMF-POCl₃ (or POBr₃) into the acrylaldehydes such as **93a** and **93b**.[161,162] Here, the effect of the nitrogen atom appears not to influence the 'chloroformylation' of an acetyl group, presumably on account of its delocalization as part of the lactam amide in its own right. The carbonyl group of oxindole behaves as a typical ketonic carbonyl, and oxindole is converted by DMF-POCl₃ into the chloroformylindole **94** (about 50% yield).[163,164] Aromatization is presumably a considerable driving force in this case, although in other cases is clearly not an overriding consideration. Thus, non-aromatic 4-dimethyl-aminomethylene-5-pyrazolones such as **95** are obtained from the corresponding 5-pyrazolones and DMF-POCl₃. No incorporation of chlorine was observed under the standard Vilsmeier conditions.[165-167] However, treatment of a pyrazolone such as **95** first with POCl₃, then with water affords the corresponding 5-chloropyrazole-5-carboxaldehyde. Derivatives of oxindole are discussed in section 2.6.2.

$$(1.41)$$

93a X=Br, 60%
93b X=Cl, 72%

94

95

An interesting vinylogous formylation of an amide takes place upon reaction of certain lactams, such as **96**, with Vilsmeier reagents. By neutralizing the reaction mixture with dimethylamine, the enamine **98** can be isolated. This can then be hydrolyzed by acetic acid to the bromopyrrole-carboxaldehyde **99**. The latter can be selectively hydrogenolized using a Pd-BaSO$_4$ catalyst, thereby affording a route to 3,4-unsymmetrically disubstituted pyrrole-2-carboxaldehydes, if different alkyl groups are located in analogs of **96**.[161,162,168]

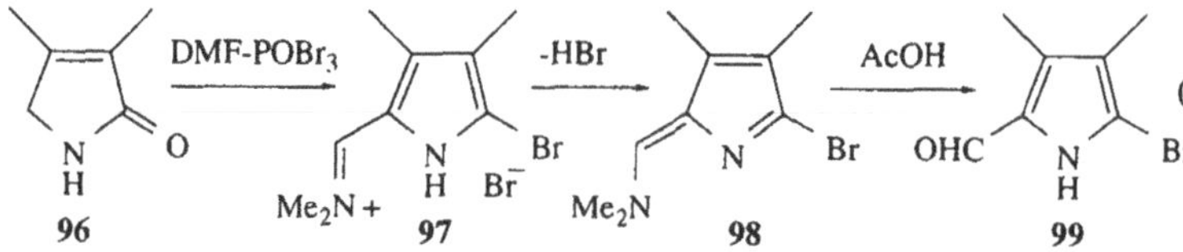

$$(1.42)$$

Yet another set of functionalities can be obtained in compounds **102** and **103**, neither of which has a heteroaromatic ring. Reaction of 2*H*-1,4-benzoxazin-3-ones **100** with DMF-POCl$_3$ affords the iminium salts **104** (R=H, 90%; R=Me, 85%).[169] In neither case is a chlorovinylaldehyde obtained upon alkaline hydrolysis; the yield of **102** is reported as 90%.

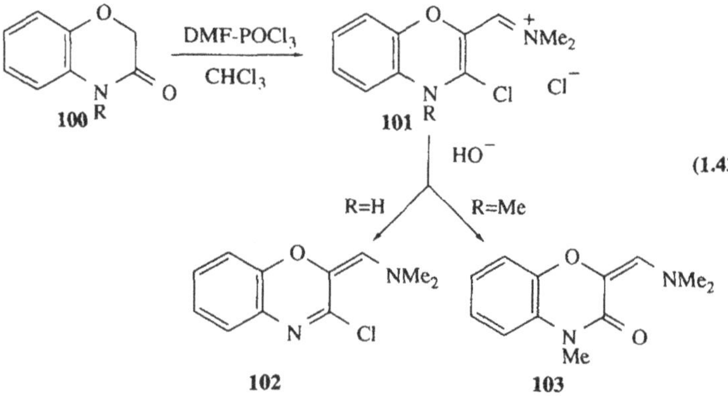

$$(1.43)$$

Enaminone **103** is an example of a Vilsmeier product that contains no chlorine. This implies a pathway involving deprotonation of iminium species such as **81** to give an enamine like **83** which subsequently undergoes iminoalkylation (at the β-position) to give **85** which is then hydrolyzed to liberate the enaminone (*e.g.* **87** or **103**). In such examples formylation occurs without displacement of the oxygen moiety by chloride.[35,163,168,170,171] However, a more usual pathway for acyclic amides, including *N,N*-dimethyl-acetamide is the right hand branch of scheme 1.37, involving rapid iminoalkylation of the α-chloroenamine **84**.

Barbituric acid and its derivatives can give either dimethylamino-methylene derivatives or chloroformylated derivatives (sections 2.8.5 and 3.2.6). Derivatives of 2-arylthiazine form the β-chlorovinylaldehydes, but the simple chlorovinyl derivatives are also formed (sections 2.11.3.7). β-Chlorovinylaldehydes are usually formed from 6-membered lactam rings that are part of a steroidally based skeleton (section 2.11.3), but for some examples dialkylaminomethylene compounds are the result of attack by DMF-POCl₃ (section 2.6). Isoquinolines can be indirectly converted into chloroformyl derivatives (section 3.2.7).

Some acetanilides react with Vilsmeier reagents to give quinolin-2-ones, by formal insertion of one carbon atom (section 4.3.3); others afford chloroformylquinolines (section 4.4.2.5). 3-Acetylthiophene undergoes ring closure to give a fused pyridine ring system (section 4.4.2.3). Some indolinones and dibenzodiazepinones react at their carbonyl group, undergoing remarkable transformations into new ring systems (chapter 5).

The acetyl group of 3-(acetylamino)benzo[*b*]thiophene behaves anomalously, being converted into an amidine (section 2.27.1).

Imides often give bis-(β-chlorovinylaldehydes), but compounds containing a chlorovinyl moiety can also be formed (section 2.11.3).

Despite this remarkable variety of functionality, the formation of most compounds derived from amides and Vilsmeier reagents can be rationalized using the pathways outlined in this chapter and subsequent ones.

References

1. O. Dimroth and R. Zoeppritz, *Ber. Dtsch. Chem. Ges.*, 1902, **35**, 995.
2. A. Vilsmeier and A. Haack, *Ber. Dtsch. Chem. Ges.*, 1927, **60**, 119.
3. A. Vilsmeier, *Chemiker-Ztg.*, 1951, **75**, 133.
4. R. Witzinger, *J. Prakt. Chem.*, 1939, **154**, 25.
5. H. Lorenz and R. Witzinger, *Helv. Chem. Acta.*, 1945, **28**, 600.
6. G. F. Smith, *J. Chem. Soc.*, 1954, 3842.
7. Z. Arnold and F. Sorm, *Collect. Czech. Chem. Commun.*, 1958, **23**, 452.
8. Z. Arnold, *Collect. Czech. Chem. Commun.*, 1961, **26**, 1723.
9. Z. Arnold, *Collect. Czech. Chem. Commun.*, 1962, **27**, 2993.
10. Z. Arnold, *Collect. Czech. Chem. Commun.*, 1963, **28**, 2047.
11. Z. Arnold, *Collect. Czech. Chem. Commun.*, 1965, **30**, 2783.
12. A. Holy and Z. Arnold, *Collect. Czech. Chem. Commun.*, 1973, **38**, 1371.
13. Z. Arnold and J. Sauliova, *Collect. Czech. Chem. Commun.*, 1973, **38**, 2641.
14. C. Jutz, in *Advances in Organic Chemistry*, vol. 9, *Iminium Salts in Organic Chemistry*, part 1, E. C. Taylor, (ed.), John Wiley, New York, 1976, pp. 225-342.
15. T. Fujisawa and T. Sato, *Org. Synth.*, 1988, **66**, 121.

16. M.-R. de Maheas, *Bull. Soc. Chim. Fr.*, 1962, 1989.
17. V. I. Minkin and G. N. Dorofeenko, *Russ. Chem. Rev.*, 1960, **29**, 599.
18. J. S. Pizey, *Synthetic Reagents* vol. 1, 1974, pp. 1-99.
19. D. Burn, *Chem. Ind. (London)*, 1973, 870.
20. S. Seshadri, *J. Sci. Ind. Res.*, 1973, **32**, 128; *Chem. Abstr.*, 1973, **79**, 104343m.
21. T. M. Bargar and C. M. Riley, *Synth. Commun.*, 1980, **10**, 479.
22. H. M. Walborsky and G. E. Niznik, *J. Org. Chem.*, 1972, **37**, 187.
23. T. Fujisawa, T. Mori, S. Tsuge, and T. Sato, *Tetrahedron Lett.*, 1983, **24**, 1543.
24. F. Effenberger, *Angew. Chem., Int Ed. Eng.*, 1980, **19**, 151.
25. G. A. Olah, L. Ohannesian, M. Arvanaghi, *Chem. Rev.*, 1987, **87**, 671.
26. M. Weissenfels and M. Pulst, *Z. Chem.*, 1973, **16**, 337.
27. O. Meth-Cohn and B. Tarnowski, *Adv. Heterocycl. Chem.*, 1982, **31**, 414.
28. Z. Arnold and J. Zemlicka, *Collect. Czech. Chem. Commun.*, 1959, **24**, 2385.
29. Z. Arnold and J. Zemlicka, *Proc. Chem. Soc.* 1958, 227.
30. H. A. Stabb and A. P. Datta, *Angew. Chem., Int. Ed. Engl.*, 1963, **3**, 132.
31. F. Wenzel and L. Bellak, *Monatsch. Chem.*, 1914, **35**, 965.
32. K. Kikugawa and T. Kawashima, *Chem. Pharm. Bull.*, 1971, **19**, 2629.
33. H. H. Bosshard, R. Mory, M. Schmid, and H. Zollinger, *Helv. Chim. Acta,* 1959, **42**, 1653.
34. H. H. Bosshard and H. Zollinger, H. *Helv. Chim. Acta,* 1959, **42**, 1659.
35. H. H. Bosshard, E. J. Jenny, and H. Zollinger, *Helv. Chim. Acta,* 1961, **44**, 1203.
36. L. F. Fieser and M. Fieser, *Reagents for Organic Synthesis*, Wiley, New York, 1967, p. 284.
37. Z. Arnold and A. Holy, *Collect. Czech. Chem. Commun.*, 1962, **27**, 2886.
38. B. A. Phillips, G. Fodor, J. Gal, F. Letourneau, and J. J. Ryan, *Tetrahedron*, 1973, **29**, 3309.
39. E. Allenstein and P. Guis, *Chem. Ber.*, 1963, **96**, 2918.
40. E. Allenstein, *Chem. Ber.*, 1963, **96**, 3230.
41. E. Allenstein and A. Schmidt, *Spectrochim. Acta.*, 1964, **20**, 1451.
42. H. Eilingsfeld, M. Seefelder, and H. Weidinger, *Angew. Chem.,* 1960, **72**, 836.
43. H. Eilingsfeld, M. Seefelder, and H. Weidinger, *Chem. Ber.*, 1963, **96**, 2671.
44. M. Grdinic and V. Hahn, *J. Org. Chem.*, 1965, **30**, 2381.

45. S. Kwon, F. Ikeda, and K. Izegawa, *Nippon Kagaku Kaishi*, 1973, 1944.
46. J. O. Wallach, *Liebigs Ann. Chem.*, 1877, **184**, 1.
47. H. Gross, J. Rusche and H. Bornowksi, *Liebigs Ann. Chem.*, 1964, **675**, 142.
48. W. Kantlehner, in *Advances in Organic Chemistry*, vol. 9, *Iminium Salts in Organic Chemistry*, part 2, E. C. Taylor, (ed.), John Wiley, New York, 1976, pp. 65-141.
49. E. Mueller, O. Orama, G. Huttner, and J. C. Jochims, *Tetrahedron*, 1985, **41**, 5901.
50. M. L. Filleux-Blanchard, M. T. Quemeneur, and G. J. Martin, *J. Chem. Soc., Chem. Commun.*, 1968, 836.
51. G. Martin and M. Martin, *Bull. Soc. Chim. Fr.*, 1963, 1637.
52. (a) G. A. Olah and S. J. Kuhn, in *Friedel-Crafts and Related Reactions*, G. A. Olah, ed., Wiley Interscience, New York, 1964, vol. 3, part 2, p. 1211; (b) H. Ulrich, *The Chemistry of Imidoyl Halides*, Plenum, New York, 1968, pp. 87-96.
53. J. C. Tebby and S. E. Willetts, *Phosphorus Sulfur*, 1987, **30**, 293.
54. W. Scheuermann, and G. McGillivray, *Proc. Int. Conf. Raman Spectrosc., 5th* 1976; *Chem. Abstr.* 1978, **88**, 104221f.
55. G. J. Martin and S. Poignant, *J. Chem. Soc., Perkin Trans. 2*, 1972, 1964.
56. G. J. Martin and S. Poignant, *J. Chem. Soc., Perkin Trans. 2*, 1974, 642.
57. A. Fratiello, D. P. Miller, and R. Schuster, *Mol. Phys.*, 1967, **12**, 111.
58. S. Clementi, F. Fringuelli, P. Linda, G. Marino, and G. Savelli, *J. Chem. Soc., Perkin Trans. 2*, 1973, 2097.
59. E. Campaigne and W. L. Archer, *J. Am. Chem. Soc.*, 1953, **75**, 989.
60. J. Schmitt, J. J. Panouse, P.-J. Cornu, H. Pluchet, A. Hallot, and P. Comoy, *Bull. Soc. Chim. Fr.*, 1964, 2760.
61. Z. Arnold and A. Holy, *Collect. Czech. Chem. Commun.*, 1961, **26**, 3059.
62. F. Klages and W. Grill, *Liebigs Ann. Chem.*, 1955, **594**, 21.
63. E. Allenstein and A. Schmidt, *Chem. Ber.*, 1964, **97**, 1863.
64. H. Bredereck, R. Gompper, K. Klemm, and H. Rempfer, *Chem. Ber.*, 1959, **92**, 837.
65. T. D. Smith, *J. Chem. Soc. (A)*, 1966, 841.
66. R. N. Silverstein, E. E. Ryskiewicz, C. Willert and R. C. Kocher, *J. Org. Chem.*, 1955, **20**, 668.
67. C. Jutz, *Chem. Ber.*, 1958, **91**, 850.
68. W. Ziegenbein and W. Franke, *Chem. Ber.*, 1960, **93**, 1681.
69. S. Alunni, P. Linda, G. Marino, S. Santini, and G. Savelli, *J. Chem. Soc., Perkin Trans. 2*, 1972, 2070.

70. P. Linda, A. Luccarelli, G. Marino, and G. Savelli, *J. Chem. Soc., Perkin Trans. 2*, 1974, 1610.

71. G. J. Martin, S. Poignant, M. L. Filleux, and M. T. Quemeneur, *Tetrahedron Lett.*, 1970, 5061.

72. H. Fritz and R. Oehl, *Liebigs Ann. Chem.*, 1971, **749**, 159.

73. H. Bredereck, R. Gompper, and K. Klemm, *Chem. Ber.*, 1959, **92**, 1456.

74. H. Bredereck and K. Bredereck, *Chem. Ber.*, 1961, **94**, 2278.

75. H. Bredereck, F. Effenberger, H. Botsch, and H. Rehn, *Chem. Ber.*, 1965, **98**, 1981.

76. C. Jutz and H. Amschler, *Chem. Ber.*, 1963, **96**, 2100.

77. J. G. Dingwall, D. H. Reid, and K. Wade, *J. Chem. Soc. (C)*, 1969, 913.

78. E. Guenther, F. Wolf, and G. Wolter, *Z. Chem.*, 1968, **8**, 63.

79. (a) G. K. Cheung, I. M. Downie, M. J. Earle, H. Heaney, M. F. S. Matough, K. F. Shuhaibar, and D. Thomas, *Syn. Lett.*, 1992, 77; (b) I. M. Downie, M. J. Earle, H. Heaney, and K. F. Shuhaibar, *Tetrahedron*, 1993, **49**, 4015.

80. F. Hallmann, *Ber. Dtsch. Chem. Ges.*, 1876, **9**, 846.

81. Z. Arnold, *Collect. Czech. Chem. Commun.*, 1959, **24**, 4048.

82. J. von Braun and C. Mueller, *Ber. Dtsch. Chem. Ges.*, 1906, **39**, 2018.

83. (a) M. H. Brown, U.S. patent, 3092637 (1963); *Chem. Abstr.*, 1963, **59**, 12764g; (b) M. H. Brown, U.S. patent, 3214428 (1965); *Chem. Abstr.*, 1966, **64**, 3501h and 3512.

84. F. S. Fawcett, C. W. Tullock, and D. D. Coffman, *J. Am. Chem. Soc.*, 1962, **84**, 4275.

85. H. Bredereck, R. Gompper, H. Rempfer, K. Klemm, and H. Keck, *Chem. Ber.*, 1959, **92**, 329.

86. F. Yoneda, M. Higuchi, T. Matsumura, and K. Senga, *Bull. Chem. Soc. Jap.*, 1973, **46**, 1837.

87. C. V. Z. Smith, R. K. Robins, and R. L. Tolman, *J. Chem. Soc., Perkin Trans. 1*, 1973, 1855.

88. (a) N. Schindler and W. Ploeger, *Chem. Ber.*, 1971, **104**, 969; (b) N. Schindler and W. Ploeger, Ger. Pat 2053840; *Chem. Abtsr.*, 1972, **77**, 34006m.

89. J. von Braun and W. Pinkernelle, *Ber. Chem. Dtsch. Ges.*, 1934, **67**, 1218.

90. (a) K. Kikugawa, M. Ichino, and T. Kawashima, *Chem. Pharm. Bull. Jap.*, 1971, **19**, 1837; (b) K. Kikugawa and T. Kawashima, *Chem. Pharm. Bull. Jap.*, 1971, **19**, 2629.

91. G. Ferre and A. L. Palomo, *Tetrahedron Lett.*, 1969, 2161.

92. E. Kuehle, *Angew. Chem.*, 1962, **74**, 861.

93. K. Harsanyl, D. Korbonits, and P. Kiss, *Acta Chim. Acad. Sci., Hung.*, 1973, **77**, 333.

94. P. A. S. Smith and N. W. Kalenda, *J. Org. Chem.*, 1958, **23**, 1599.

95. T. Kojtscheff, F. Wolf, and G. Wolter, *Z. Chem.*, 1966, **6**, 148.

96. H. K. Hall, *J. Am. Chem. Soc.*, 1956, **78**, 2717.

97. H. Bredereck, R. Gompper, and H. Seiz, *Chem. Ber.*, 1957, **90**, 1837.

98. H. Bredereck, F. Effenberger, and G. Simchen, *Chem. Ber.*, 1963, **96**, 1350.

99. K. Ikawa, F. Takami, Y. Fukui, and K. Tokuyama, *Tetrahedron Lett.*, 1969, 3279.

100. F. H. Suydam, W. E. Greth, and N. R. Langermann, *J. Org. Chem.*, 1969, **34**, 292.

101. M. Itoh, *Chem. Pharm. Bull. Jap.*, 1970, **18**, 784.

102. R. R. Koganty, M. B. Shambhue, and G. A. Digenis, *Tetrahedron Lett.*, 1973, 4511.

103. V. A. Pattison, J. G. Solson, R. L. K. Carr, *J. Org. Chem.*, 1968, **33**, 1084.

104. A. J. Speziale, L. R. Smith, and J. E. Fedder, *J. Org. Chem.*, 1965, **30**, 4303.

105. D. E. Horning and J. M. Muchowski, *Can. J. Chem.*, 1967, **45**, 1247.

106. D. Davidson and H. Skovronek, *J. Am. Chem. Soc.*, 1958, **80**, 376.

107. K. Dickore and F. Kroehnke, *Chem. Ber.*, 1960, **93**, 1068.

108. K. Dickore and F. Kroehnke, *Chem. Ber.*, 1960, **93**, 2479.

109. (a) H. Nordmann and F. Kroehnke, *Angew. Chem.*, 1969, **81**, 747; (b) H. Nordmann and F. Kroehnke, *Liebigs Ann. Chem.*, 1970, **731**, 80.

110. A. G. Martinez, R. M. Alvarez, J. O. Barcina, S. de la Moya Cerero, E. T. Vilar, A. G. Fraile, M. Hanack, and L. R. Subramanian, *J. Chem. Soc., Chem. Commun.*, 1990, 1571.

111. E. Allenstein and A. Schmidt, *Z. Anorg. Allgem. Chem.*, 1966, **344**, 113.

112. D. R. Hepburn and H. R. Hudson, *Chem. Ind.*, 1974, 664.

113. S. Christopherson and P. Carlsen, *Tetrahedron Lett.*, 1973, 211.

114. H. Meerwein, P. Laasch, R. Mersch, and J. Spille, *Chem. Ber.*, 1956, **89**, 209.

115. G. Seitz, H. Morck, K. Mann, and R. Schmiedel, *Chem.-Ztg.*, 1974, **98**, 459.

116. G. Seitz and H. Morck, *Chimia*, 1972, **26**, 386.

117. H. Eilingsfeld, M. Seefelder, and H. Weidinger, *Chem. Ber.*, 1963, **96**, 2899.

118. R. M. Wagner, Dissertation, Tec. Univ., Munich, 1972.

119. F. Dallacker and F.-E. Eschelbach, *Liebigs Ann. Chem.*, 1971, **689**, 171.

120. A. R. Katritzky and C. M. Marson, *J. Org. Chem.*, 1987, **52**, 2726.

121. A. R. Katritzky, I. V. Shcherbakova, R. D. Tack, and P. J. Steel, *Can J. Chem.*, 1992, **70**, 2040.

122. P. Linda, G. Marino, and S. Santini, *Tetrahedron Lett.*, 1970, 4223.

123. C. Grundmann and J. M. Dean, *Angew. Chem.*, 1965, **77**, 966.

124. C. Grundmann and H. Hooks, *Angew. Chem.*, 1966, **78**, 747.

125. K. Hafner and C. Bernhard, *Angew. Chem.*, 1957, **69**, 533.

126. K. Hafner and C. Bernhard, *Liebigs Ann. Chem.*, 1959, **625**, 108.

127. F. Dallacker, D. Bernabei, R. Katzke, and P.-H. Benders, *Chem. Ber.*, 1971, **104**, 2526.

128. S. Akabori, *Bull. Chem. Soc. Jap.*, 1926, **1**, 96.

129. A. Streitwieser, Jr., *Molecular Orbital Theory for Organic Chemists*, Wiley, New York, 1961, p. 335.

130. C. H. Tilford, W. J. Hudak, and R. E. Lewis, *J. Med. Chem.*, 1971, **14**, 328.

131. C. F. Candy, R. A. Jones, and P. H. Wright, *J. Chem. Soc. (C)*, 1970, 2563.

132. N. Y. Koshelev, A. V. Reznichenko, L. S. Efros, and I. Y. Kvitko, *Zh. Org. Khim.*, 1973, 2201.

133. S. McKenzie and D. H. Reid, *J. Chem. Soc. (C)*, 1970, 145.

134. G. Duguay, D. H. Reid, K. O. Wade, and R. G. Webster, *J. Chem. Soc. (C)*, 1971, 2829.

135. J. Bignebat and H. Quiniou, *C. R. Hebd. Seances Acad. Sci., Ser. C*, 1971, **269**, 1129.

136. J. Bignebat and H. Quiniou, *Bull. Soc. Chem. Fr.*, 1972, 4181.

137. C. J. Schmidle and P. G. Barnett, *J. Am. Chem. Soc.*, 1956, **78**, 3209.

138. C. Jutz and R. Heinicke, *Chem. Ber.*, 1969, **102**, 623.

139. H. Hartmann, *J. Prakt. Chem.*, 1970, **312**, 1194.

140. A. W. Nichol, *J. Chem. Soc. (C)*, 1970, 903.

141. D. T. Witiak, D. R. Williams, S. V. Kakodkar, G. Hite, and M.-S. Shen, *J. Org. Chem.*, 1970, **39**, 1242.

142. E. J. Seus, *J. Org. Chem.*, 1965, **30**, 2818.

143. C. Jutz and W. Mueller, *Chem. Ber.*, 1967, **100**, 1536.

144. M. J. Grimwade and M. G. Lester, *Tetrahedron*, 1969, **25**, 4535.

145. Z. Arnold and A. Holy, *Collect. Czech. Chem. Commun.*, 1963, **28**, 2040.

146. C. Jutz, W. Mueller, and E. Mueller, *Chem. Ber.*, 1966, **99**, 2479.

147. J. Zemlicka and Z. Arnold, *Collect. Czech. Chem. Commun.*, 1961, **26**, 2852.

148. G. Kalischer, A. Scheyer and K. Keller, Ger. Pat. 514415 (1927); *Chem. Abstr.*, 1927, **25**, 1536.

149. S. Pennanen, *Acta. Chem. Scand.*, 1973, **27**, 3133.

150. K. Bodendorf and R. Mayer, *Chem. Ber.*, 1965, **98**, 3554.

151. K. Bodendorf and P. Kloss, *Angew. Chem.*, 1963, **75**, 139.

152. H. Laurent, G. Schulz, and R. Wiechert, *Chem. Ber.*, 1966, **99**, 3057.

153. H. Laurent and R. Wiechert, *Chem. Ber.*, 1966, **101**, 2393.

154. G. W. Moersch and W. A. Neuklis, *J. Chem. Soc.*, 1965, 788.

155. A. Consonni, F. Mancini, U. Pallini, B. Patelli, and R. Sciaky, *Gazz. Chim. Ital.*, 1970, **100**, 244.
156. J. O. Karlsson and T. Frejd, *J. Org. Chem.*, 1983, **48**, 1921.
157. P. R. Giles and C. M. Marson, *Tetrahedron Lett.*, 1990, **31**, 5227.
158. C. M. Marson, *Tetrahedron*, 1992, **48**, 3659.
159. J. Liebscher, A. Knoll, H. Hartmann, and S. Anders, *Collect. Czech. Chem. Commun.*, 1987, **52**, 761.
160. A. Kirschner, Dissertation, Tec. Univ., Munich, 1969.
161. H. von Dobeneck and F. Schmierle, *Tetrahedron Lett.*, 1966, **bb**, 5327.
162. F. Schmierle, H. Reinhard, N. Dieter, E. Lippacher, and H. von Dobeneck, *Liebigs Ann. Chem.*, 1968, **715**, 90.
163. K. E. Schulte, J. Reisch, and U. Stoess, *Angew. Chem.*, 1965, **77**, 1141.
164. K. E. Schulte, J. Reisch, and U. Stoess, *Arch. Pharm.*, 1972, **305**, 523.
165. Yu. N. Koshelev, I. Ya. Kvitko, and L. S. Éfros, *J. Org. Chem., USSR* 1972, **8**, 1789.
166. B. A. Porai-Koshits, I. Ya. Kvitko, and E. A. Shutkova, *Pharm. Chem. 1. (New York)*, 1970, 138.
167. M. A. Kira and W. A. Bruckner, *Acta Chim. Acad. Sci. Hung.*, 1968, **56**, 47.
168. B. Hansen and H. von Dobeneck, *Chem. Ber.*, 1972, **105**, 3630.
169. M. Mazharuddin and G. Thyagarajan, *Tetrahedron Lett.*, 1971, 307.
170. H. Bredereck, G. Simchen, H. Wagner, and A. A. Santos, *Liebigs Ann. Chem.*, 1972, **766**, 73.
171. T. Messerschmitt, U. von Specht, and H. von Dobeneck, *Liebigs Ann. Chem.*, 1971, **751**, 50.

chapter two

Synthesis of Functional Groups using Vilsmeier Reagents

2.1 Bromo Compounds

2.1.1 Alkyl Bromides

Alcohols are converted into the corresponding bromides by halomethyleneiminium bromides[1-3] and also by salts such as **1**.[4,5]

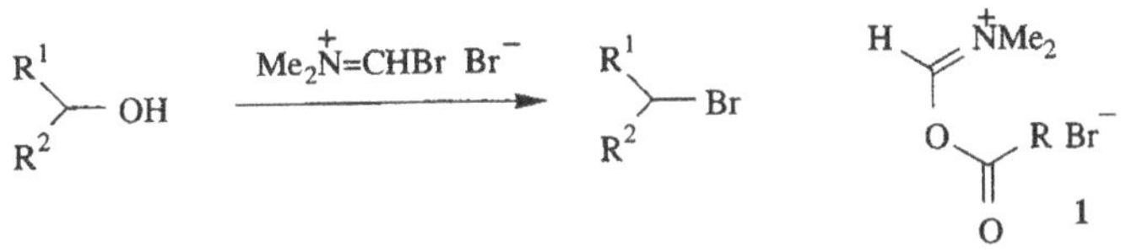

(2.1)

Halomethyleneiminium bromides have also been used to convert alcohols in sugars, and nucleosides into the corresponding bromides.[6,7]

DMF-PBr$_3$ has also been described in patents for the conversion of hydroxy groups into bromo.[8] Other methyleneiminium bromides have been used to convert alcohols into alkyl bromides.[9]

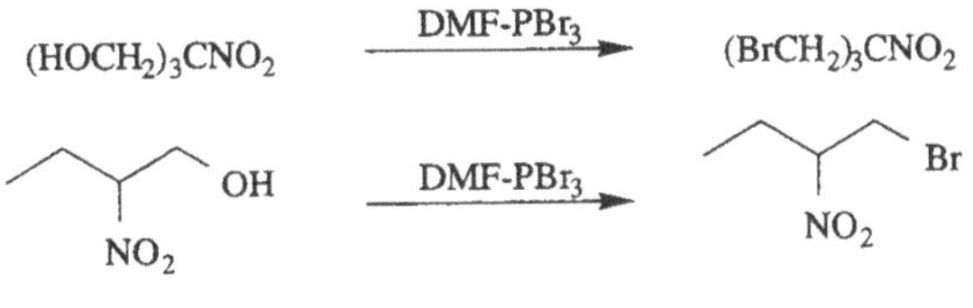

(2.2)

2.1.2 Vinyl Bromides

4-(Dimethylamino)-α-bromostilbenes **3** are formed by the reaction of β-4-(dimethylamino)desoxybenzoins **2** with DMF-POCl$_3$.[10] The anticipated bromoformyl stilbenes (section 2.11.2) were not isolated.

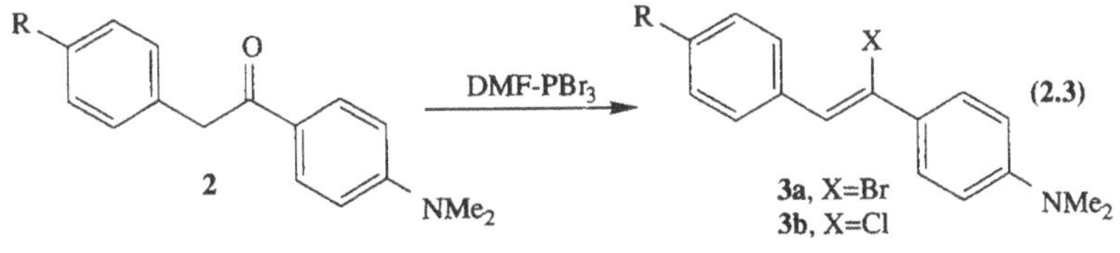

3a, X=Br
3b, X=Cl

(2.3)

An indirect route to 1-bromo-1,3-dienes using Vilsmeier reagents employs the bromoformylation of ketones (section 2.11.2), followed by a Wittig reaction which does not affect the bromovinyl moiety.[11] The bromo group is then available for metalation followed by reaction with electrophiles. For example, reaction of **4** with butyllithium and zinc bromide afforded bromozinc intermediates that undergo palladium (0) catalyzed coupling with vinyl and aryl iodides to give conjugated trienes.[12]

$$R^2 \diagdown \quad \xrightarrow{\text{DMF-POCl}_3} \quad R^2 \diagdown CHO \quad \xrightarrow{\text{Ph}_3\text{P}=\text{CR}^3\text{R}^4} \quad R^2 \diagdown = CR^3R^4 \qquad \textbf{(2.4)}$$

$$R^1 \diagup O \qquad\qquad R^1 \diagup Br \qquad\qquad R^1 \diagup Br$$
$$\textbf{4}$$

2.1.3 Heteroaryl Bromides

Hydroxy groups in various heterocycles can be converted into bromo by DMF-POBr$_3$.[13]

$$\xrightarrow{\text{DMF-POBr}_3} \qquad\qquad \textbf{(2.5)}$$

2.2 Chloro Compounds

2.2.1 Alkyl Chlorides, including Chloromethyl Compounds

A simple procedure for converting alcohols into chlorides employs chloromethyleneiminium chloride.[1,2] The reaction has been shown to proceed with inversion of configuration; the formation of 2-chlorooctane from 2-octanol proceeds with 100% inversion.[14]

$$R^1 \diagdown \!\!-OH \quad \xrightarrow{\text{Me}_2\overset{+}{\text{N}}=\text{CHCl} \ \ \text{Cl}^-} \quad R^1 \diagdown \!\!-Cl \qquad \textbf{(2.6)}$$

$$R^2 \qquad\qquad\qquad\qquad\qquad R^2$$

The fact that only two chlorine atoms of the presumed 1:1 complex derived from DMF-POCl$_3$ can be used to convert ROH into RCl suggests that the adduct could either be **5** or **6**.[15]

$$\underset{\text{Me}}{\overset{\text{Me}}{\diagdown}}\overset{+}{\text{N}}=\!\!\!<\!\!\overset{\text{OPOCl}_2}{\underset{\text{H Cl}^-}{}} \qquad\qquad \underset{\text{Me}}{\overset{\text{Me}}{\diagdown}}\overset{+}{\text{N}}=\!\!\!<\!\!\overset{\text{Cl}}{\underset{\text{H PO}_2\text{Cl}_2^-}{}} \qquad \textbf{(2.7)}$$

$$\textbf{5} \qquad\qquad\qquad\qquad\qquad \textbf{6}$$

Chloromethyleneiminium chlorides convert straight chain,[16,17] branched, and polyfunctionalized alcohols[18,19] into the corresponding chlorides (scheme 2.8).

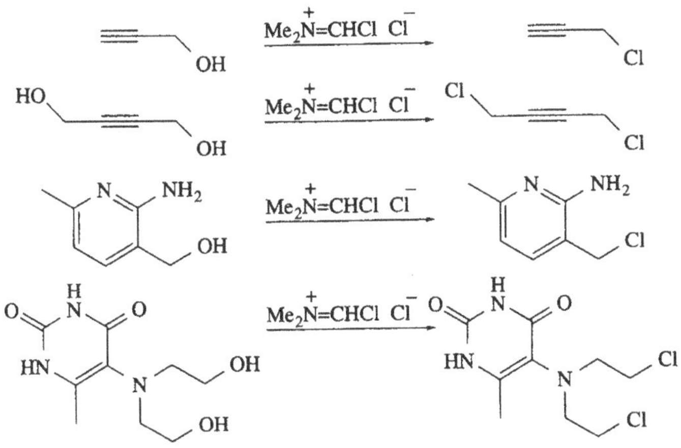

(2.8)

In none of the above examples need the iminium salt be isolated; a catalytic quantity of DMF is sufficient.[17] The halogenating agent may be $COCl_2$ or $SOCl_2$.

Ziegenbein and Franke[20] showed that the epoxides of alkenes (and cyclo-alkenes) underwent ring opening to the corresponding 1-acyloxy-2-chloro-alkanes (scheme 2.9).

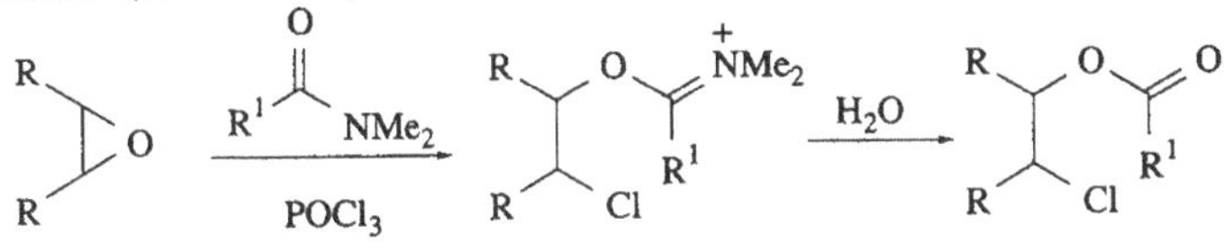

(2.9)

The use of $DMF-COCl_2$ converts propene oxide into 1,2-dichloro-propane. The same reagent effects cleavage of a variety of epoxides and ethers to the corresponding 1,2-dichloro compounds;[20-22] a catalytic quantity of DMF can be used. Regioselective ether cleavage is also possible.[21]

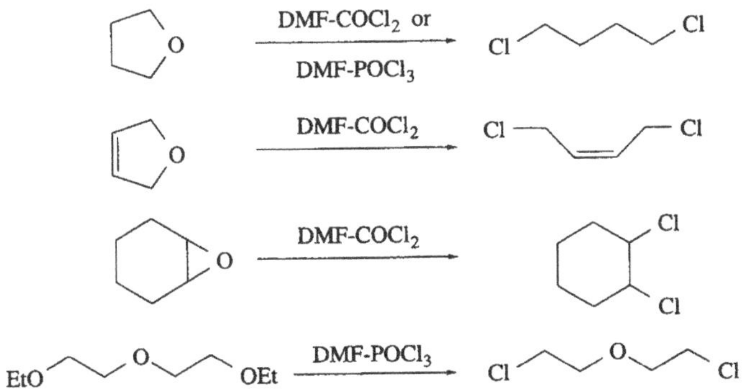

(2.10)

The dimethyl acetal of methoxyethanal has been reported to react with DMF-COCl$_2$ to give CH$_3$OCH$_2$CHCl$_2$.[23]

Diazomethane reacts with dimethylaminomethyleneiminium chloride to give 1,3-dichloro-2-dimethylaminopropane.[24] This behavior differs markedly from other diazoalkanes (section 2.10).

The conversion of an hydroxy group to chloro by dimethylchloromethyl-eneiminium chloride may give competing reactions, *e.g.* chlorination of alkenes or afford poor yields. Fujisawa reported that *N,N*-diphenylchloro-phenylmethyleneiminium chloride (prepared from *N,N*-diphenylbenzamide and COCl$_2$) in the presence of triethylamine, converted a wide range of saturated and unsaturated alcohols cleanly into the corresponding chlorides, in high yields. Reaction at a chiral centre proceeds with complete inversion.[25]

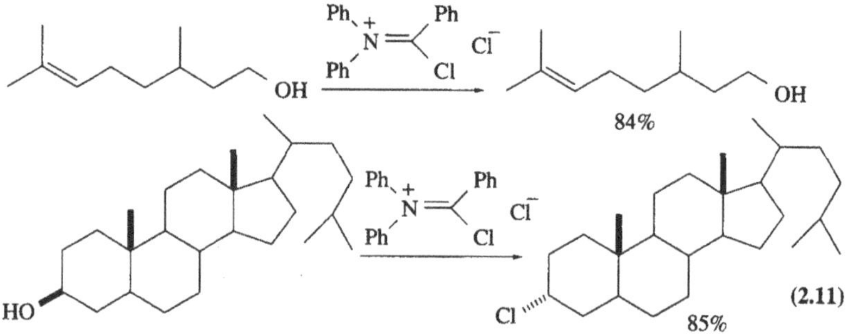

(2.11)

Other chloroiminium salts have been used to convert alcohols into alkyl chlorides, particularly 6-chloro-1,3-dimethyl-4,5-dihydropyridazinium phosphorodichloridate, but the isolated yields were lower than for the DMF-POCl$_3$ adduct.[9]

ArCHO reacts with DMF-SOCl$_2$ to give ArCHCl$_2$;[26] some vinylogous aldehydes react similarly. The yields of the dichloro compounds from the following aldehydes are: benzaldehyde (95%), 1-naphthaldehyde (91%) and cinnamaldehyde (90%). The study provides evidence for [Me$_2$N=CHOSOCl]$^+$ Cl$^-$ as the structure of the active species, and a cyclic mechanism is proposed involving the delivery of a chlorine atom to the cationic benzylic site.

Unprotected pentitols and hexitols are converted into 1,5- and 1,6-dichloro derivatives by DMF-POCl$_3$.[27] For example, after acetylation, the dichlorides 7 (77%) and 8 (50%) were isolated.

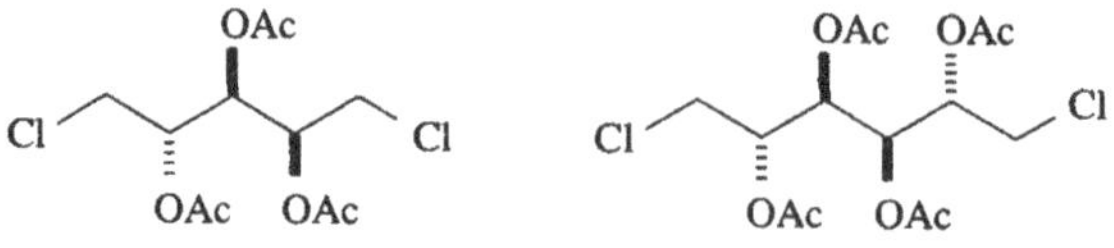

(2.12)

Glycosyl chlorides can also be prepared from the corresponding glycosides and a Vilsmeier reagent; the anomeric hydroxy group of a protected sugar generally undergoes replacement by chloro.[28]

2.2.2 Vinyl Chlorides

β-Chlorovinylaldehydes react with Wittig reagents to give 1-chloro-1,3-dienes; thus 1-chloro-2-formylcyclopentene reacts with methylenetriphenylphosphorane to give 1-chloro-2-vinylcyclopentene.[29]

4-(Dimethylamino)-α-chlorostilbenes are formed by the reaction of β-4-(dimethylamino)desoxybenzoins with DMF-POCl₃.[10]   The anticipated chloroformyl stilbenes were not isolated.

The vinyl chloride **9** is obtained by heating the corresponding 3-acetyl pyrrole with a chlorine-containing Vilsmeier reagent.[30]   However, other 3-acetylpyrroles afford the corresponding 3-alkynyl derivatives.[31]

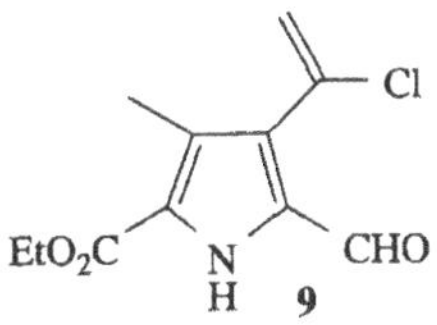

(2.13)

Chlorovinyl compounds possessing an hydroxy or alkoxy group at the allylic position are obtained in moderate yields, along with β-chlorovinylaldehydes (section 2.11.3), by reacting 2-alkyl-2-cyclohexen-1-ones with *N*-formylmorpholine-POCl₃.   The presence of a cationic allylic species, possibly **10** or its equivalent is confirmed by the fact that quenching the reaction mixture with alkoxide afforded the corresponding alkoxy ethers **11** (R¹=alkyl).[32]

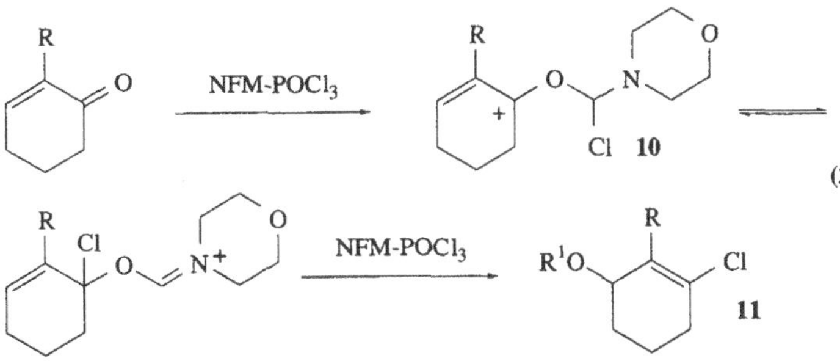

(2.14)

Reaction of conjugated enones that are part of the steroidal ring system can lead to vinyl chlorides (section 2.11.3).

N-Acyl-2,3-dihydro-4-pyridinones react with one equivalent of DMF-POCl₃ to give 1-acyl-4-chloro-1,2-dihydropyridines in excellent yields.[33] The presumed key intermediate **12** is subject to nucleophilic attack by chloride, and subsequent loss of DMF. No chloroformylated products were observed.

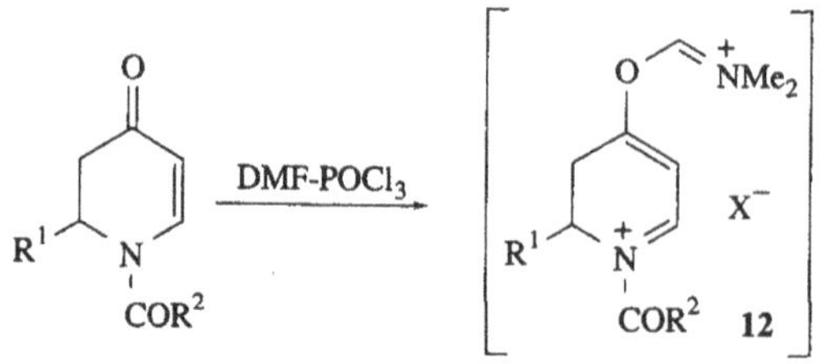

$$\text{DMF-POCl}_3 \qquad (2.15)$$

Treatment of oxathiadiazapentalenes with DMF-POCl$_3$ afforded chloro-vinylthiadiazolium salts (scheme 2.16; R = dimethyl or cyclohexyl).[34]

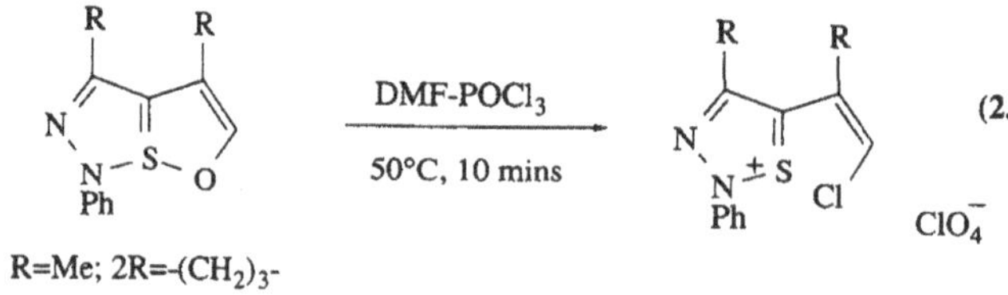

$$\xrightarrow[\text{50°C, 10 mins}]{\text{DMF-POCl}_3} \qquad (2.16)$$

R=Me; 2R=-(CH$_2$)$_3$-

2.2.3 Aryl Chlorides

Replacement of phenolic OH by Cl can be achieved with iminium chlorides if the aromatic ring is sufficently susceptible to nucleophilic attack, as it is in picric acid.[35-37]

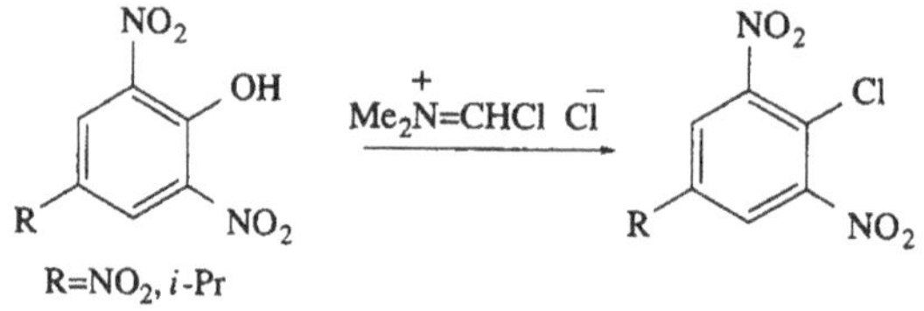

$$\xrightarrow{\text{Me}_2\text{N=CHCl Cl}^-} \qquad (2.17)$$

R=NO$_2$, *i*-Pr

2.2.4 Heteroaryl Chlorides

Hydroxy groups attached to heterocyclic aromatic rings can be converted into chloro groups (scheme 2.18).[16,38] Catalytic amounts of DMF suffice.

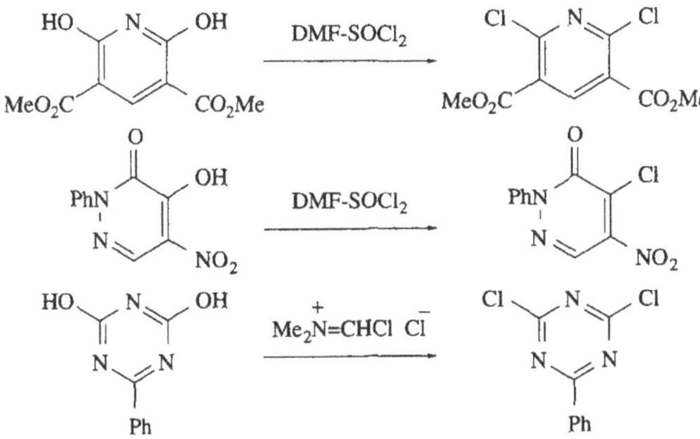

(2.18)

Substituted 1-formyl-1,5-benzodiazepines were prepared (scheme 2.19) in improved yield by decreasing the amount of Vilsmeier reagent used (*e.g.* 10 g of amide **13**, and 50 ml of both DMF and $POCl_3$).[39]

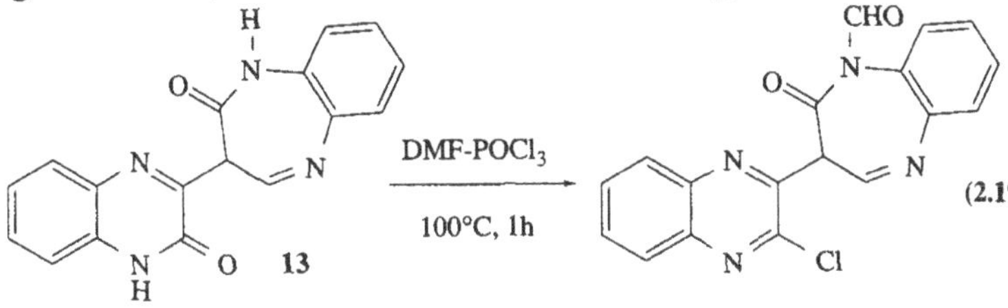

(2.19)

A number of 2-methoxypyridines containing bromo and chloro substituents has been reacted with $DMF-POCl_3$ to give the corresponding 2-chloropyridines.[40]

2.3 Alkynes

The pentamethinium salts **14** and **15**,[41] obtained respectively by the action of $DMF-POCl_3$ on acetone and methyl ethyl ketone (section 2.29.4) are hydrolyzed with fragmentation to give highly functionalized alkynes, albeit in low yields (scheme 2.20).[42]

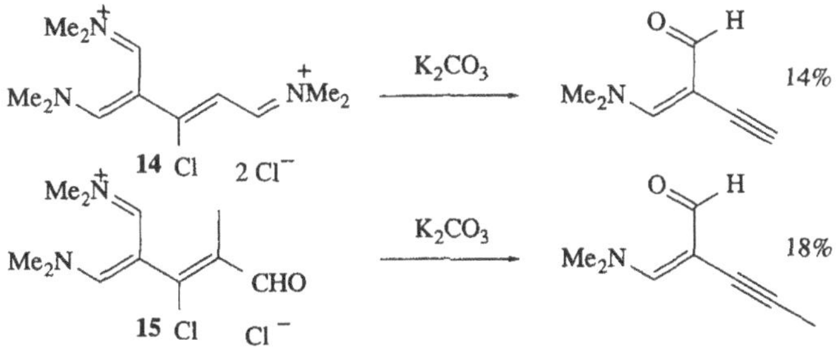

(2.20)

The alkynic aldehyde **16** (34%) was obtained by the action of a Vilsmeier reagent upon 3-acetyl-4-methylpyrrole.[31]

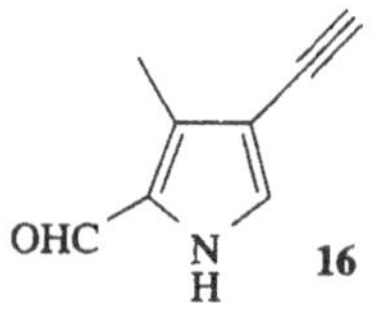

(2.21)

Chloroacrylaldehydes (and their iminium salts) which are not substituted at the α-position undergo fragmentation induced by alkali to give a valuable synthetic route to terminal alkynes (scheme 2.22).[43-46] *p*-Diethynylbenzene has been prepared by heating β,β'-dichloro-*p*-benzenediacrolein.[47]

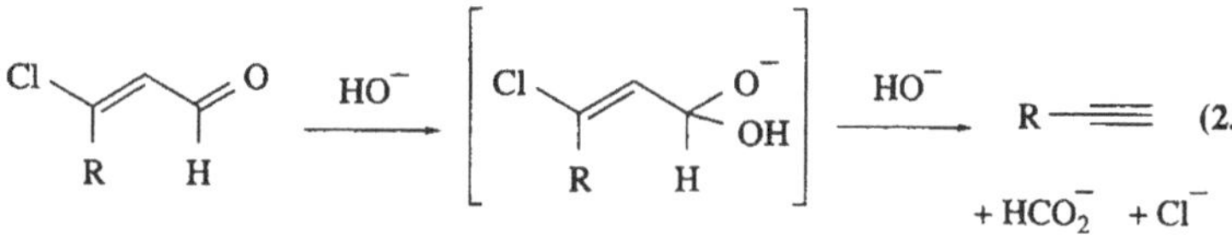

(2.22)

2.4 Ethers

Aryl chloroformates react with DMF to give iminium salts of type **17** which react with alcohols to give alkyl aryl ethers.[48]

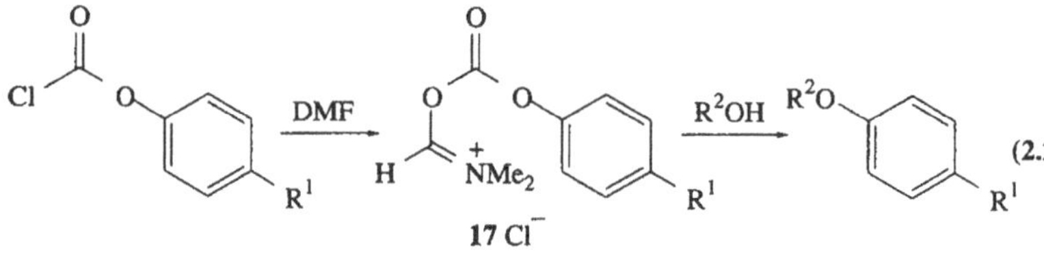

(2.23)

2.5 Dialkylaminomethyl Compounds

An interesting aromatization of 1-ketotetrahydrocarbazoles **18** that also introduces a dimethylaminomethyl group has been observed (scheme 2.24).[49]

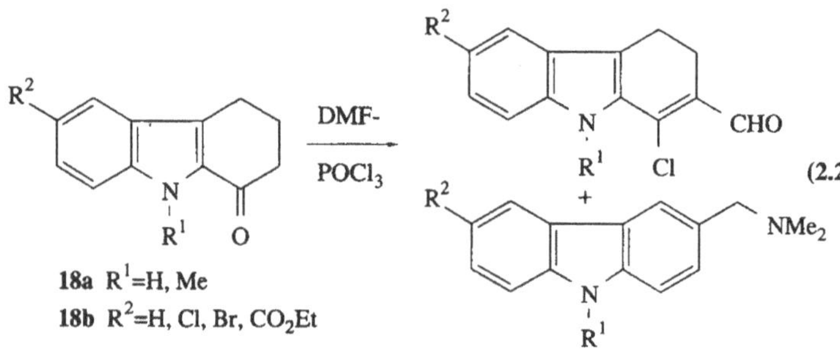

(2.24)

18a R^1=H, Me
18b R^2=H, Cl, Br, CO$_2$Et

N-Benzyl-*N*-cyanoethyl-4-methylanilines **19** are acted upon by DMF-POCl$_3$ to give not only the product **20** of *ortho*-formylation, but the dimethylaminomethyl compounds **21** which evidently arise by a [1,5]-shift of hydride.[50]

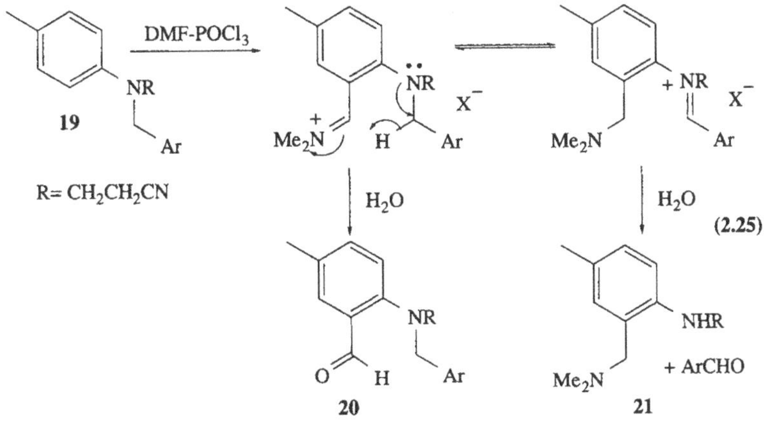

(2.25)

2.6 α-Dialkylaminomethylene Compounds

2.6.1 α-Dialkylaminomethylene Heterocycles

A lactam, upon treatment with DMF-POCl$_3$ affords an iminium salt that reacts with alkali (during work-up) to give an unusual vinylogous amidine (scheme 2.26).[51]

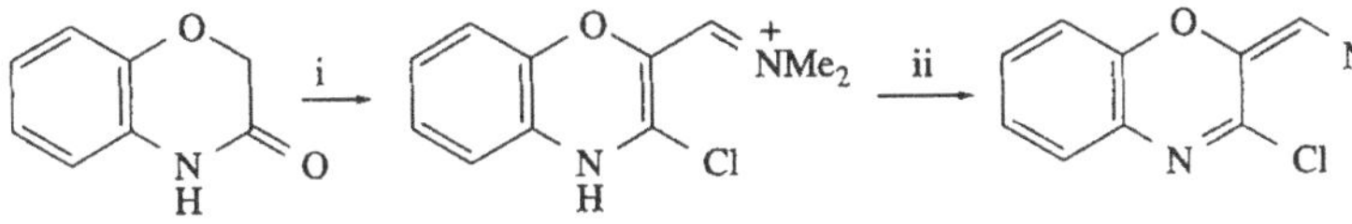

i, DMF-POCl$_3$; ii, HO$^-$

4*H*-1,4-Benzothiazin-3-ones **22** are converted into the corresponding enamines (scheme 2.27), but these decomposed during a period of two days.[52]

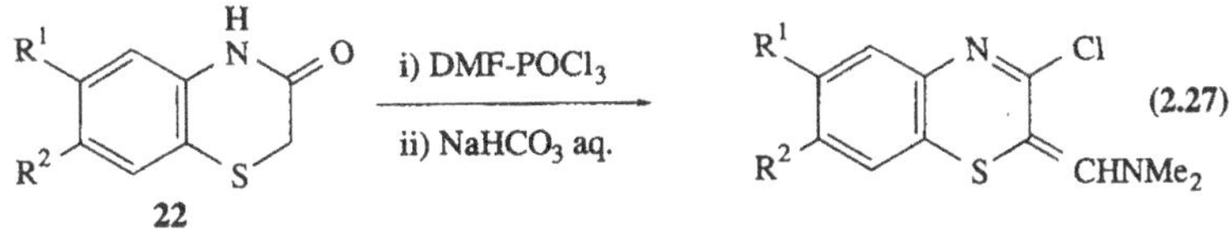

(2.27)

2.6.2 α-Dialkylaminomethylene Lactams

DMF-Acetic anhydride can be used to introduce a dimethylamino-methylene group into active methylene compounds.[53]

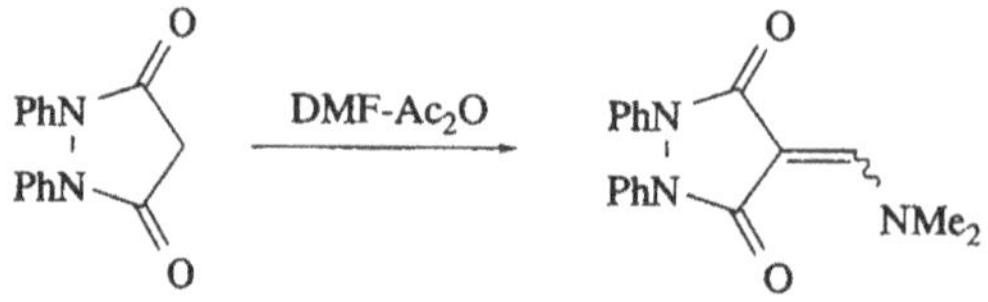

(2.28)

The methyl group of some 4-methylpyrylium salts can also be simliarly attacked by Vilsmeier reagents, which afford an enamine group at the 4-position.[54]

α-Pyrrolones are converted by Vilsmeier reagents into either the 2-chloro-3-formyl derivatives[55] or the 2-halo-5-formyl derivatives,[56] depending upon the substitution in the ring. In some cases, the intermediate enamines may be isolated. The 2-bromopyrrole-5-aldehydes so obtained are useful compounds for the synthesis of pyrromethanes. In some other cases α-pyrrolones afforded the dimethylaminomethylene derivatives **23**.[57] The lactams **24** and **25** are obtained[57b] from *N*-methyl-δ-valerolactam and *N*-methyl-ε-caprolactam respectively, but were accompanied by other products.

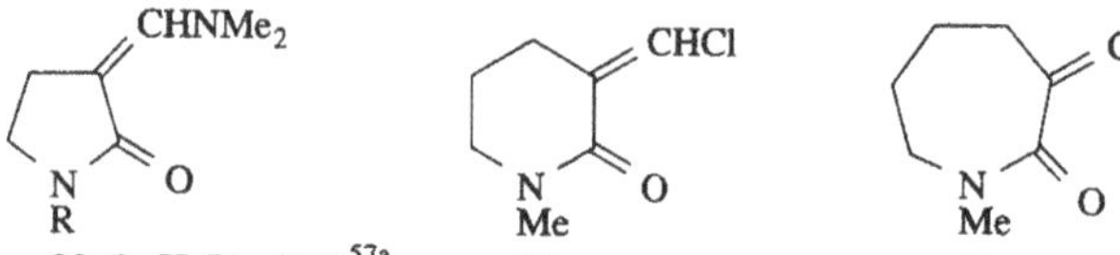

23 R=H, Ph, CHO[57a]   **24**   **25**
R =Me, Bu[57b]

(2.29)

The behavior of oxindoles towards DMF-POCl$_3$ depends upon temperature; at about 80°C, α-(dimethylaminomethylene) compounds are obtained in excellent yields after hydrolytic work-up (scheme 2.30).[58]

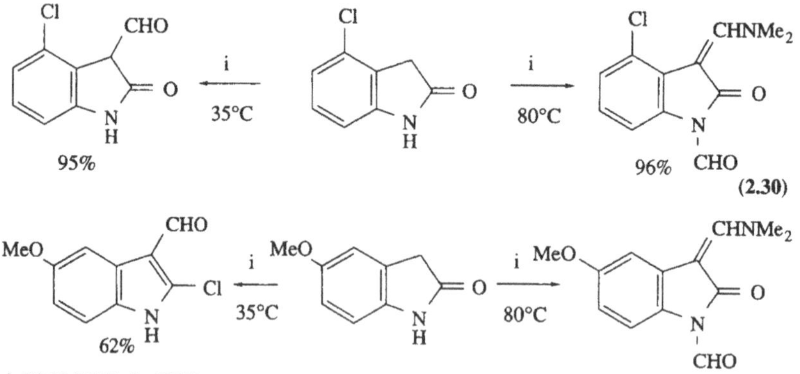

(2.30)

i, DMF-POCl$_3$ in CHCl$_3$

Vilsmeier reaction of benzodiazepin-2-ones **26** afforded the corresponding aminomethylene derivatives (scheme 2.31), rather than the corresponding chloroformylated compounds.[59,60]

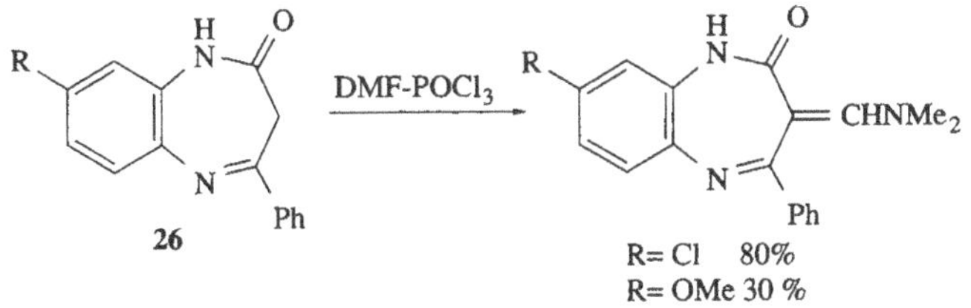

26

R= Cl   80%
R= OMe 30 %

(2.31)

2.7 Epoxides

cis-Diols, as part of a cyclohexane or pyran ring, are converted into epoxides by reaction with phosgeneiminium chloride (Viehe's salt) and subsequent ring closure with methoxide.[61]

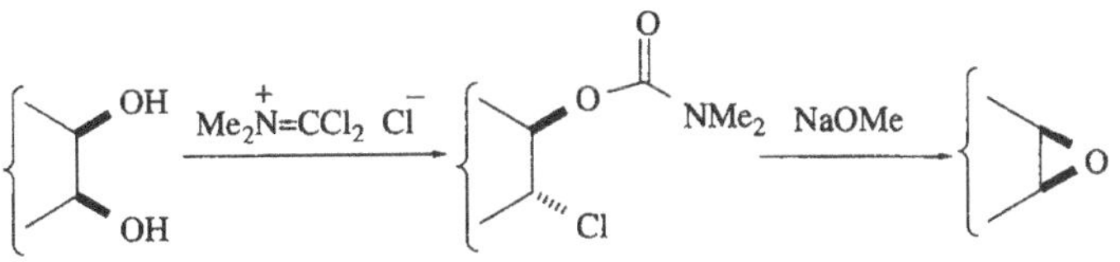

(2.32)

Reaction of hydrogen peroxide with DMF-POCl$_3$ affords adduct **27**, which is capable of epoxidizing double bonds. Unfortunately, the selectivity of this adduct is poor and yields were low; chlorination of the starting material was also noted.[62] However, by using the adduct **28**, generated *in situ* from the *N*-methylpyrrolidinone-POCl$_3$ adduct **29** and hydrogen peroxide, the yield and selectivity of the reaction was markedly improved.[63]

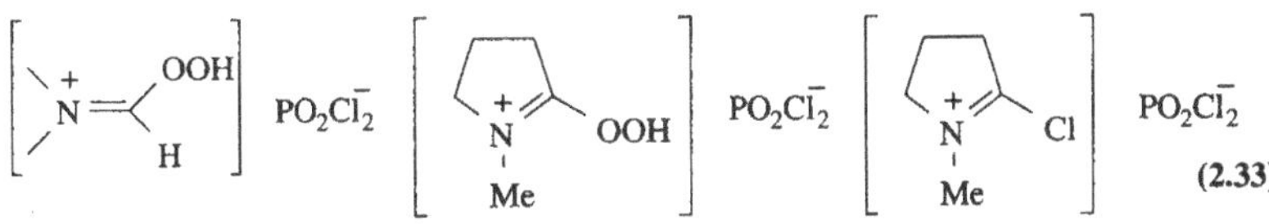

27   28   29

(2.33)

Epoxidation occurs at the most substituted carbon-carbon double bond; sensitive groups, *e.g.* acetals are not cleaved. No Baeyer-Villiger reaction was noted with cyclohexanone (scheme 2.34).

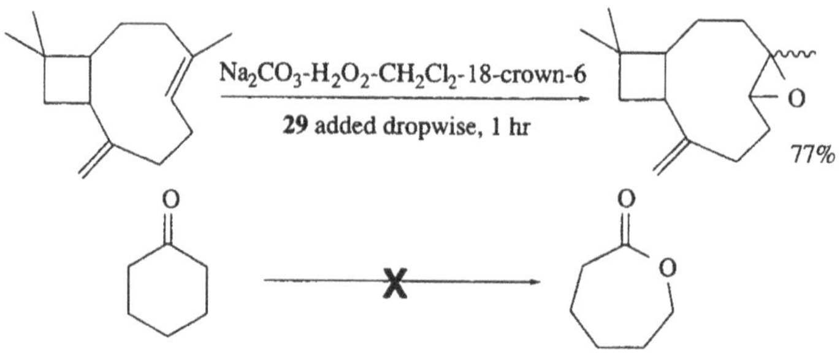

$$(2.34)$$

The mechanism and scope of such epoxidations is similar to *m*-CPBA.

2.8 Aldehydes

2.8.1 Aliphatic Aldehydes

A direct and chemoselective conversion of carboxylic acids into aldehydes is achieved by *O*-alkylation with a chloroiminium salt followed by reduction.[64,65] The reaction is widely applicable; α,β-unsaturated aldehydes result from the corresponding carboxylic acids, and aryl- and heteroaryl-aldehydes can also be prepared, in good yields. Other functionality is tolerated (scheme 2.35).

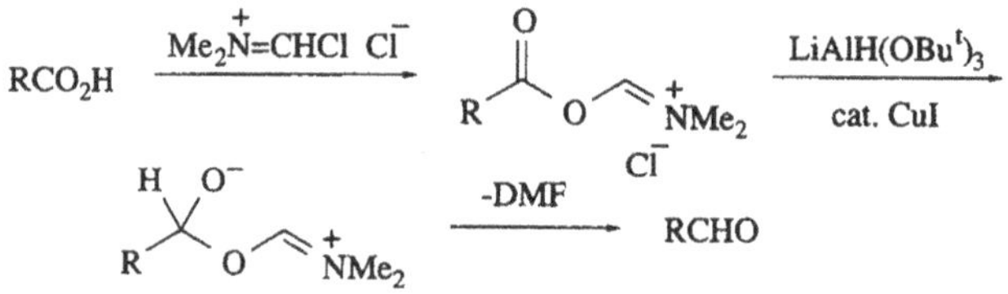

$$(2.35)$$

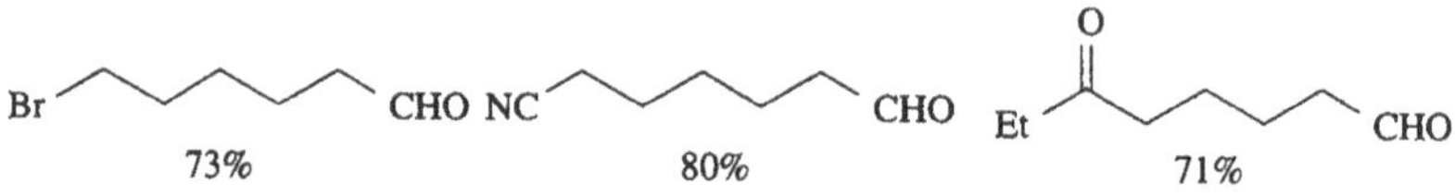

An alternative to the traditional condensation of carboxylic esters with alkyl formates to give α-formyl esters is a Vilsmeier formylation of the *O*-silylated enolate of the ester.[66] The procedure is satisfactory for short chain aliphatic α-formyl esters.

$$(2.36)$$

A two-step procedure involves conversion of a cycloalkanone into the corresponding β-chlorovinylaldehyde, followed by dissolving metal reduction of the latter to give an anion which may be quenched with an alkyl iodide to give an aldehyde containing no α-hydrogen atoms.[67,68]

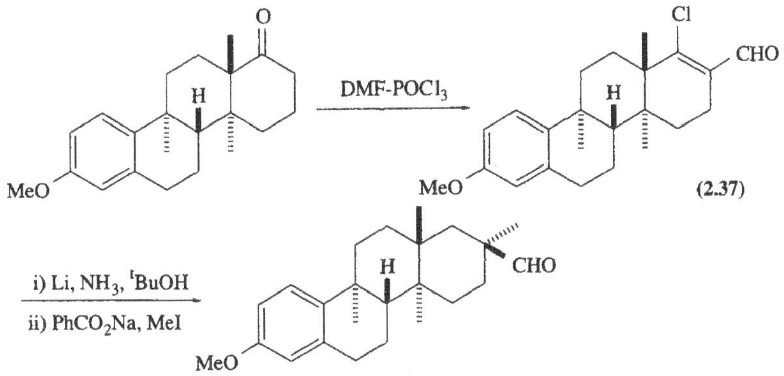

(2.37)

i) Li, NH$_3$, ^{t}BuOH

ii) PhCO$_2$Na, MeI

The monoacetals of β-ketoaldehydes can be prepared *via* the β-chlorovinylaldehydes of the corresponding ketones (scheme 2.38).[69]

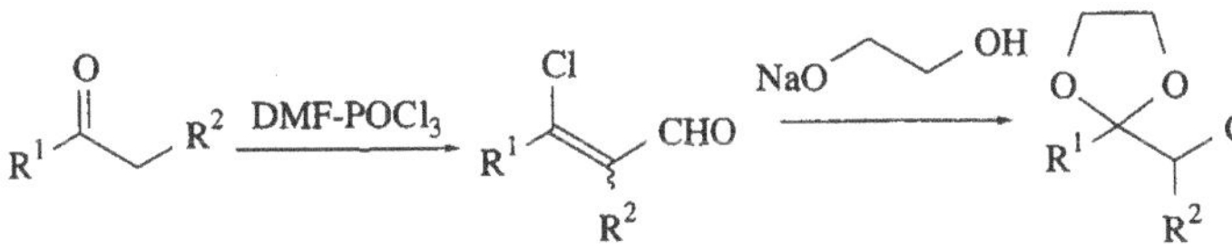

(2.38)

2.8.2 α,β-Unsaturated Aldehydes

2.8.2.1 α,β-Unsaturated Aldehydes with no Functional Group at the β-Position

From Ketones

A two-step procedure of reasonable generality involves the conversion of a ketone into a β-chlorovinylaldehyde, and hydrodechlorination of the latter (scheme 2.39). The procedure has been shown to be effective for a variety of cycloalkanes (n=1 to 4).[70] A similar conversion may be achieved using zinc dust in ethanol.[71]

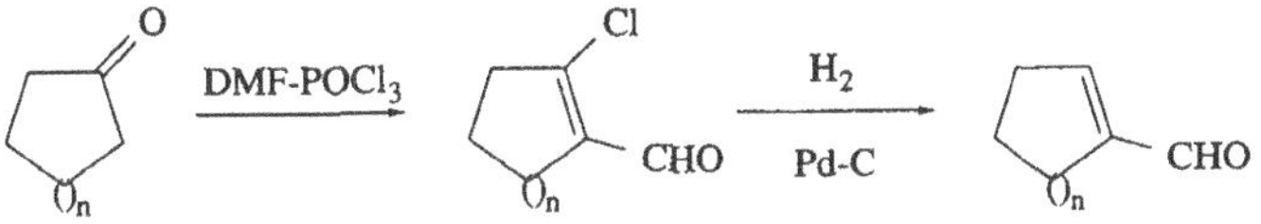

(2.39)

Direct formylation of trisubstituted alkenes $R^1R^2C=CHR^3$ is frequently possible using a disubstituted formamide-POCl$_3$ reagent.[72,73]

By Formylation of Alkenes

There are relatively few examples of efficient monoformylation of alkenic double bonds, but an interesting paper by Katritzky and co-workers[74] shows that *N*-formylmorpholine-POCl$_3$ reacts with a number of alkenic compounds in this way, whereas DMF-POCl$_3$ and *N*-methylformanilide-POCl$_3$ complexes have generally failed to effect monoformylation, with the exception of exocyclic methylene groups.[75,76] Thus, 3,3-dimethylbut-1-ene affords enal **31** (80%) and cyclohexene gives the corresponding aldehyde **32** (35%). Both alkenes **33** and **34** afforded an (*E*)/(*Z*) mixture of enals **35a** and **35b**; the acidity of the reaction medium provides an equilibrium concentration of alkene **33** from alkene **34**. Norbornene, however, did not give the corresponding enal, but instead a complex product derived by iminoalkylation, a [1,5]-hydride shift, and subsequent hetero Diels-Alder reaction.[74]

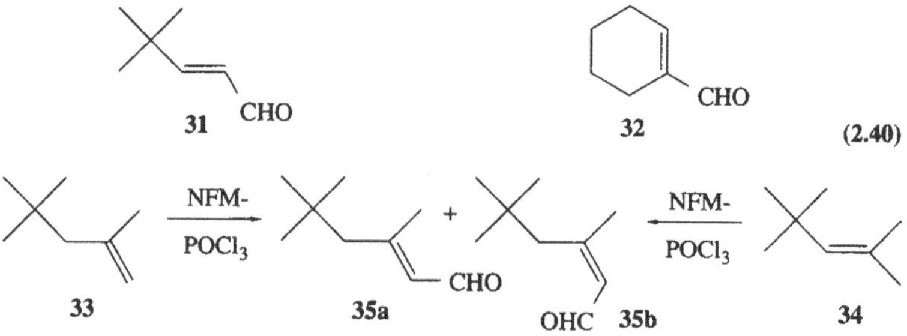

(2.40)

An extensive study of the reaction of Vilsmeier reagents with vinylcyclopropanes showed that in many cases, simple monoformylation of the alkenic double bond can occur, leaving intact the cyclopropane ring;[77] enals **36** and **37** are representative. However, more complex products can arise.

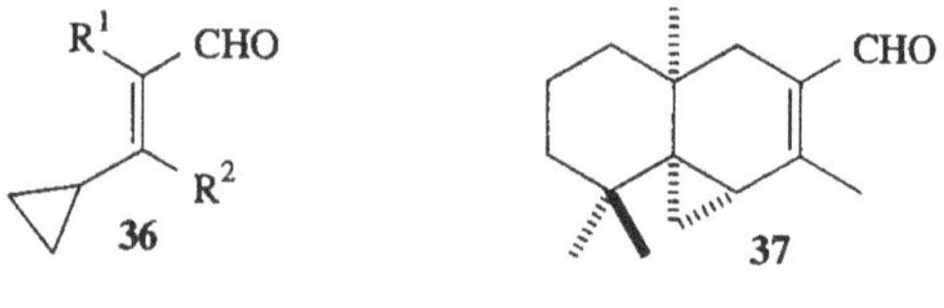

(2.41)

In a study of the formylation of conjugated acyclic trienes using DMF-POCl$_3$, mechanistic studies for products were provided. Trienes are usually formylated at C-1, but the relative stability of the carbocations determines the products (scheme 2.42). Yields of polyenals are lowered by competing cyclization of polyunsaturated enamines to benzene rings.[78]

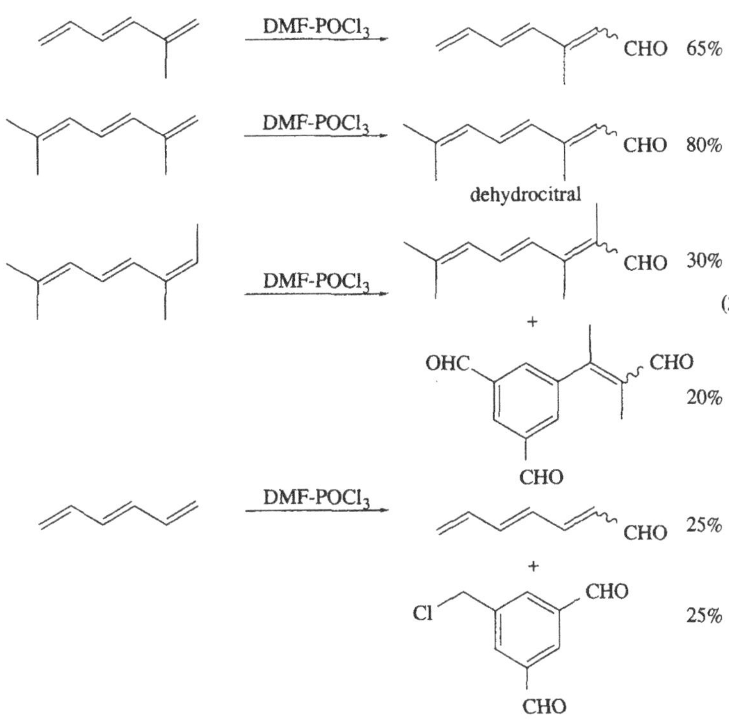

A simple situation arises for alkenes such as camphene (scheme 2.43) that do not possess a hydrogen atom in the allylic position that can be lost as a proton; an alkyliminium salt is formed that can often be isolated (*e.g.* **38** (69%) by the addition of perchloric acid). Alkaline hydrolysis of the alkyliminium salt leads to an α,β-unsaturated aldehyde in the example below.[75]

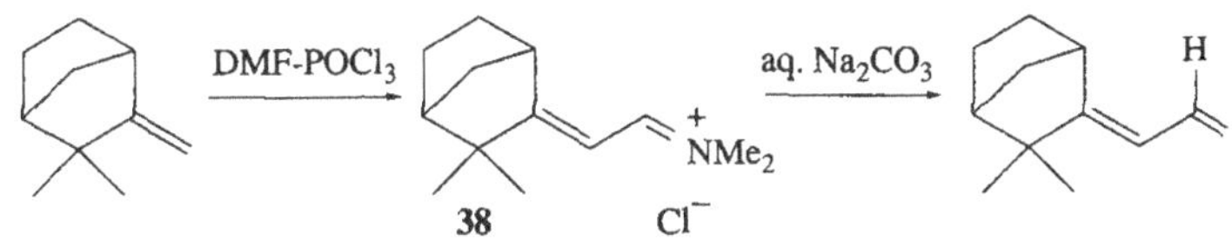

Indene **39** is an example of where the corresponding α,β-unsaturated aldehyde **41** is formed at ambient temperatures, but where iminoalkylation proceeds further at 80°C, in the presence of excess Vilsmeier reagent (scheme 2.44). Deprotonation of **40** to give the unusual intermediate **42**, 2,3-benzo-6-dimethylaminofulvene, presumably occurs, followed by electrophilic attack at the unsubstituted and electron-rich positions of the fulvene ring. Work-up with aqueous perchloric acid affords **43** (66%).[79]

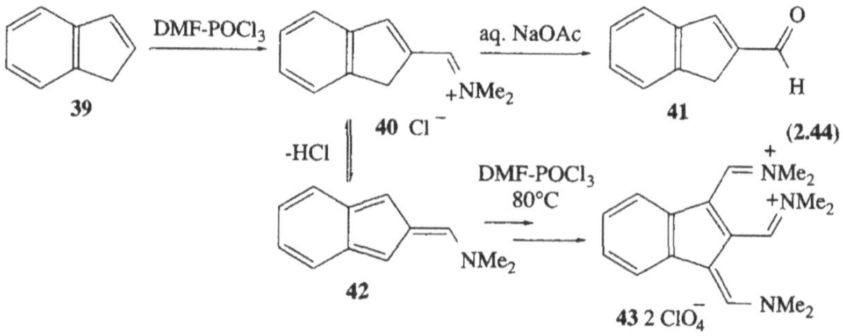

(2.44)

An interesting and regioselective formylation of limonene using DMF-POCl$_3$ affords the α,β-unsaturated aldehyde **44** which with CH$_2$=C(CH$_3$)CH$_2$MgCl was converted into atlantone.[80]

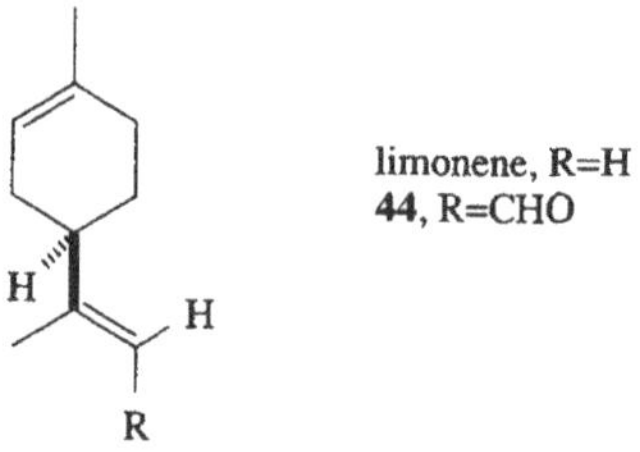

limonene, R=H
44, R=CHO   (2.45)

Cyclopentadiene

The general sequence of iminoalkylation, deprotonation and further iminoalkylation is followed when cyclopentadiene **45** reacts with Vilsmeier reagents. Thus, although 6-dimethylaminofulvene **46** is evidently an intermediate (which has been isolated from other reactions), it reacts further, depending upon the reaction conditions,[81-86] to give the useful unsaturated aldehydes **47-51**. The intermediate iminium salts can also be isolated, especially as the perchlorates.[87,88]

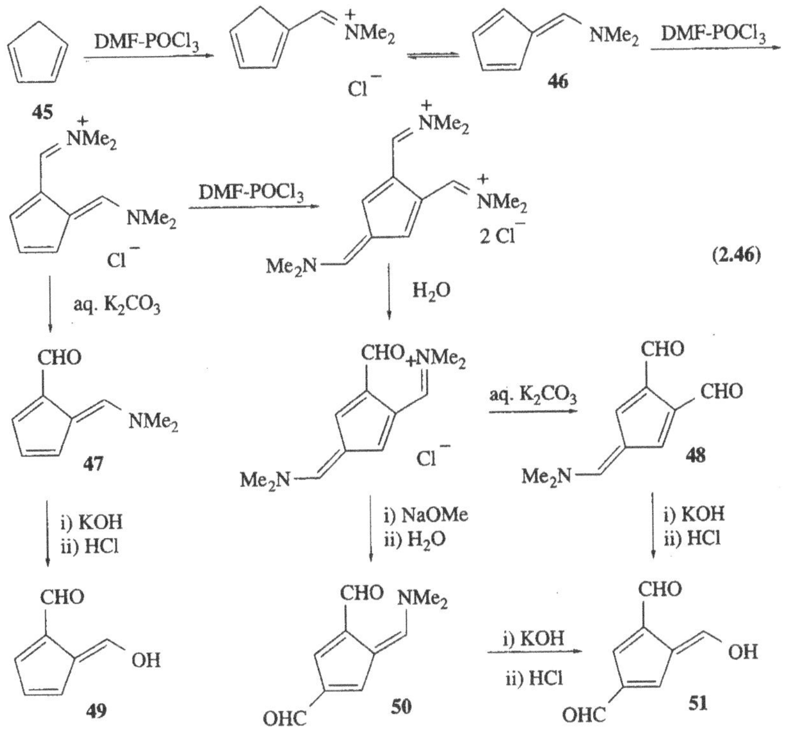

(2.46)

Fulvenes

The 1-position of the fulvenes **52a** and **52b** is attacked to give, after hydrolytic work-up, the corresponding 1-formylfulvenes **53a** and **53b**.[82-86]

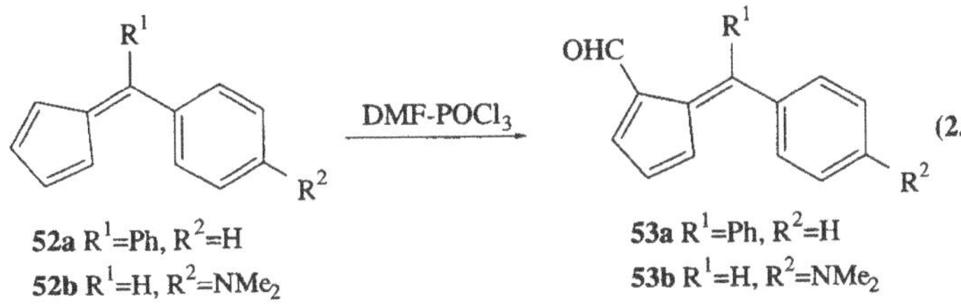

(2.47)

52a R^1=Ph, R^2=H
52b R^1=H, R^2=NMe$_2$

53a R^1=Ph, R^2=H
53b R^1=H, R^2=NMe$_2$

Cycloheptatriene

1-Formylcycloheptatriene has been obtained in yields not greater than 30% by the action of Vilsmeier reagent upon cycloheptatriene.[89]

Steroidal Alkenes

The location of an exocyclic methylene group on a steroidal framework determines the extent of reaction with Vilsmeier reagents. Thus,

17-methylene-5α-androstan-3β-ol acetate reacts with DMF-POCl$_3$ (14 days, 20°C) to give the aldehyde **54** in moderate yield. However, 3-methylene-5α-androstan-17β-ol acetate **55** did not react with Vilsmeier reagents.[90]

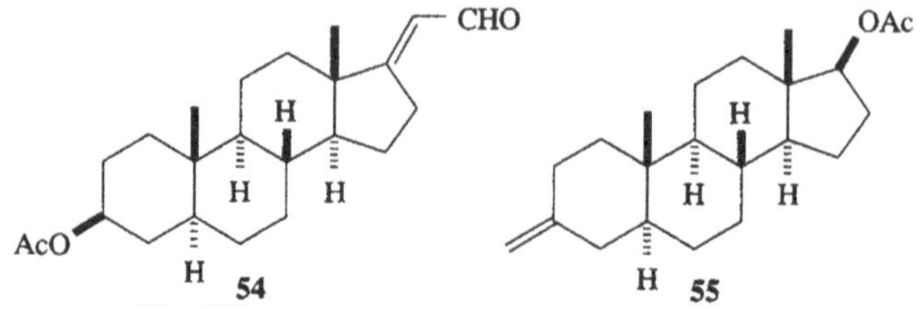

(2.48)

Steroidal dienes that can isomerize to give one exocyclic methylene group such as 3-methyl-3,5-androstadiene-17β-ol acetate **56**, can show differing, but regioselective, formylation depending upon the reaction temperature. Thus diene **56** affords only aldehyde **59** (as an (*E*)/(*Z*) mixture) at 20°C, presumably *via* the acid-catalyzed isomerization of **56** into **58**. At higher temperatures, attack at the 6-position is evidently faster than the rate of formation of **58**, and the 6-formyl derivative **57** is obtained as the major product.[90]

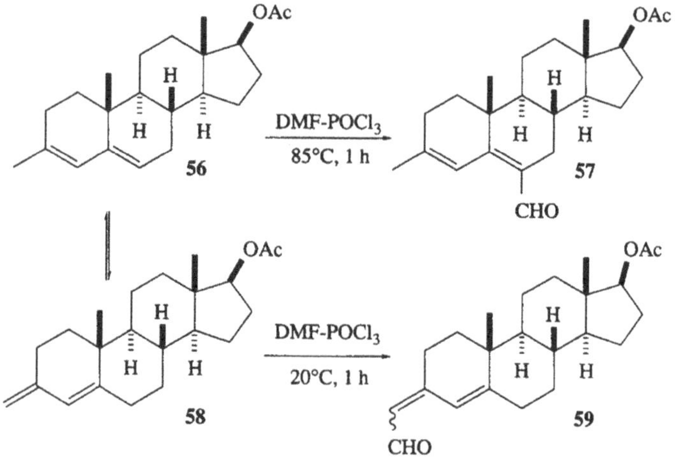

(2.49)

A combination of steric and electronic factors influence the site for formation of conjugated steroidal dienes and trienes. 3,5-Androstadiene-17β-ol acetate reacts exclusively at the 3-position to give aldehyde **61** together with its isomer **62**.[90]

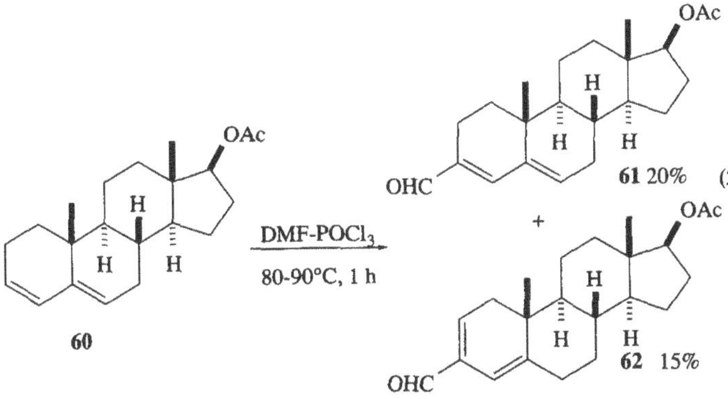

$$61 \ 20\% \quad (2.50)$$

$$62 \ 15\%$$

2,4,6-Androstatriene-17β-ol benzoate **63** undergoes formylation at the 2-position; the almost equally reactive 7-position is not attacked, being more sterically hindered. 2-Formyl-6-methyl-3,5-pregnadiene-17α-ol-2-one acetate **65** was formed from the alcohol **64**, presumably by dehydration of the alcohol and subsequent isomerisation to the 6-methyl-2,4,6-triene prior to iminoalkylation.[90]

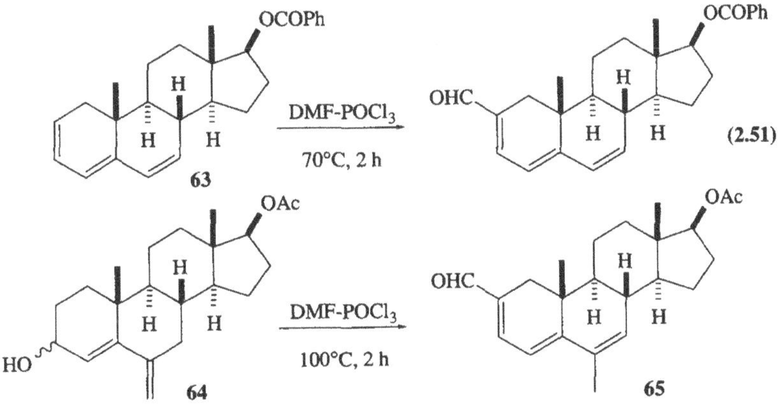

$$(2.51)$$

Other steroidal alkenes have been formylated using DMF-COCl$_2$.[91]

Cinnamaldehyde can be prepared in excellent yield from styrene by formylation using a 3:1 mixture of DMF-POCl$_3$.[92] Styrene-divinylbenzene copolymers have been formylated by a Vilsmeier reaction, where other methods failed.[93]

Treatment of **66** with DMF-POCl$_3$ in pyridine afforded only aldehyde **67** when the reaction was conducted at 40°C for 15 min. Heating the reaction at 60°C for 40 min afforded a mixture of **67** and **68** (40 and 35% respectively). The *cis*-relation of the aryl substituents was confirmed by ^{1}H and ^{13}C NMR studies.[94]

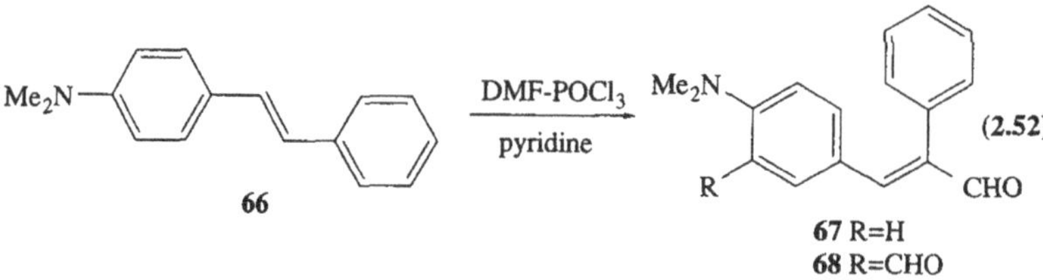

(2.52)

66

67 R=H
68 R=CHO

Formylation of Benzylic Alcohols

Benzylic and allylic alcohols are dehydrated by DMF-POCl$_3$ (scheme 2.53), presumably with formation of the alkene which then undergoes formylation to give α,β-unsaturated aldehydes.[95]

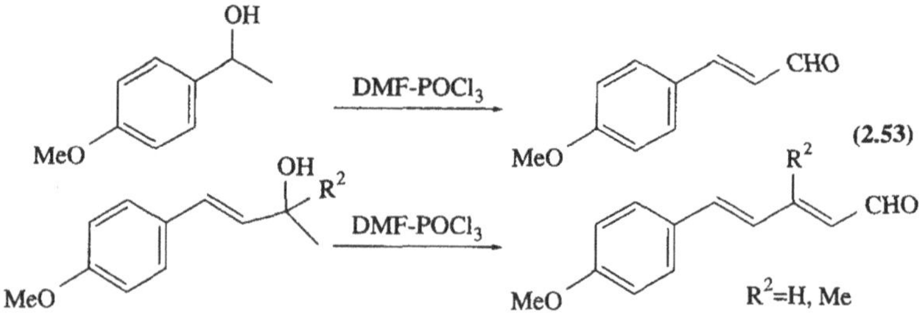

(2.53)

The alcohol **69** was converted into the conjugated α,β-unsaturated aldehyde **70** in 73% yield.[96]

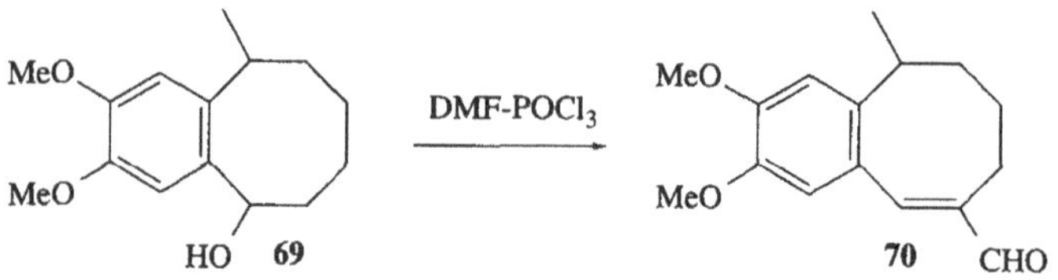

(2.54)

Substituted α-tetralols usually react similarly to the alcohols above, although the site at which formylation occurs is evidently finely balanced.[97-100] The location of a p-methoxy group may have a profound effect upon the regioselectivity of the reaction, owing to activation of the double bond when para to it (scheme 2.55).[101] The α-tetralols may be subsequently aromatized using DDQ.[101,102] Such a procedure has been employed in the synthesis of a substituted benz[a]anthracene.[102]

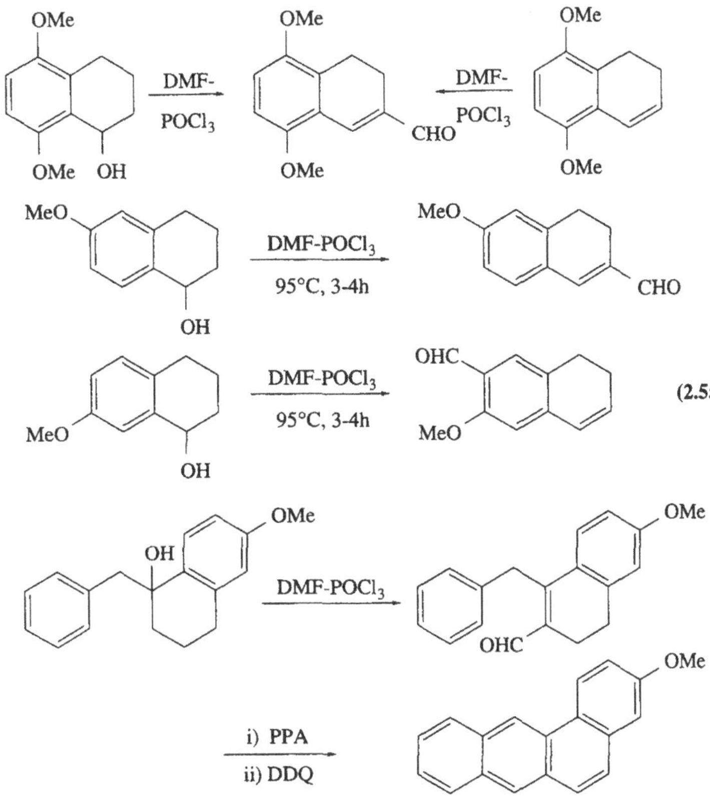

Formylation of a double bond as part of a heterocyclic ring is often efficient, and the site of reaction is that of greatest electron density, which can be determined by the influence of the heteroatom as for **71** and **72**. Figure 2.56 illustrates the sites of formylation (R=H to R=CHO). For systems **71** and **72**, in which the nitrogen atom activates the β-position of the alkene, formylation can be effected at 0-5°C, using DMF-POCl₃.[103,104] The aldehydes **73** were prepared from the corresponding 2*H*-chromenes.[105] The aldehyde **74** (63%) was obtained by formylation of the corresponding dihydropyran, ancubin hexacetate with DMF-POCl₃.[106] By using [¹³CHO]-DMF a related [11-¹³C]-iridotriglucoside was prepared. A number of C-2-formyl glycals including **75** (80%) has been similarly prepared.[107] *N-O*-Nitrobenzoyldihydropyrrole derivatives undergo similar formylation to **71** when treated with Vilsmeier reagents.[108]

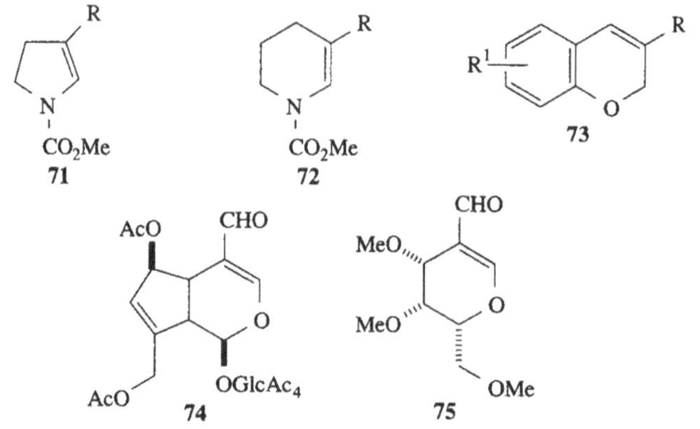

71   72   73   (2.56)

74   75

Vinylogous Formylations

Activated aromatic and heteroaromatic rings undergo conversions into the (E)-α,β-unsaturated aldehydes upon treatment with a vinylogous Vilsmeier reagent (scheme 2.57).[109]   Typical substrates include *N,N*-dimethylaniline; pyrrole and 1-alkyl pyrroles react at the 2-position.

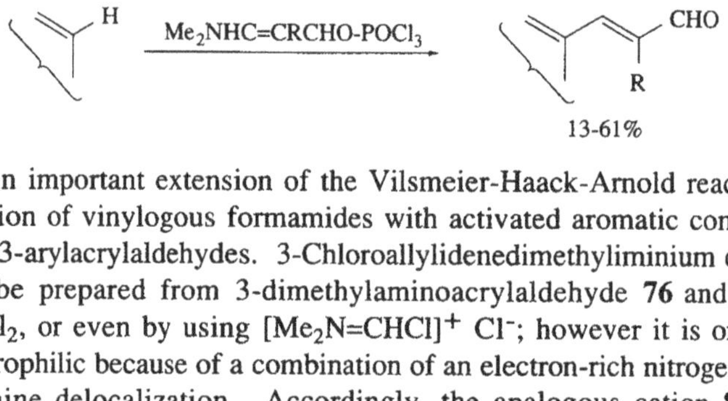

$$\text{Me}_2\text{NHC=CRCHO-POCl}_3$$

(2.57)

13-61%

An important extension of the Vilsmeier-Haack-Arnold reaction is the reaction of vinylogous formamides with activated aromatic compounds to give 3-arylacrylaldehydes. 3-Chloroallylidenedimethyliminium chloride **79** can be prepared from 3-dimethylaminoacrylaldehyde **76** and POCl₃ or COCl₂, or even by using [Me₂N=CHCl]⁺ Cl⁻; however it is only weakly electrophilic because of a combination of an electron-rich nitrogen atom and enamine delocalization. Accordingly, the analogous cation **80** is more electrophilic; it can be prepared from 3-dimethylanilinoacrylaldehyde **77** and POCl₃. The next highest vinylogue **81**, preparable from 5-methyl-anilinopenta-2,4-dienal **78** (Zincke aldehyde) and POCl₃, also reacts with activated arenes, giving 5-arylpenta-2,4-dienals.

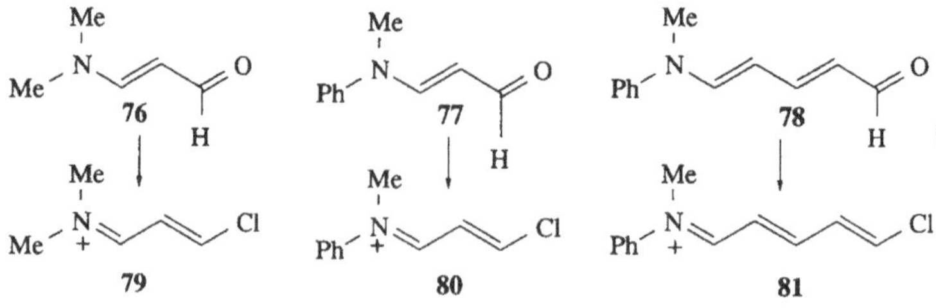

76   77   78   (2.58)

79   80   81

N,N-Dimethylaniline reacts with **76**-POCl$_3$ to give **82a** (R=Me, 70%);[110,111] alternatively **76** and N$_3$P$_3$Cl$_6$ or **76** and PhCOBr may be used, but the yields of **82a** are lower (54[112] and 15%[112-116] respectively). The aldehyde **82b** can be prepared in 84% yield from *N,N*-diethylaniline and the reagent **76**-POCl$_3$.[110,111] The same reagent affords **85** (85%) from 1-phenyl-1-(4′-dimethylaminophenyl)ethene;[110,111] **84** (90%) from rescorcin dimethyl ether;[110,111] and **86** (97%) from azulene.[110,111] The reagent Zincke aldehyde-POCl$_3$ converts *N,N*-dimethylaniline into **83** (18%), and azulene into **87** (90%).[110,111] Other azulenepolyenals have been similarly obtained in excellent yields.[82-86,110,111]

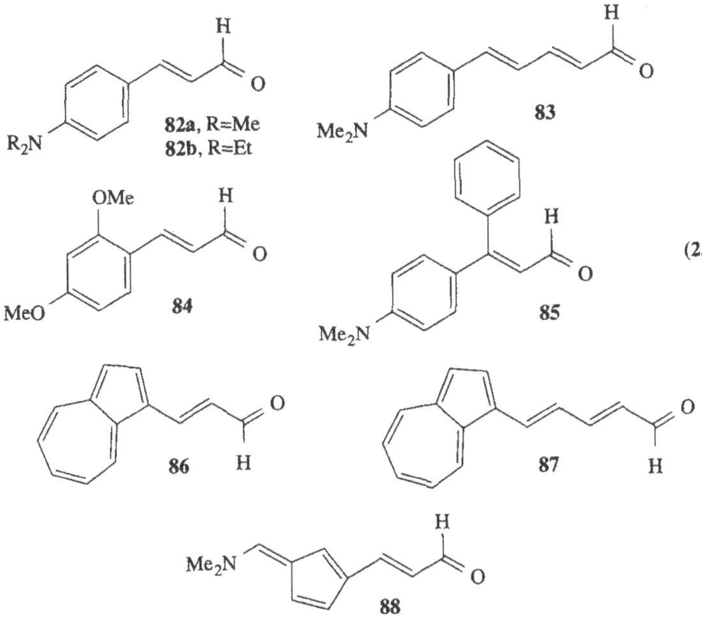

(2.59)

The extensively conjugated system **88** has been obtained from 6-dimethylaminofulvene and the reagent **76**-POCl$_3$.[82-86] The aldehydes **77** and **78** have also been used with either acetic anhydride[110,111,117,118] or acyl bromides.[113-115] In all cases, intensely coloured polyeneiminium salts are the first isolable products.[82-86,110,111]

2.8.2.2 β-Hydroxy- and β-Alkoxy-α,β-Unsaturated Aldehydes

If the substrate functionality or structure does not induce an alternative pathway, vinyl ethers undergo iminoalkylation at the β-position, and 1-alkoxy-1,3-dienes at the 4-position. After hydrolysis, the products are usually alkoxy unsaturated aldehydes. Examples include the pyridinone **89**[119] and the alkoxydiene **90**.[120] Many functional groups are tolerated with the exception of certain epoxides and hydroxy groups.

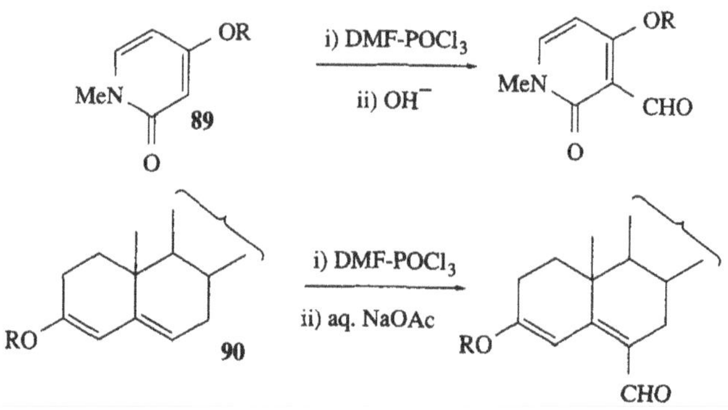

(2.60)

β-Chlorovinylaldehydes have been converted into the corresponding β–alkoxyvinylaldehydes by reaction with an alcohol under basic conditions. These conversions include acyclic β-chlorovinylaldehydes,[121] chloroformyl-steroids[122] and chloroformylcodeine derivatives.[123]

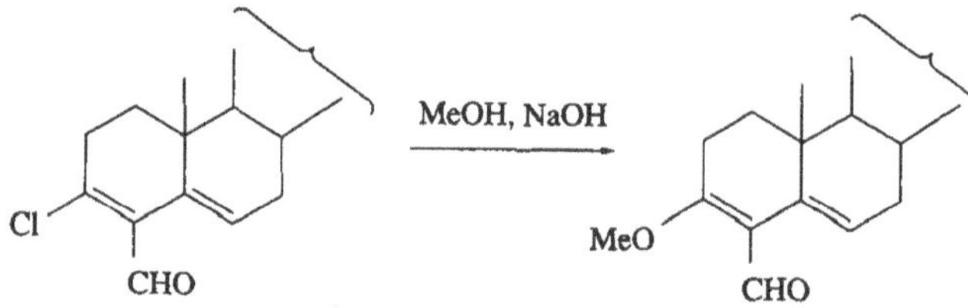

(2.61)

Methyl groups of electron-deficient heteroaromatic compounds undergo double iminoalkylation. 4-Methylpyridine reacts with DMF-POCl$_3$ to give the aldehyde **91** after alkaline work-up;[124] *N*-alkyl-4-picolinium salts react faster than the corresponding pyridines.

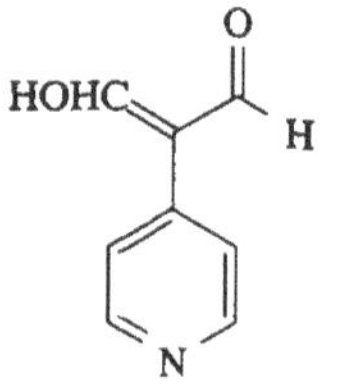

(2.62)

91

A general mechanism consistent with these observations that also applies to 4-methylpyrimidine (scheme 2.63) has been proposed by Bredereck and his group.[125]

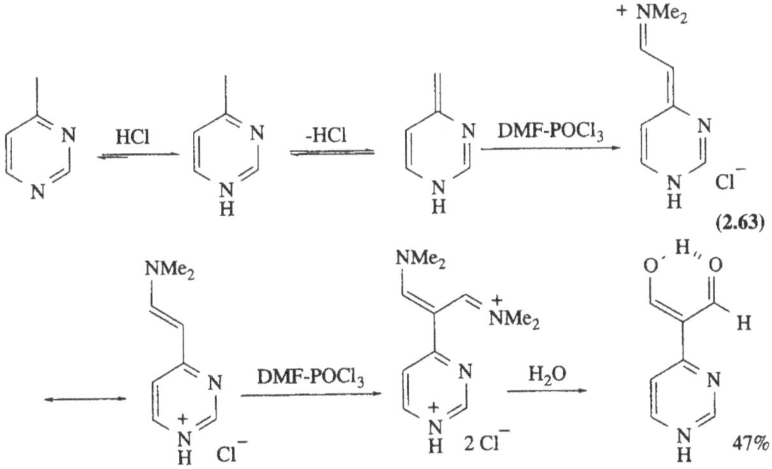

(2.63)

Malondialdehyde derivatives have been similarly obtained by reacting DMF-POCl$_3$ with the following heterocycles: 4-methyl-2-phenylpyrimidine and 2-methyl-4-phenylpyrimidine;[126] 2-methylbenzoxazoles;[127,128] 2-methylbenzothiazoles;[128] 2-methylselenazole;[128] and quinaldine and lepidene.[128] The action of Vilsmeier reagents on 2-hydroxyacetophenone oximes affords 2-methyloxazoles by Beckmann rearrangement, so the former can be converted in one pot into the malondialdehydes of 2-methyloxazoles.[129] Mixtures of 4-dimethylamino-2-methylpyrimidin-2-malondialdehyde and 4-dimethylamino-2-methylpyrimidin-6-malondialdehyde are obtained from 2,6-dimethyl-4-pyrimidinone and 4-chloro-2,6-dimethylpyrimidine and Vilsmeier reagents.[126]

4-Quinazolone-2-malondialdehyde is obtained from 2-methyl-3-phenyl-4-quinazolone and DMF-POCl$_3$. That Vilsmeier reagent also converts 6-methylpurine into 3-dimethylamino-2-(6-purinyl)allylidenedimethyliminium chloride; alkaline hydrolysis affords purine-6-malonaldehyde (82%).[130] However, no identifiable products were obtained by treating purine, hypoxanthine, adenine, 8-methyladenine, guanine, or guanosine with DMF-POCl$_3$.[130]

In contrast to quinaldine, 2-methylpyridine gave no isolable formylated product.[124] However, pyridine-2-ethyl acetate and pyridine-2-acetonitrile did react with DMF-POCl$_3$, although the corresponding dimethyl dimethylaminovinyl derivative was obtained.[124]

The interesting compound 'tetraformylethene' **93** can be prepared by double iminoalkylation of 1,4-bis(dimethylamino)-1,3-butadiene **92**, followed by stepwise hydrolysis (scheme 2.64).[131]

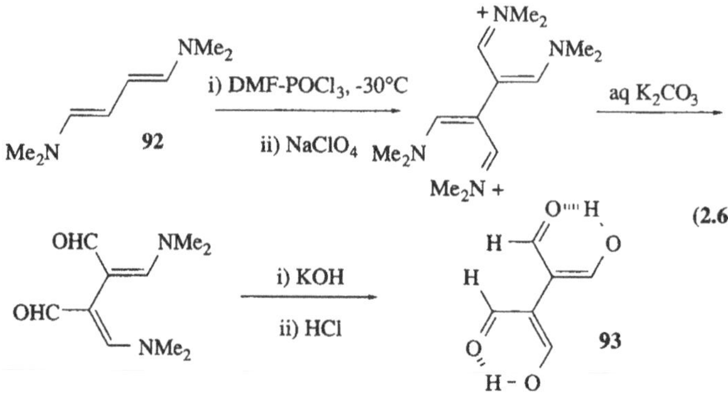

(2.64)

Reaction of several enol ethers of 14-hydroxydihydrodeoxycodeinone with DMF-POCl$_3$ in dichloroethane afforded 7-formylated products; the 14-hydroxy group also underwent O-formylation, but the formate group could be hydrolyzed to give **94b**.[123]

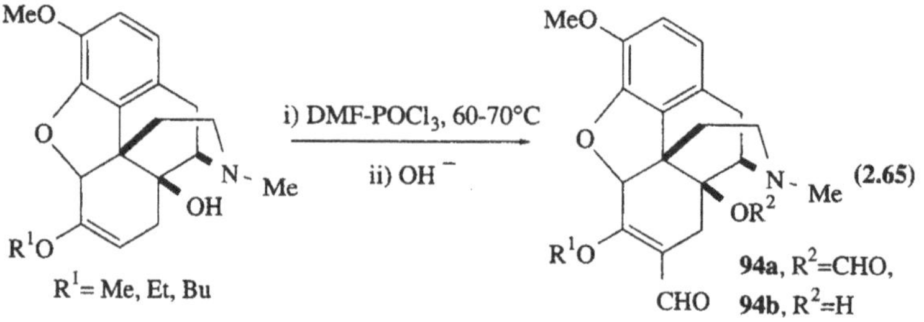

(2.65)

By conversion of aldehydes **94a** into the corresponding oximes, semicarbazones, hydrazones, and anils, those were cyclized to give the corresponding derivatives of oxazole, pyrazole, and quinoline.

Isopropyl methyl ketone reacts with Vilsmeier reagents to give the cation **95** which by deprotonation gives a dienamine that has only one site that can be further iminoalkylated, the δ-position being doubly substituted. The final compound is isobutyrylmalondialdehyde **97** (scheme 2.66). The sequence is limited by the poor yields of **96**.[132,133]

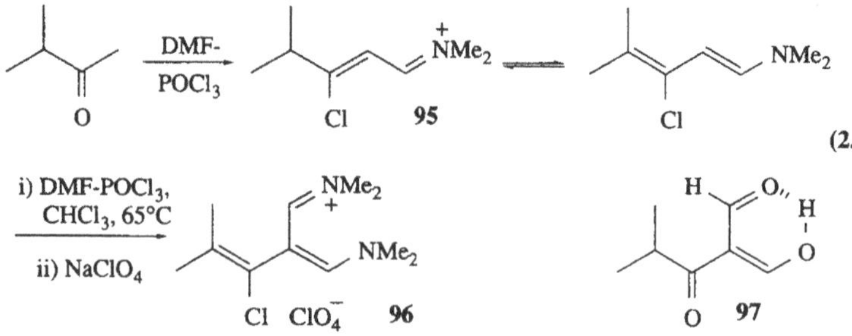

(2.66)

Triformylmethane has been prepared by reaction of β-dimethylamino-acrylaldehyde with DMF-COCl$_2$ in chloroform.[134]

Aryl malondialdehydes **99** can be obtained by the iminoalkylation of the trimethinium salts **98**, followed by alkaline hydrolysis. The aldehydes **99** (R=H, 84%[134]), (R=^{t}Bu, 80%[132]), (R=Ph, 85%[132]) have been so prepared. The procedure can be applied to activated aromatic rings, although formylation also occurs on the aromatic rings.[132]

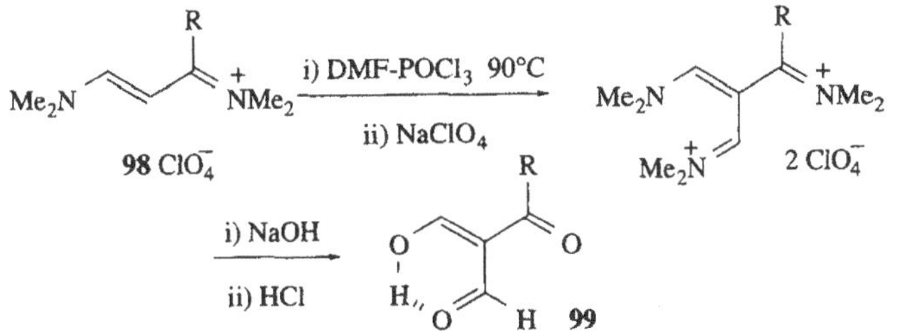

$$(2.67)$$

A related sequence can be performed on 3,4-dimethoxyacetophenone; the iminium salt **100** is further iminoalkylated, and the product of hydrolysis, **101**, can be indirectly hydrolyzed to give the acyl malondialdehyde **102**.[135]

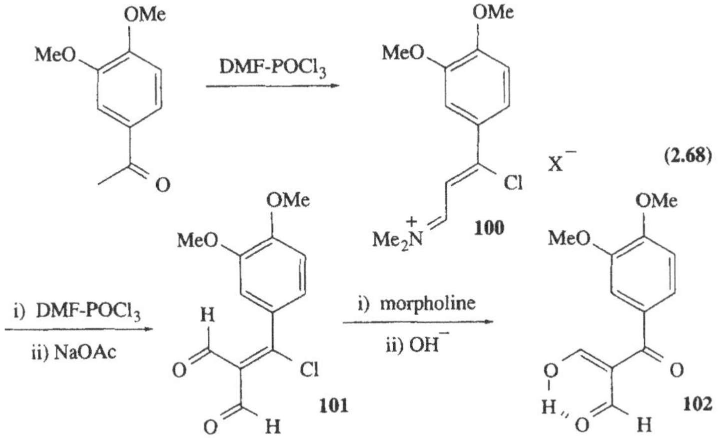

$$(2.68)$$

Malondialdehydes having a ketoaryl group at the 2-position can be prepared by analogous reactions (scheme 2.69).

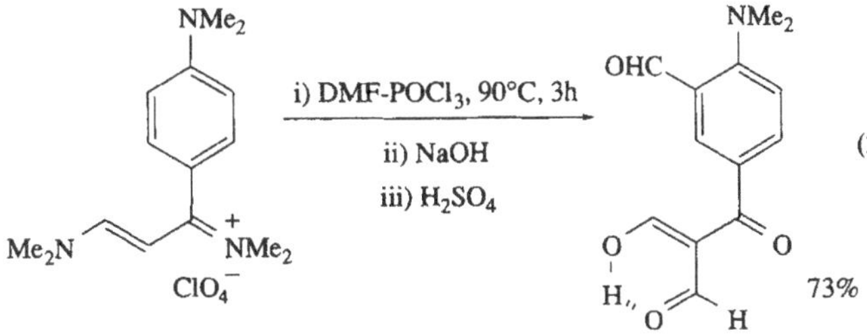

$$(2.69)$$

73%

The alkoxy malondialdehydes **103**, ethers of triose reductone, can be prepared by reaction of 1,2-dialkoxyethenes with *N*-methylformanilide-$POCl_3$ to give the corresponding β-methylanilinoacrylaldehydes that are then hydrolyzed (scheme 2.70).[136]

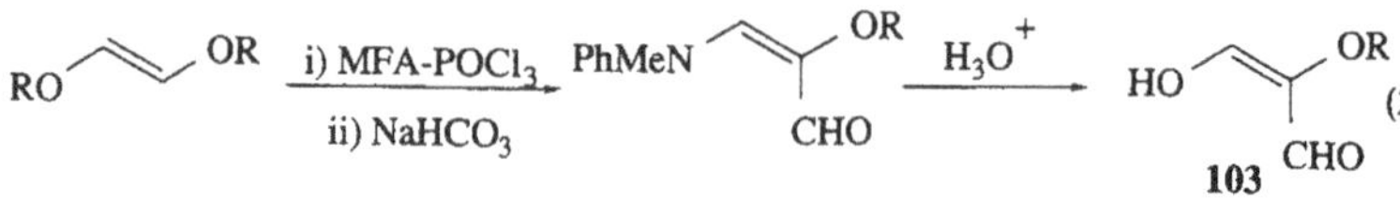

Ketals react with Vilsmeier reagents in dichloroethane at 40°C, in the same manner as do acetals.[137] The ethoxyiminium salts **104** can be reacted with Me_2NH to give trimethinium salts, from which β-(dialkylamino)acrylaldehydes can be obtained by further nucleophilic substitution. However, hydrolysis of salts **104** with alkali affords good yields of the β-alkoxyacrylaldehydes **105** (R^1=Me, R^2=Ph 92%; R^1=H, R^2=^{t}Bu 82%; R^1= R^2=(CH$_2$)$_4$ 59%).

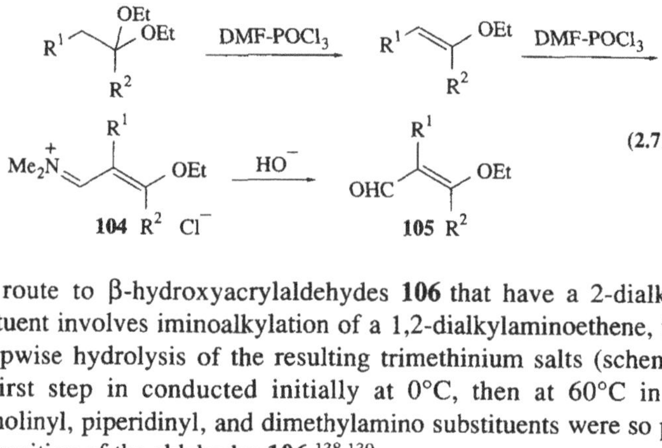

A route to β-hydroxyacrylaldehydes **106** that have a 2-dialkylamino substituent involves iminoalkylation of a 1,2-dialkylaminoethene, followed by stepwise hydrolysis of the resulting trimethinium salts (scheme 2.72). The first step in conducted initially at 0°C, then at 60°C in $CHCl_3$. Morpholinyl, piperidinyl, and dimethylamino substituents were so placed at the 2-position of the aldehydes **106**.[138,139]

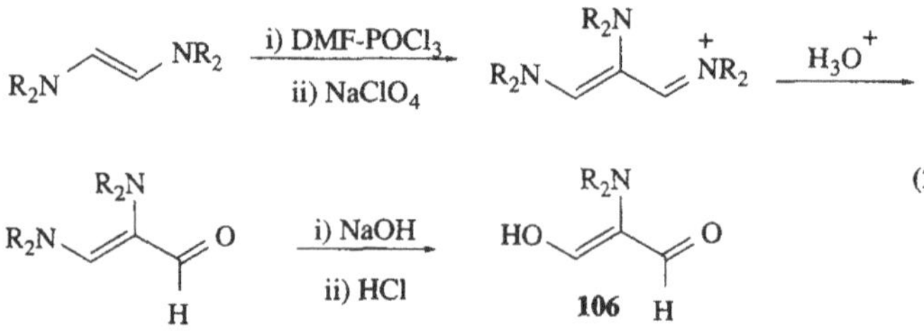

Reaction of the morpholinyl enamines **107** with *N*-formylmorpholine-oxalyl chloride, followed by hydrolysis of the vinamidinium chlorides **108**, afforded malondialdehydes **109** in yields of 42-66%.[140]

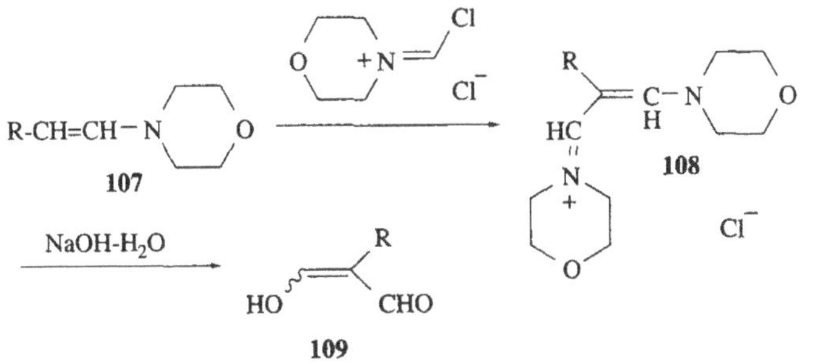

$$(2.73)$$

R=Ph$_3$C, phenyl, 2-thienyl, *t*-butyl, adamantyl

Aliphatic dialdehydes, upon reaction with Vilsmeier reagents, hydrolysis of the intermediates, and reaction with arylamines, lead to malondialdehyde anils.[141]

2.8.2.3 α,β-Unsaturated Aldehydes with Other Functional Groups at the β-Position

The displacement of a halide from the β-position of β-halovinylaldehydes provides a useful route to a variety of β-substituted α,β-unsaturated aldehydes. Thus, cyclic β-chlorovinylaldehydes react with sodium azide to give β-azidovinylaldehydes; those are reduced by hydrogen sulfide in methanol to the corresponding β-aminoaldehydes.[142]

Arylhydrazones undergo enaminic reactions with DMF-POCl$_3$ to give 1,4,5-triaza-1,3-pentadienium salts **110** which can be isolated and hydrolyzed to give the hydrazones **111**. The latter undergo acid catalyzed rearrangements to the hydrazones **112**.[143]

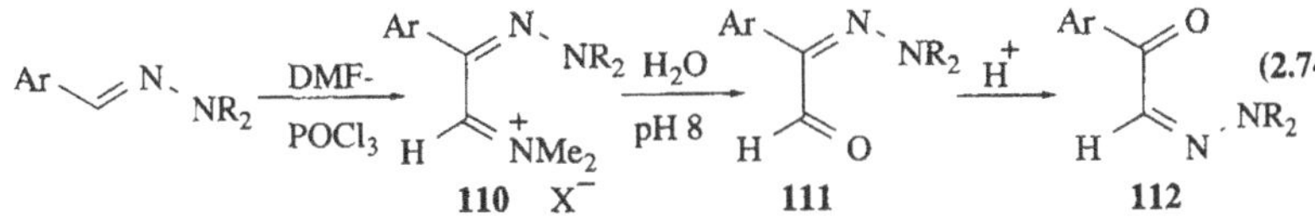

$$(2.74)$$

Reaction of alkali metal sulfides with β-chlorovinylaldehydes in methanolic solution has been used to prepare a series of β-formylvinylthiols (scheme 2.75).[144-148]

$$Cl\text{-}CR\text{:}CR\text{-}CHO \xrightarrow{Na_2S} NaS\text{-}CR\text{:}CR\text{-}CHO \xrightarrow{H^+} HS\text{-}CR\text{:}CR\text{-}CHO$$

$$(2.75)$$

β-Mercaptovinylaldehydes can be prepared by reacting chloroalkanes and β-formylvinylthiols under basic conditions.[146]

β-Chlorovinylaldehydes react with ammonium thiocyanate below 40°C to afford β-thiocyanovinylaldehydes. Heating the β-thiocyanovinylaldehydes at temperatures above 40°C has enabled the synthesis of 1,2-thiazoles (isothiazoles, see section 4.4.1.9.).[149-151]

2.8.2.4 α,β-Unsaturated Aldehydes with a β-Amino Group (Vinylogous Formamides)

The literature concerning the reaction of aldehydes with Vilsmeier reagents is not extensive. Arnold and Zemlicka described the preparation of 2-(dimethylaminomethylene)butenal from DMF-POCl$_3$ and butanal,[152,153] Barton and co-workers described an improved procedure (51% yield) using 1,1-diethoxybutane.[154]

Displacement of the chloride atom from β-chlorovinylaldehydes allows the synthesis of β-aminovinylaldehydes (scheme 2.76). Klimko[155] reacted ethylamine, methylamine, and ammonia with α-alkyl- and α-aryl-β-chlorovinylaldehydes to give the β-aminovinylaldehydes. Arnold and Zemlicka[152] achieved similar transformations using secondary amines. However, they noted that at higher temperatures and pressures, the (dimethylaminomethylene)cyclobutanone was isolated; whereas at ambient temperatures and pressures, the β-aminovinylaldehyde was formed.

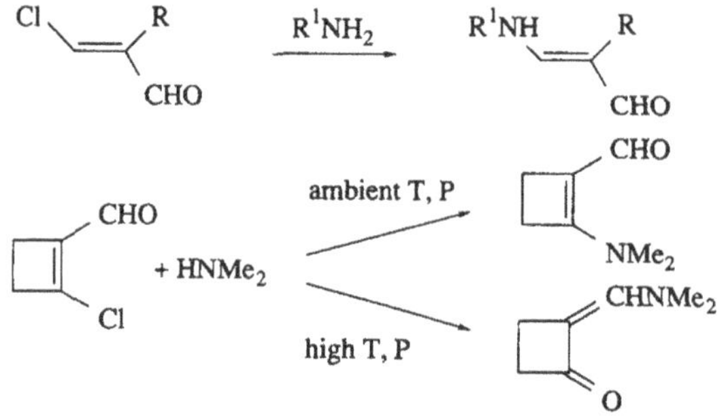

(2.76)

The action of DMF-POCl$_3$ on sodium trifluoroacetate, followed by addition of Et$_3$N, affords 2-fluoro-3-(dimethylamino)acrylaldehyde which can be converted into fluoromalondialdehyde.[156]

The acryaldehydes **114** were prepared by Vilsmeier reaction of the corresponding acetic acids **113**.[157]

(2.77)

Acetals, or vinyl ethers (into which Vilsmeier reagents convert acetals) react with DMF-POCl$_3$ to give alkoxyiminium salts such as **115** that are hydrolyzed by aqueous potassium carbonate to give β-(dimethylamino)-acrylaldehydes **116**. Further hydrolysis affords malondialdehydes (section 2.8.2.2). Arnold[153] was the first to report the formylation of acetals; later workers showed that vinyl ethers could also be used to the same effect.[158] The yields of **116** are good to excellent when 2.5 mole equivalents of Vilsmeier reagent is used with either the acetal or the vinyl ether.

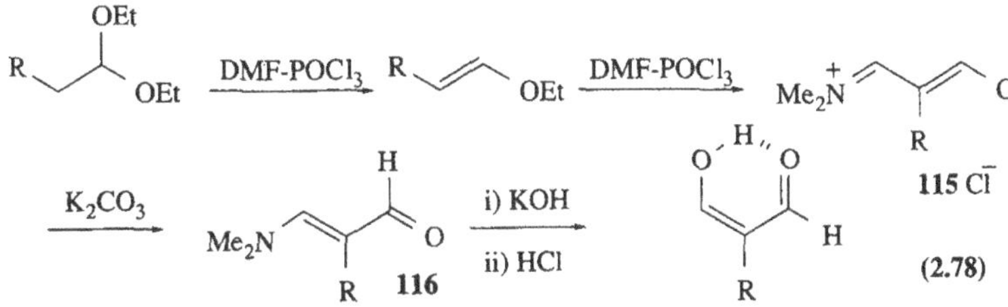

$$(2.78)$$

2-Acyl-3-(dimethylamino)acrylaldehydes **119** have been prepared by hydrolysis of the perchlorate salts **118** with aqueous potassium carbonate (*e.g.* R=NMe$_2$ in 76%[159]).

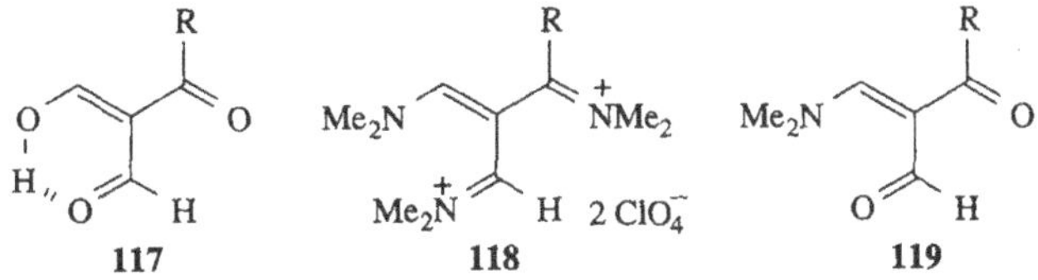

$$(2.79)$$

The salts **119** were prepared by iminoalkylation of the corresponding trimethinium perchlorates (section 2.29.3). The aldehydes **119** (R=^{t}Bu) and **119** (R=Ph) were obtained by reacting copper chelates of **117** with Me$_2$NH;[132] the preparation of aldehyde **119** from trimethinium salts is described in section 2.29.3. These and other procedures provide access to triformylmethane **119** (R=H).

Where the enaminic unit cannot undergo subsequent hydrolysis, for example because a nitrogen atom is part of a ring, enamines are formylated in the β-position, and dienamines at the δ-position, usually in high yields.[72,73,160-163] The enamides **121**, or their precursors **120** (cyclic tautomers of 2-acetylbenzamides that dehydrate to give **121**) undergo formylation to give the 3-formylmethylene derivatives **122** which can be condensed with active methylene compounds to give cyanine dyes.[164]

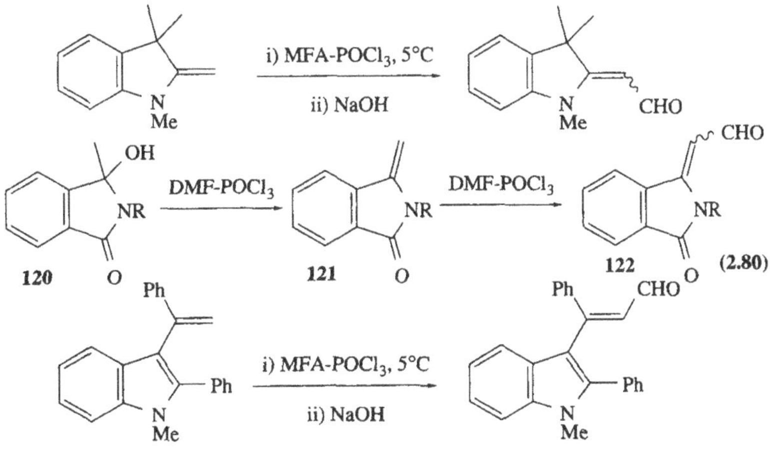

$$(2.80)$$

A 1,4-dihydropyridine that contains methyl groups at the 2- and 5-positions reacts with DMF-POCl$_3$ at one of the methyl groups to give a vinylogous formamide.[165]

3-Methylquinoxalinones afford the dimethylaminoacrylaldehydes **123**. Hydrolysis of the dimethylamino group with aqueous sodium hydroxide gave malondialdehydes (section 2.8.2.2).[166]

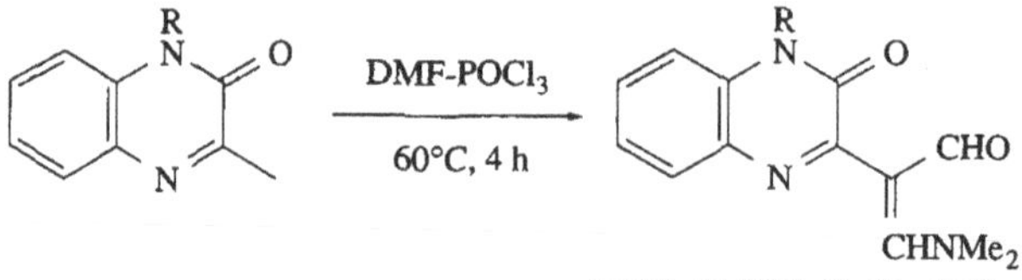

$$(2.81)$$

123 R=H, 72%; R=Me, 75%

2.8.2.5 *Other Vinylogous Formamides*

5-Dimethylamino-2,4-pentadienals are obtained (42-50%) by reacting 1-ethoxydienes with DMF-POCl$_3$, first in dichloroethane at 0-15°C (30 min), followed by heating at 55-60°C (30 min), then hydrolysis with aqueous potassium carbonate.

$$(2.82)$$

The reactions were shown to proceed for R^1=R^2=R^3=H, and for several mono- and di-methyl analogs.[167] This procedure has been applied to extensively conjugated vinyl ethers, the corresponding vinylogous formamides **124a**,[168] **124b**,[169] and **124c**[169] being obtained in fair yields.

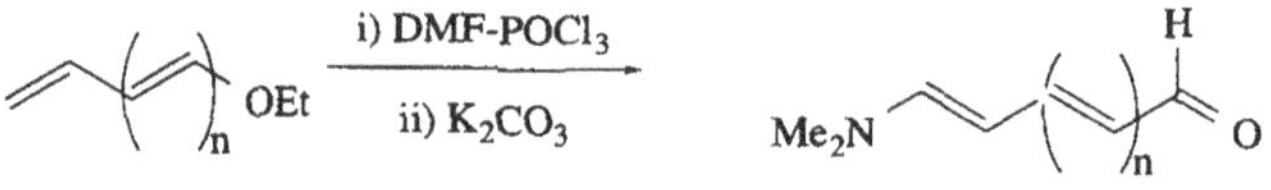

$$(2.83)$$

124a, n=1; **124b**, n=2; **124c**, n=3

The aldehyde **125** (68%) is obtained by reacting DMF-POCl$_3$ in THF with 1,2,3,4-tetrachloro-1,3-cyclopentadiene followed by an alkaline work-up.[170]

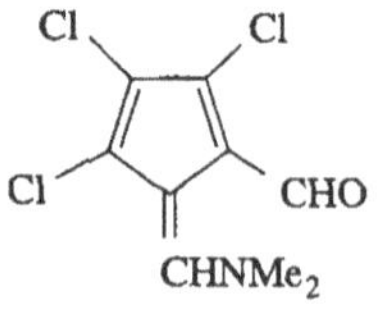

$$(2.84)$$

125

2.8.3 Aliphatic Thioaldehydes

Some β-chlorovinylaldehydes containing at least one phenyl group have been converted by sodium sulfide into the corresponding monothiomalondialdehydes.[171] Those react with aniline to give vinylogous thioformamides.

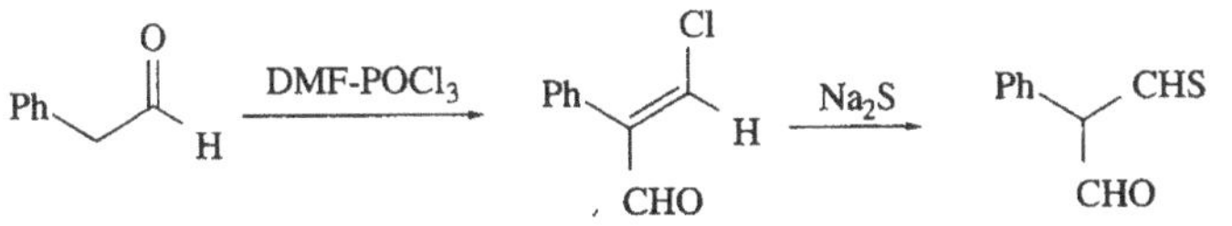

$$\text{Ph}\underset{H}{\overset{O}{\diagdown\diagup}}\xrightarrow{\text{DMF-POCl}_3}\underset{\text{, CHO}}{\text{Ph}\overset{\text{Cl}}{\diagup\diagdown}\text{H}}\xrightarrow{\text{Na}_2\text{S}}\underset{\text{CHO}}{\text{Ph}\diagdown\text{CHS}}\qquad (2.85)$$

Enamino thioaldehydes can be prepared by a reaction of enamines with Vilsmeier reagents followed by *in situ* solvolysis with NaSH.[172]

$$\underset{\text{Ph}}{\text{R}_2\text{N}}\diagdown\xrightarrow[\text{ii) NaSH}]{\text{i) DMF-POCl}_3}\underset{\text{Ph}}{\text{R}_2\text{N}}\diagdown\overset{\text{H}}{\diagup}=\text{S}\qquad (2.86)$$

Improvements to the original procedure have led to the preparation of simple enamino thioaldehydes that need not have an aryl substituent; for example 3-amino-2-cyanothiocrotonaldehyde (60%).[173]

2.8.4 Aromatic aldehydes

The introduction of a formyl group into aromatic aldehydes is well-documented. A formyl group can be introduced by a wide range of methods, for example, Reimer-Tiemann,[174] Gattermann,[175-178] Gattermann-Koch,[177] Karrer,[176] and Duff[178,179] reactions, though each is of of limited scope. For example, the Reimer-Tiemann reaction is suitable for phenolic or heterocyclic substrates only and seldom gives yields above 50%. Some substrates give dichlorocarbene insertion products as well as or instead of, the expected product. The Duff reaction, closely related to the Reimer-Tiemann reaction, employs hexamethylenetetramine [(CH$_2$)$_6$N$_4$] as the alkylating agent instead of chloroform. The Duff reaction is applicable only to phenols and amines, although the yields are low. This reaction has some application to alkylbenzenes if trifluoroacetic acid is used. The yields are usually higher for the Reimer-Tiemann reactions. The Gattermann-Koch reaction has been used to formylate benzene and alkylbenzenes and is essentially limited to these arenes; the requirement of a mixture of carbon monoxide and hydrogen chloride, in the presence of aluminium chloride and copper chloride is often inconvenient. Of wider scope is the Gattermann reaction, which can be used to formylate phenols, phenolic ethers, and many heterocyclic compounds. Zinc cyanide and hydrogen chloride are the most convenient reagents to carry out this reaction.

As a convenient alternative to the above reactions, the Vilsmeier-Haack reagents give smooth reactions with many substrates, in high yields, and with the use of less toxic reagents. One of the early papers on this subject was by

Fischer[180] who found that on heating N-methylacetanilide with $POCl_3$, an acetyl migration occurred. Vilsmeier and Haack[181] continued the investigation and established, in principle, a general method for the introduction of an aldehyde group into an aromatic ring by reaction with a substituted formamide.

Further studies have shown the Vilsmeier-Haack method to be a convenient way of introducing an aldehyde group into many classes of aromatic compounds; many such formylated compounds are difficult to obtain by other methods.

Aromatic hydrocarbons which are much more electron-rich than benzene (e.g. azulene and ferrocenes), as well as arylamines (127, X=NR^{3}R^4) and phenols (127, X=OH) can be formylated using either an arylalkylamide or a dialkylamide with $POCl_3$, or other halogenating agent. Ketones can sometimes be prepared by using other amides. The attacking species is the cation 126,[4] as has been confirmed by NMR spectroscopy,[182] kinetic studies,[183] and deuterium labelling studies.[184] Such cations had been invoked by earlier workers.[185,186] Either the formation of 126 or its reaction with the substrate can be rate-determining, depending upon the reactivity of the substrate.[183] In any case, the product of the reaction prior to hydrolytic work-up is the iminium salt 128.

$$R^1R^2N-CHO \ + \ POCl_3 \ \longrightarrow \ R^1R^2N-\overset{Cl}{\underset{OPOCl_2}{\overset{|}{C}}}-H \ \rightleftharpoons \ \underset{\underset{126}{}}{\overset{R^1}{\underset{R^2}{}}N\overset{+}{=}\overset{Cl}{\underset{H}{C}} \ \ PO_2Cl_2^-}$$

$$\text{(2.87)}$$

127   128

Monocyclic, aromatic hydrocarbons such as benzene, toluene, and their derivatives do not generally react with Vilsmeier-Haack reagents even under the most forcing conditions, and have been used as inert solvents in the formylation reaction. Naphthalene does not react; however anthracene and other polynuclear hydrocarbons are easy to formylate and give good yields of aldehydes. In recent work, an adduct of DMF-trifluoroacetic anhydride has been reported to formylate 1,3,5-trimethylbenzene and naphthalene.[187] A summary of some of the compounds formylated is contained in Table 2.1.

Attack by Vilsmeier reagents that leads to diformylated products is rare, and only occurs in highly activated systems. Azulene undergoes iminoalkylation to give a yellow dication that may be regarded as a combination of a tropylium cation with a pentamethinium unit. Hydrolysis affords azulene-1,3-dicarboxaldehyde (43%).[188,189]

Starting Material	Product	Conditions	Yield (%)	References
Anthracene	Anthracene-9-aldehyde	A, 20 min, 90-95°C	77-92	190-192
Azulene	Azulene-1-aldehyde	B, 24 h, 20°C	85	193
Azulene	Azulene-1-aldehyde	A, 10 min, 0°C	90-95	188, 189, 193
Azulene	Azulene-1,3-dialdehyde	A, 45 min, 20°C	43	193
Benz[a]anthracene	Benz[a]anthracene-10-aldehyde	A, 2 h, 95°C	64	191
Ferrocene	Ferrocenaldehyde	A, 72 h, 20°C	77	194
Guiazulene	Guiazulene-3-aldehyde	A or B, 5-30 min, 0°C	90-95	188, 193
Pyrene	Pyrene-3-aldehyde	A or B, 2-6 h, 100°C	53	195
Perylene	Perylene-3-aldehyde	A, 30 min, 95°C	63	196
1,3,5-Trimethyl-benzene	1,3,5-Trimethyl-benzaldehyde	C, 48 h, 20°C	60	187

CONDITIONS. A; $DMF-POCl_3$; B; $MFA-POCl_3$; C; $DMF-Tf_2O$

Table 2.1. Aromatic Aldehydes prepared using Vilsmeier Reagents

$$(2.88)$$

Other 1,3-diformylazulenes have been prepared by Vilsmeier formylation (*e.g.* of 6-octylazulene) as intermediates for drugs and sensitizers for electrographic photoconductors.[197] Vilsmeier reactions on azulene using ω-carbethoxy-*N*,*N*-dialkylamides and $POCl_3$ afforded 1-(α-oxo-ω-carbethoxyalkyl)azulenes.

1-(Dimethylamino)naphthalene has also been diformylated at the 2- and 4-positions in 70% yield using $DMF-POCl_3$ (1:3:2 ratio).[198]

Vilsmeier formylation of aromatic amines is an early example of the synthesis of aldehydes by the reaction of organic compounds with substituted formamides.[181] The formylation of secondary and tertiary aromatic amines is well-documented, and Vilsmeier reactions can be used to formylate *N*-alkyl- and *N*,*N*-dialkylanilines,[181] toluidines,[181] diarylamines,[199] and triarylamines.[200] $MFA-POCl_3$ readily formylates *N*,*N*-dimethylaniline at 0-10°C giving *p*-dimethylaminobenzaldehyde.[181] With $DMF-POCl_3$, the yield is increased, but heating is required. Formylation usually occurs at the *para* position, but when this is blocked *ortho* substitution can occur.

Starting Material	Product	Conditions	Yield (%)	Reference
Phenol	4-Hydroxbenz-aldehyde	A, 2 h, reflux	low	201
Resorcinol	2,4-Dihydroxy-benzaldehyde	A, 15 h, reflux	46	202
Anisole	Anisaldehyde	A, 15 h, reflux	70	201
1,3-Dimethoxy-benzene	2,4-Dimethoxy-benzaldehyde	B, 1-2 h, reflux	85	203
2-Naphthol	2-Hydroxy-1-naphththaldehyde	B, 1 h, reflux	-	204
1-Methoxy-naphthalene	4-methoxy-1-naphthaldehyde	A, 3 h, reflux	90	204

CONDITIONS. A; DMF-POCl$_3$; B; MFA-POCl$_3$;

Table 2.2. Phenolic Aldehydes prepared using Vilsmeier Reagents.

Phenols, naphthols and their derivatives readily undergo formylation by Vilsmeier reagents (Table 2.2, above).

The influence of steric factors on aromatic formylation has been investigated. Whereas 129 (R^1=Me, R^2=H; Ar=1-anthraquinonyl) failed to react with DMF-POCl$_3$, under the same conditions 130 (R^1=H, R^2=Me) underwent formylation at the 4-position (37%), the remainder being chiefly the unreacted diarylamine.[205] It is argued that cation 130 is inactivated by the geometrical requirements at the nitrogen atom.

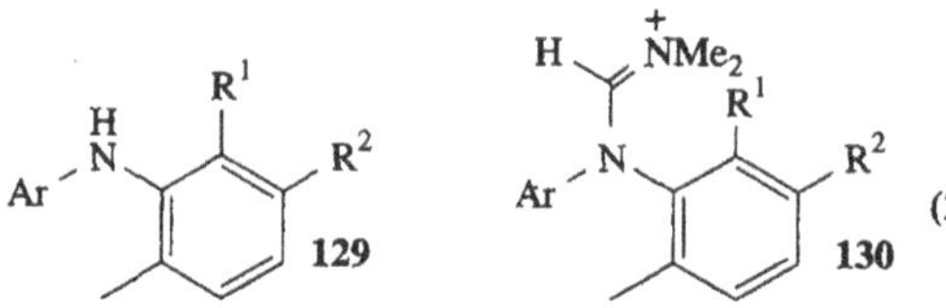

(2.89)

Three moles of *N,N*-dimethylaniline undergo condensation with DMF-POCl$_3$ at 90-100°C to give, after hydrolysis, the triarylmethane 131 (93%). The degree of involvement of iminium salts in this reaction has been questioned. When the *p*-position to an activating alkoxy group on an aromatic ring is blocked, *o*-formylation can often be effected; this led to aldehyde 132, from which several phenolic sesquiterpenoids were synthesized.[206]

(2.90)

Formylation of dibenzo[a,l]pyrene using MFA-POCl$_3$ occurs only in the 10-position; however, dibenzo[a,e]fluoranthene gives a mixture of mono-formyl derivatives.[207]

3-Bromoazulene-1-carboxaldehyde (92%) is formed from 1,3-dibromo-azulene by DMF-POCl$_3$ at 150°C; reaction of 3-bromoazulene-1-carboxaldehyde with DMF-POCl$_3$ induces a second displacement, giving azulene-1,3-dicarboxaldehyde (23%).[208] Reaction of benzyl-1-azulenyl ketone with DMF-POCl$_3$ affords the corresponding 3-formyl derivative (38%) and also 3-(azulenyl-1)-3-chloro-2-phenylacrylaldehyde (54%), the latter by reaction at the carbonyl group.[209]

A series of porphyrin metal complexes[210] (of copper, magnesium, zinc, nickel, cobalt, and manganese) underwent monoformylation at a methine position, as did aethioporphyrin-I-Cu(II).[211]

Formylation of octamethylferrocene using DMF-POCl$_3$ in chloroform at 60-70°C affords octamethylformylferrocene (95%)[212] which on reduction (LiAlH$_4$) and treatment with HBF$_4$ affords the (octamethylferrocenyl)methyl carbocation tetrafluroborate in excellent yield. Since the latter carbocation is attacked by nucleophiles, it can be used to functionalize surfaces of silica, platinum, and indium tin oxide. Vilsmeier formylation has also been employed in order to prepare 1′-chloro and 1′-bromoferrocenecarbox-aldehydes.[213]

A small but interesting change in selectivity was observed in the formylation of 2,3-methylenedioxyanisole when reacted with DMF-POCl$_3$ in the presence and absence of potassium iodide (scheme 2.91). With KI, the formylation occurred in a 3:1 ratio in the 2- and 4-positions in 80% yield. In the absence of KI, the formylation was 5:3 at the 2:4-positions, in 83% yield.[214]

(2.91)

Dihydrobenzopyran and dihydrobenzofuran thiazolin-2,4-diones, potential anti-hypoglycemic agents,[215] were synthesized via the formyl derivatives **133**, which were obtained from substituted benzopyrans and furans using Vilsmeier techniques.

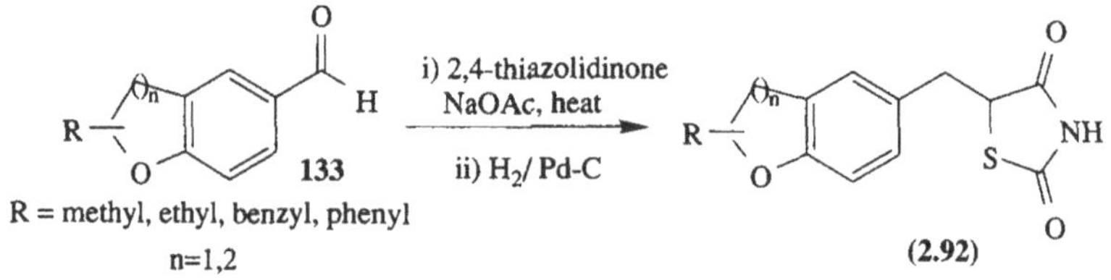

R = methyl, ethyl, benzyl, phenyl

n=1,2

(2.92)

2.8.5 Heterocyclic Aldehydes

Formylation using DMF or MFA and POCl$_3$ can provide a wide variety of heteroaromatic aldehydes. This application is of particular importance, since the alternative formylation reactions outlined in section 2.8.4 give poor results owing to the acid-sensitivity of certain heteroaromatic compounds.

2.8.5.1 Indoles, Pyrroles and Porphyrins

Indole-3-aldehyde was the first heteroaromatic aldehyde to be prepared using Vilsmeier reagents (72%).[216,217] Formylation of indole at the 2-position has been reported in 83% yield.[218] A mixture of 3:1 DMF:BCl$_3$ afforded an excellent yield of 3-formylindole.[92] 3-Arylindoles undergo Vilsmeier formylation to give the corresponding 2-formyl derivatives.[219]

4,6-Dimethoxyindoles react with DMF-POCl$_3$ at 0°C to give the corresponding 7-formylindoles; at 5°C the 3-isomer is also obtained.[220] 7-Formylindoles react with *o*-phenylenediamine to give 2-(7-indolyl)benzimidazoles, and with ethyl acetate in the presence of NaOEt afford 4-oxo-4*H*-pyrrolo[3,2,1-*i,j*]quinolines.[221] 1*H*,10*H*-Benzo[*e*]pyrrolo-[3,2-*g*] indole **134** was formylated at 50°C by DMF-POCl$_3$ to give the formyl derivative **135** (50%). However, with a five-fold excess of DMF-POCl$_3$ the chlorinated formyl derivative **136** was formed (64%).[222]

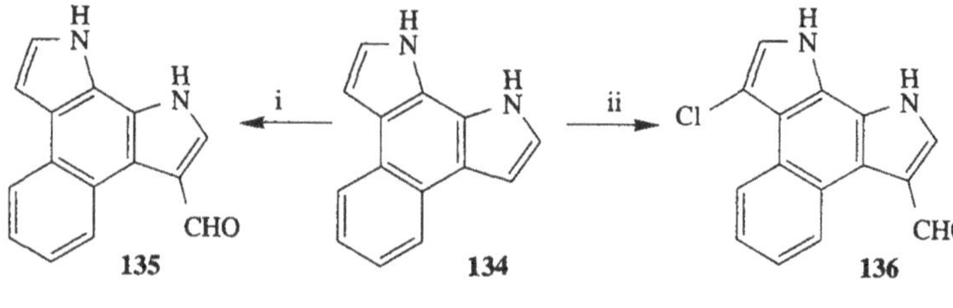

i, 3 eq DMF- 1 eq DMF-POCl$_3$, 50°C; ii, 5 eq DMF-POCl$_3$, 50°C

Formylation using DMF-POCl$_3$ gave indole **137** (60%).[223] Formylation of 2-methyl-2-pyrido[3,2-*e*]pyrrolo[1,2-*a*]pyrazine gave the 9-formyl derivative **138** (47%).[224] The indolizine **139** (50%) was obtained by formylation using DMF-POCl$_3$.

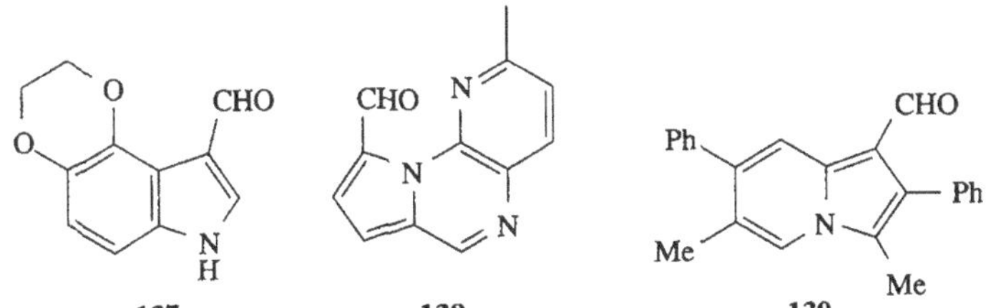

137   138   139   (2.94)

Pyrrole reacts with DMF-POCl$_3$ at 0°C giving pyrrole-2-aldehyde.[218] Strong electron-withdrawing groups on the nitrogen atom influence the susceptibility of the pyrrole nucleus towards electrophilic substitution, and lead exclusively to the 2-formyl derivatives in good yields.[225] Other related aldehydes were obtained in a similiar fashion.[226]

Formylation of substituted pyrroles with DMF-POCl$_3$ afforded 2,5-dimethyl- and 1,2,5-trimethylpyrrole-3,4-dialdehydes that were used in a new route to 2-azaazulenes.[227]

3-Octadecylpyrrole reacts with DMF-POCl$_3$ to give an 85% yield of a 3:1 mixture of the corresponding 2-formyl and 5-formyl derivatives.[228]

The rates of formylation of substituted pyrroles by variously substituted formamides has been studied by White.[229] *N,N*-Dimethylbenzamide-POCl$_3$ reacted with the pyrrole **140** to give >98% of **141** (by UV analysis) in 5 minutes. The cyclohexyl benzamide adduct took 45 minutes to achieve the same conversion. The ketone **142** was isolated in excellent yield.

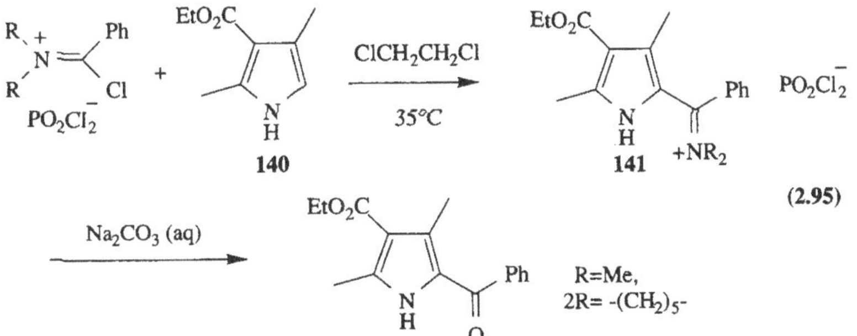

(2.95)

1-Alkyl-5-chloropyrrole-2-carboxaldehydes and 1-alkyl-5-chloropyrrole-2,5-dicarboxaldehydes have been prepared by reacting the appropriate *N*-alkylsuccinamidal with DMF-POCl$_3$ in the respective molar equivalents of 3:3, and 5:10.[230]

Nitropyrroles are important components of pharmaceuticals, fungicides, and herbicides. They are readily formylated in the activated positions. In this way Ono[231] formylated a series of 3-aryl-4-nitropyrroles in good yields.

No formylation of the aromatic ring was noted unless an excess of DMF-POCl$_3$ was used.

Acylation of pyrrole with Vilsmeier reagents gave the iminium salt **143**. Reaction of the iminium species with excess sodium cyanide afforded α-dialkylaminonitriles **144** in good yields. The α-dialkylaminonitriles (R^1=H) could be subsequently converted into a methyl ester or aldehyde depending on the conditions used.[232]

$$\text{143} \xrightarrow[\text{MeCN}]{\text{NaCN}} \text{144} \qquad \textbf{(2.96)}$$

Several 5,5′-diformyldipyrrylmethanes have been prepared by Vilsmeier formylation using DMF-PhCOCl.[233]

In 1966, Inhoffen and co-workers reported the formylation of porphyrins using Vilsmeier reagents.[234] Since then, it has been routinely used in order to introduce substituents into the meso-position of numerous copper (II) and nickel(II) porphyrins and chlorins. A vinylogous formylation applied to the pyrrole ring has been described (scheme 2.97).[235]

$$\xrightarrow[\text{POCl}_3]{\text{Me}_2\text{NCH=CHCHO-}} \qquad \textbf{(2.97)}$$

This vinylogous formylation has been recently applied to porphyrins and chlorins.[236] For example, reaction of nickel(II) octaethylporphyrin with 3-(dimethylamino)acrylaldehyde-POCl$_3$ gives a disubstituted derivative in which the acrolein substituents are on adjacent rather than opposite meso positions. Such acrolein substituents can be cyclized with acid, a reaction in which ethyl migration followed by aromatization occurs, but only if a centrally chelated metal is present.

Several porphyrin metal complexes have been shown to undergo formylation.[210,237] Some metal porphyrins that have all meso-positions substituted by aryl groups undergo Vilsmeier monoformylation at the 2-position.[238]

Steric factors concerning the formylation of porphyrins by Vilsmeier reagents has been examined.[239] Meso-positions of chiral porphyrins has been formylated using DMF-POCl$_3$.[240] Methylporphyrins have also been formylated at the meso-position, and Vilsmeier methodology enabled the synthesis of the first porphyrin fused with two cyclopentane rings.[241]

2.8.5.2 Furans and Thiophenes

Furan undergoes formylation at the 2-position by DMF-POCl$_3$ (64%).[242] 3,5-Disubstituted furans give 5-formyl derivatives in moderate yield.[243]

One of the most satisfactory methods of preparing furan-2,5- and thiophene-2,5-dicarboxaldehydes is not by Vilsmeier formylation, but *via* metalation of the parent heterocycles and formylation using DMF.[244]

Thiophene and its derivatives have been studied in detail by a number of workers. King and Nord[245-247] showed that the formyl group enters the *alpha*-position unless both *alpha*-positions are blocked, in which case it enters the *beta*-position. Thiophene dialdehydes were not obtained. Kato reported the conversion of 3-methoxythiophene into the corresponding 2-formyl derivative using DMF-COCl$_2$ in 98.8% yield.[248] Reaction of $2\lambda^4\delta^2$-thieno[3,4-*c*] thiophene **145a** with 10 eq DMF-POCl$_3$ afforded only the monoformyl derivative **146a** in 40% yield. Reaction of **145b** under the same conditions afforded **147** in addition to **146b** (33%). The structure of **147** was confirmed by X-ray crystallography.[249]

$$(2.98)$$

145a, R=*t*-butyl
145b, R=*i*-propyl

146a, b   **147**

2.8.5.3 Pyridines

The 2-methyl group of a 2,6-dimethyl-1,4-dihydropyridine was formylated by a Vilsmeier reagent, leaving intact the ester groups at the 3- and 5-positions.[165]

Formylation of 3-substituted-1-(phenoxycarbonyl)-1,2-dihydropyridines with DMF-POCl$_3$ afforded the corresponding 5-formyl derivatives **148** (R=Br, Cl, OMe, Me, Ph).[250] Aromatization with sulfur gave the corresponding 5-substituted-pyridine-3-carboxaldehydes. Comins has also shown that 1-(phenoxycarbonyl)-1,2-dihydropyridines afford the 5-formyl derivatives (81%) with DMF-POCl$_3$.[251] The dihydropyridines **149** have been similarly prepared.[252] Earlier work had shown that 1,4-dihydroquinolines underwent Vilsmeier formylation at the 3-position.[253]

148   **149** (2.99)

Reaction of 4-arylglutarimides with Vilsmeier reagents and alkaline work-up has been shown to lead exclusively to conjugated aldehydes **150**. The latter can be efficiently oxidized to the pyridine-3,5-dicarboxaldehydes **151**.[254]

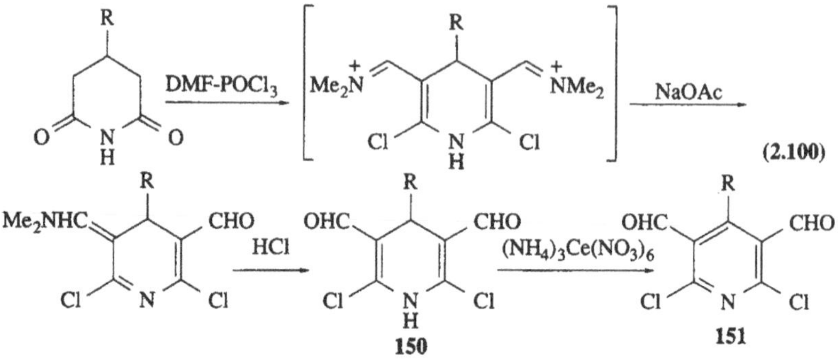

(2.100)

150   **151**

2.8.5.4 Pyrimidine Derivatives (including Barbituric Acid)

Unactivated pyrimidines such as 4,6-dichloropyrimidine do not generally react with Vilsmeier reagents. However, an unsubstituted 5-position of certain derivatives of barbituric acid, uracil, and 4-hydroxy-6-oxodihydropyrimidines can be formylated and 4-hydroxy-6-oxopyrimidines undergo formylation, in accordance with their reactivity as β-enamides.[255]

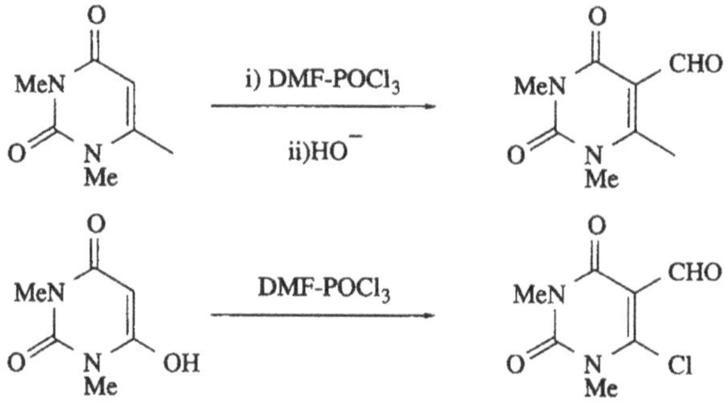

(2.101)

Barbituric acid derivatives **152** afforded either **153** or **155** depending upon the solvent employed (scheme 2.102). The 5-dimethylamino-methylenebarbituric acid derivatives were hydrolyzed by alkali to the 5-(hydroxymethylene)barbituric acids **154**. The Vilsmeier reaction of 1,3-disubstituted uracil derivatives afforded 1,3-disubstituted 5-formyluracil derivatives.[255,256] Aldehyde **156** was obtained from the N,N-dimethyl derivative of barbituric acid and DMF-POCl₃.[257]

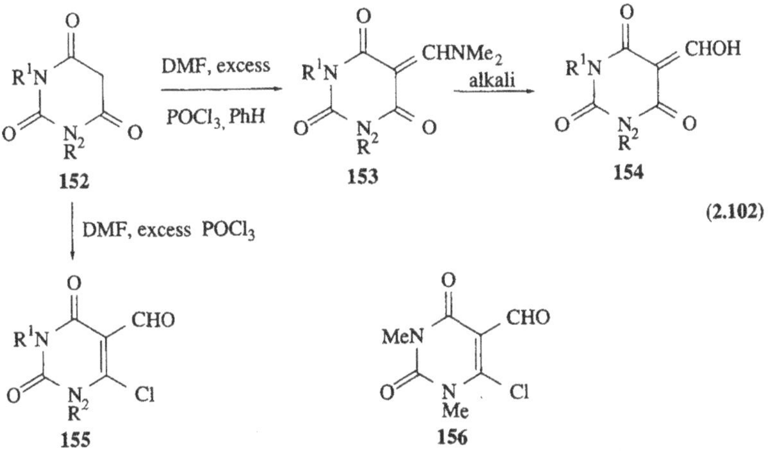

(2.102)

When such heterocycles are reacted with DMF-POCl$_3$, only iminoalkylation occurs, as established by hydrolysis to the aldehyde. Conversion of the oxygen atoms into chloro can only be accomplished by a separate reaction with POCl$_3$ (scheme 2.103).[258]

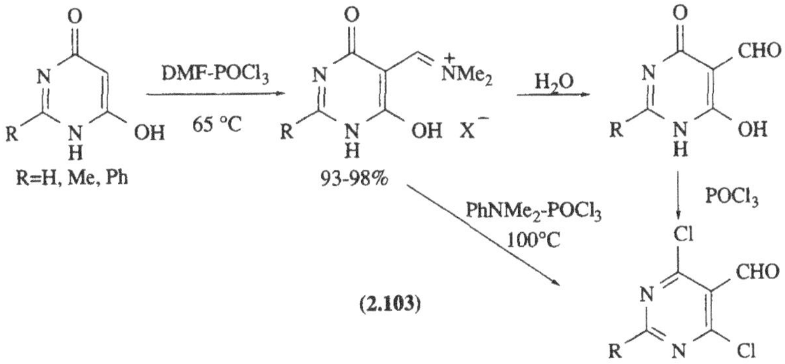

(2.103)

Formylation occurs at the 5-position of 4-(*O*-acetyl-β-D-glycopyranosyl-amino)-6-oxopyrimidines upon treatment with DMF-POCl$_3$.[259]

2.8.5.5 Miscellaneous

Heterocyclic mesomeric betaines can be formylated by DMF-POCl$_3$, the means by which aldehydes such as **157** have been prepared.[260]

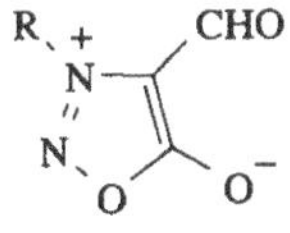

(2.104)

Some remarkable formylations of carbazoles and some derivatives have been reported (*e.g.* R=H to R=CHO in **158**).[261] The pathway is purported to be more reasonable than a mechanism previously proposed.[262]

(2.105)

5-Amino-3-methyl-1-phenylpyrazole reacted with DMF-POCl$_3$ at the 4-position, the amino group being converted into a dimethylaminomethylen-amino group.[263] The reactions of pyrazole and N-benzoylpyrazole with Vilsmeier-Haack reagents were unsuccessful,[264,265] but N-methylpyrazole and N-phenylpyrazole gave the corresponding 4-aldehydes.[265]

1,3,5-Triphenyl and 1,5-diphenyl-3-styryl-Δ^4 pyrazolines react with DMF-POCl$_3$ to give the 1-p-formylphenyl derivatives; the 4-position of the pyrazole is not attacked.[266]

The 3-formyl derivative is formed from 2-(2′-thienyl)indole and DMF-POCl$_3$.[267] A number of indoles, including 1-substituted-5-acetoxyindoles,[268] 2-substituted-benz[e]indoles and 2-substituted-benz[g]indoles underwent formylation by DMF-POCl$_3$ to give the corresponding 3-carboxaldehydes.[237]

Imidazo[1,5-a]pyridine was shown to resemble indolizine in its reactivity towards DMF-POCl$_3$; the 1-formyl derivative was isolated as the major product, along with the 3-formyl isomer.[269] Quinoline does not react with Vilsmeier reagents, but N-methyl-1,2,3,4-tetrahydroquinoline is formylated at the 6-position in 46% yield.[270]

Systems of formally antiaromatic properties have been reacted with DMF-POCl$_3$. 1,3-Di(ethoxycarbonyl)pyrido[2,1,6-d,e]quinolizine (cycl-[3,3,3]azine) afforded a mixture of the 4- and 6-monoformyl derivatives.[271] 3a-Aza-4-azulenone reacted like a 1-acylpyrrole, giving the 3-carbox-aldehyde (73%).[272]

1-Pyrrolo-2-pyrazine was formylated at the 2-position with DMF-POCl$_3$ to give **159** in 56% yield. No chlorination of either ring was noted; however, when 1-pyrrolyl-2-pyridine **160** was reacted under similar conditions, 2-chloro-3-(2-formylpyrrolyl)pyridine **161** was isolated in 83% yield.[273]

(2.106)

159   160   161

The pyrrolopyrrole **162** was obtained in 90% yield by the reaction of the parent compound with DMF-POCl$_3$.[274]

2-Substituted 3-formyl-4*H*-pyrido[1,2-*a*]pyrimidin-4-ones have been prepared by direct formylation.[275] Comparison of the reactivities of the pyrimido[4,5-*b*]quinolines **163** and **164** towards Vilsmeier reagents showed that **164** did not react significantly, even under severe conditions, whereas **163** gave high yields of the 3-formyl derivatives [R^1=Me, R^2=Et, and R^1, R^2= (CH$_2$)$_4$].[276]

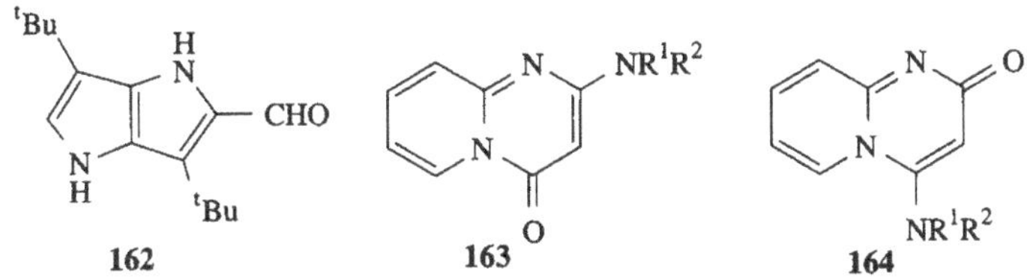

(2.107)

162   163   164

2.8.6 Heterocyclic Thioaldehydes

The thioformylation of enamines by reaction first with a Vilsmeier reagent, then with NaSH was referred to in section 2.8.3. The same procedure can be used to introduce a thioformyl group at a site of high electron density in a heterocycle. In this way, 2-methylindolizine has been converted into 2-methyl-3-thioformylindolizine (86%);[277] and pyrrolo-[2,1-*b*]thiazole thioaldehydes have also been prepared.[278]

2.9 Ketones

2- and 3-Chloroacetylpyrroles result by the acylation of pyrroles with *N,N*-dimethylacetamide. The absence of a substituent on nitrogen leads exclusively to 2-chloroacetyl derivatives, but an *N*-substituent leads to a mixture of the 2- and 3-isomers which was 1:9 for *N*-phenylpyrrole or *N*-benzylpyrrole.[279]

Aroylation of pyrroles by a Vilsmeier reagent is usually superior to other methods of aroylation (scheme 2.108).[280]

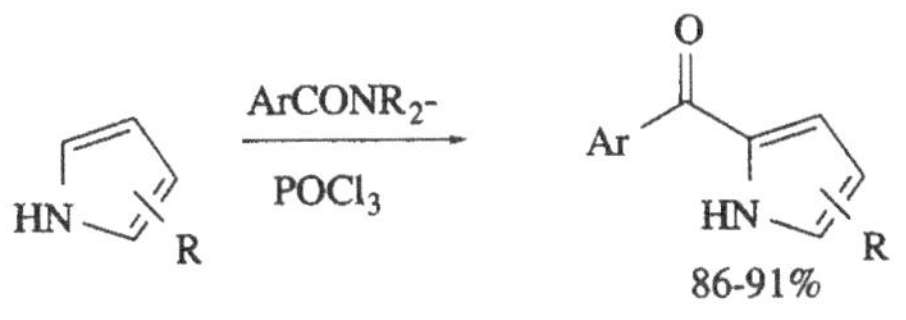

(2.108)

86-91%

Indole-2-acetonitrile undergoes aroylation at the 3-position by *N,N*-dimethylacetamide-POCl$_3$ to give 2-acetylindole-2-acetonitrile (24%).[281]

Some thioketones that have an α-carbonyl group have been prepared by reaction of the corresponding β-chlorovinylcarbonyl compounds with sodium sulfide (see also section 2.8.3).[171]

2.9.1 2-Chloroketones

Only the hydroxy group of benzoin is converted into a chloro group by DMF-POCl$_3$.[282] Analogously, 5-chlorofurfuryl 2′-furyl ketone is one of the products obtained by the action of DMF-POCl$_3$ upon furoin.[282]

$$\text{(2.109)}$$

2.9.2 β-Amino-α,β-Unsaturated Ketones

A route to iminium salts **165** from ketones has been outlined in section 2.8.2.2. Two sequential nucleophilic displacements afford a route to a few β-(dialkylamino)-α,β-unsaturated ketones, albeit in low yields. The conversion of **165** into **166** is better for cyclic compounds [*e.g.* R^1, R^2= (CH$_2$)$_3$] than for acyclic ones.[137]

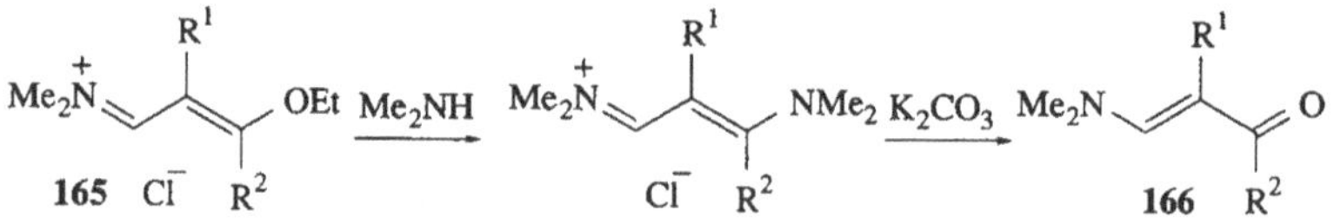

$$\text{(2.110)}$$

2.10 Other α,β-Unsaturated Carbonyl Compounds

Formylation of an acetal or ketal provides a route to α-ketoaldehydes.[283]

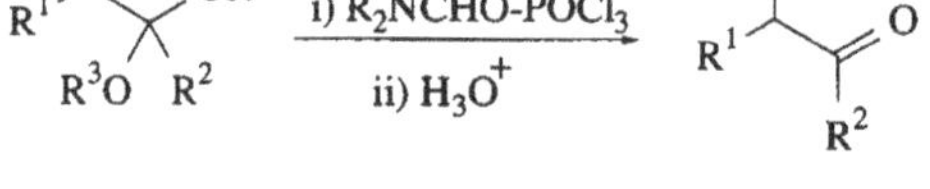

$$\text{(2.111)}$$

Enamines of cycloalkanones and α-tetralone, particularly the morpholinoenamines, react with DMF-POCl$_3$ in CHCl$_3$, followed by hydrolysis with alkali to give the respective 1,3-dicarbonyl compounds **167** and **168** that probably exist as the hydroxymethylene tautomers.[284]

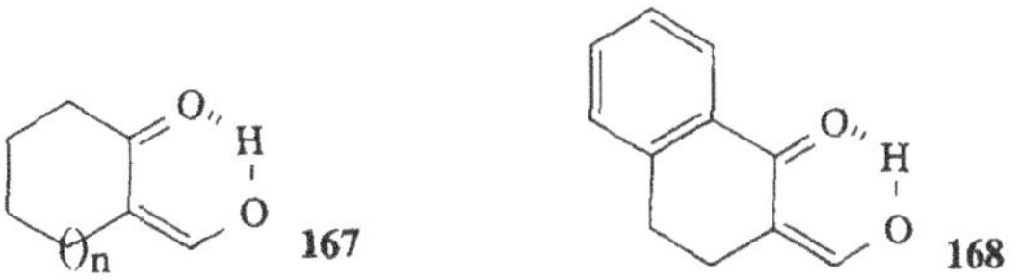

(2.112)

Some 2,3-disubstituted-2-cyclohexene-1-ones are converted into 6-hydroxymethylene derivatives by Vilsmeier reagents. Those derivatives have been aromatized to substituted salicylaldehydes.[285]

The pentamethinium salt **169**, which is formed by the reaction of acetone with DMF-POCl$_3$ (section 2.29.4) is converted into the ketoaldehyde **170** by nucleophilic displacements with Me$_2$NH, and hydrolysis.[41,42]

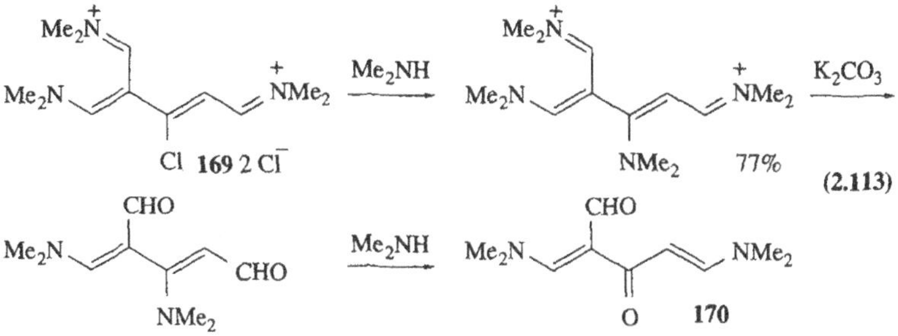

(2.113)

Diazoacetophenone **171a** and ethyl diazoacetate **171b** are powerful nucleophiles, and they react with [Me$_2$N=CHCl]$^+$ Cl$^-$ to give the corresponding diazaiminium chlorides **172** which are converted into the formyldiazo compounds **173** by careful hydrolysis.[24] Interestingly, diazomethane reacts quite differently, giving 1,3-dichloro-2-dimethylamino-propane.

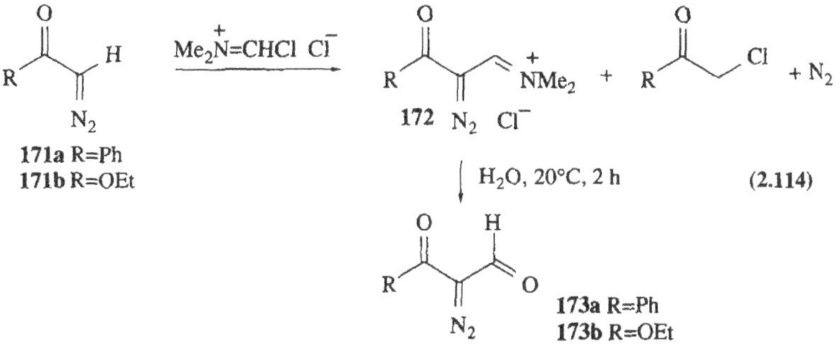

(2.114)

2.11 β-Halovinylaldehydes

2.11.1 Formation of β-Halovinylaldehydes

The usual reaction of ketones with Vilsmeier reagents to give β-halo-vinylaldehydes, and the mechanistic considerations have been outlined in section 1.3.5. β-Halovinylaldehydes are the products of monoformylation; further attack of the intially formed halomethyleneiminium cation can proceed, but usually only if deprotonation is possible, giving a dienamine. The dienamine may then be attacked by one or more mole equivalents of Vilsmeier reagent; this leads to polyformylated products upon hydrolysis. However, even in cases where further iminoalkylation is possible, avoiding an excess of Vilsmeier reagent and performing the reaction at relatively low temperatures (*e.g.* 20-50°C) allows a good yield of the β-halovinylaldehyde to be obtained. The regioselectivity of the reaction has been briefly discussed in section 1.3.5, and will be referred to again in the present section.

Many methyl ketones, as well as acyclic and cyclic methylene ketones, are converted by Vilsmeier reagents into 3-haloacrylaldehydes (scheme 2.115); these reactions have been extensively surveyed in previous reviews.[3,286,287] The usual procedure involves slow addition of the ketone to the Vilsmeier reagent (2.5-5 eq.), with cooling. Solvent is usually employed to control the exothermic, and sometimes violent reaction. After the initial reaction has subsided, the mixture may be heated, prior to being quenched with ice and neutralized by cold aqueous sodium acetate or sodium carbonate. β-Haloacrylaldehydes of low molecular mass are lachrymatory oils which decompose spontaneously within a few hours, sometimes violently.

(2.115)

In the simplest cases, monochloroformylation occurs, and usually with the formation of a mixture of the *(Z)*-isomer and the *(E)*-isomer. Acetone afforded 3-chlorobut-2-enal in 39% yield.[133]

2.11.2 β-Bromovinylaldehydes

DMF does not react with carbonyl bromide. However, cyclic β-bromo-vinylaldehydes can usually be prepared using complexes of DMF-POBr₃,[288] or of DMF-PBr₃.[288,289] The chemical reactivity of those adducts is the same as dimethylbromomethyleneiminium bromide **174**, which can be obtained by the treatment of dimethylchloromethyleneiminium chloride with HBr. The structure **175** has been suggested for the structure of the DMF-PBr₃ adduct (scheme 2.116).[289]

Ketone	Aldehyde		Reagent	Reaction	Reaction	Yield
	R^1	R^2		Temp. (°C)	Time (h)	%
Acetone	Me	H	**175**	20	12	20
Ethyl methyl ketone	Me	Me	**175**	60	7	36
Benzyl methyl	Me	Ph	**175**	60	2	25
ketone			**174**	60	4	56
Pinacolone	CMe₃	H	**175**	60	4	75
Cyclopentanone	-(CH₂)₃-		**175**	20	12	45
			174	65	1.5	31
Cyclohexanone	-(CH₂)₄-		**175**	20	12	54
Cycloheptanone	-(CH₂)₅-		**175**	70	3	45
			174	65	1.5	67
Cyclooctanone	-(CH₂)₆-		**175**	60	12	37
			174	60	1.5	63
Acetophenone	Ph	H	**175**	60	2	45
			174	60	1.5	68
Propiophenone	Ph	Me	**175**	60	5	71
			174	70	3	85
Desoxybenzoin	Ph	Ph	**175**	60	3	75
17β-Acetoxy-5α-androstan-3-one	17β-Acetoxy-3-bromo-2-formyl-5α-androst-2-ene		**175**	75	1	40

Table 2.3. Preparation of β-Bromoacrylaldehydes $R^1(Br)C=CR^2CHO$ from Ketones

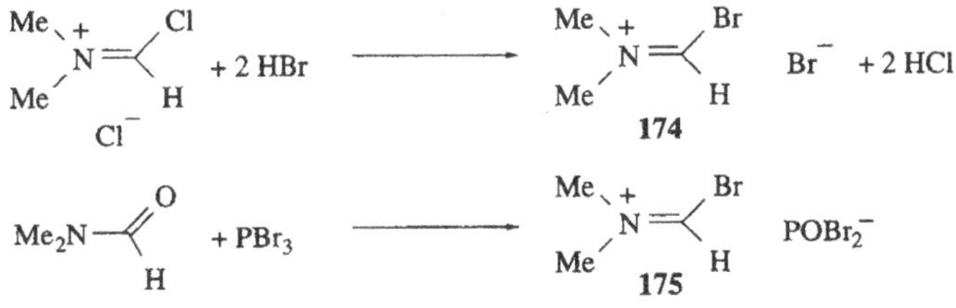

$$(2.116)$$

Whatever the relative importance of the electronegativity of the halogen atom in halomethyleniminium salts **174** and **175**, *versus* any steric factors, the bromomethyleneiminium salts **174** and **175** are found to be less reactive than the corresponding chloromethyleneiminium salts.

Table 2.3 gives the preparation of various β-bromoacrylaldehydes using either of the salts **174** or **175**, conducted in chloroform, with the exception of entry 12, for which trichloroethane was used. There appears to be little general indication of when salt **175** should be preferred to salt **174**. 3-Bromoacrylaldehydes are unstable, and many have been characterized as their oximes or semicarbazones.

Cyclohexanone has been reported as giving a product **176** derived from *two* iminoalkylations, albeit in low yield.[288]

Br   H

HO   O   (2.117)

176

3-Bromoacrylaldehydes have been exploited as bifunctional electrophilic reagents, for example as key intermediates in the synthesis of some 11-*cis*-retinoids.[290] The aldehyde group was converted into 2-hydroxyethyl by Wittig methylenation and subsequent oxidative hydroboration. The bromo substituent was metalated and coupled with a copper acetylide in order to assemble the desired polyene chain.

2.11.3 β-Chlorovinylaldehydes

2.11.3.1 Introduction
β-Chlorovinylaldehydes can be prepared by the chloroformylation of certain alkynes with Vilsmeier reagents.[291]

Interestingly, reaction of thiophenes **177** with MFA-POCl$_3$ affords predominantly the (*E*)-isomer of the corresponding β-chlorovinylaldehyde, and this was established conclusively by an X-ray structure of the 2,4-dinitrophenylhydrazone of the aldehdye **178**. Formylation on the ring to give junipal occurred only to a minor extent. 3-Aryl-3-chlorovinylaldehydes have also been prepared by the reaction of Vilsmeier reagents upon arylalkynes.[292]

OHC   S   S   S   CHO

junipal   **177**   **178** Cl   (2.118)

MFA-POCl$_3$

Many ketones that bear either a methyl group, or a methylene group (that are acyclic or cyclic), can be converted into the corresponding β-chlorovinyl-aldehydes, *via* the iminium species (scheme 2.119). Representative examples are given in Table 2.4. The general procedure[293] involves addition of the ketone to an excess of the Vilsmeier reagent (either dimethylchloro-methyleneiminium chloride or DMF-POCl$_3$, typically 2.5-5 mol equiv.) with cooling. The reactions are exothermic, and on a large scale without solvent can be violent. Accordingly, although DMF-POCl$_3$ can be used with only a small excess of DMF in most cases, and in all large scale preparations, the use of a solvent such as dichloroethane or trichloroethene is advisable. After the intial reaction has subsided, the mixture is heated (between 35-70°C) for a further period (0.5-4 hours). The precise conditions depend upon the ketone; a higher temperature is usually appropriate where a solvent is used.

Entry	R^1	R^2	Reagent	Yield %	Reference
1	Me	H	A	39	152
			A	32	43
2	Me	Me	A	67	152
	Et	H	A	78	43
3	CHMe$_2$	H	B	80	152
4	^{t}Bu	H	A	14	43
5	Et	Me	A	77	152
6	Me	CHMe$_2$	A	20	43
	Me$_2$CHCH$_2$	H			294
7	Me	Et	A	59	43
8	-(CH$_2$)$_2$-		A	33	133
9	-(CH$_2$)$_3$-		A	66	152
			A	82	295, 296
10	-(CH$_2$)$_4$-		A	54	152
			A	83	295, 296
11	-(CH$_2$)$_5$-		A	65	152
			B	88	295,296
12	4-Br-C$_6$H$_4$	H	A	24	297
13	4-Ph-C$_6$H$_4$	H	A	36	297
14	Ph	Ph	A	60	297, 298
15	Ph	Me	A	91	152
			B	60	152
16	4-MeO-C$_6$H$_4$	H	A	24	297
17	4-NO$_2$-C$_6$H$_4$	H	A	71	43
18	3,4-(MeO)$_2$-C$_6$H$_4$	Me	A	56	43

Reagent A; DMF-POCl$_3$; B; [Me$_2$N=CHCl]$^+$ Cl$^-$

Table 2.4. Preparation of β-Chloroacrylaldehydes from Ketones

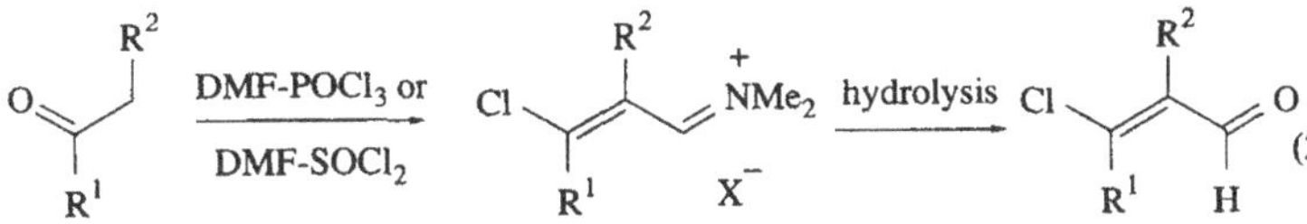

$$\text{(2.119)}$$

The mixture is quenched with ice and then neutralized with cold aqueous sodium acetate (or sodium carbonate).   However, strongly alkaline conditions during work-up must be avoided.   Aliphatic β-chlorovinylaldehydes of low molecular mass are oils that can be distilled, but are powerful lachrymators and can decompose within a few hours.   The decomposition is catalyzed by traces of alkali, can be spontaneous and sometimes violent.   Aryl groups confer much greater stablity.   A typical procedure is the preparation of 2-chloro-1-formyl-1-cyclohexene.[293]

β-Chloroacrylaldehydes can exist as the (E)- and (Z)-isomers.[294,298,299] In some cases, chromatographic separation of the 2,4-dinitrophenylhydrazones of those isomers has been effected.[299]

β-Chloroacrylaldehydes have many synthetic applications, including their alkali-induced fragmentation to give terminal alkynes[43] (section 2.3). Their use in forming a heterocyclic ring is extensively described in Chapter 4. In that regard, it is of particular value that the ketone used can have other functionality that can be carried through to give an appropriately substituted heterocycle, such as **179**.[300]

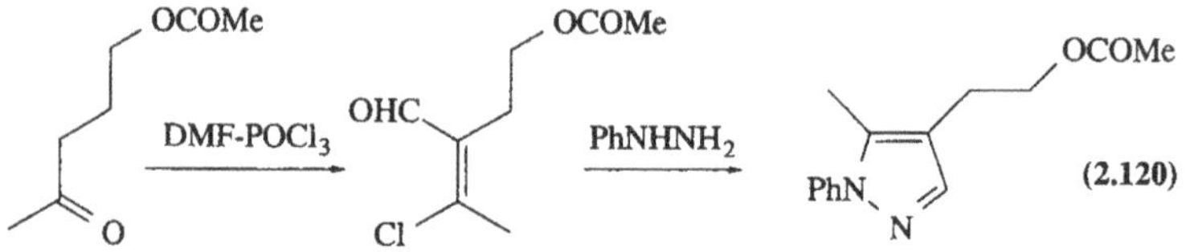

(2.120)

Alkyltrifluoromethyl ketones react with DMF-POCl$_3$ to give (E) and (Z)-mixtures of the corresponding fluorinated β-chlorovinylaldehydes.[301]

The usual pattern of regioselectivity is illustrated by butan-2-one, the more substituted alkene **180** being obtained.[152] However, where the 2-substituent is particularly bulky, the less substituted alkene [(E)- and/or (Z)-isomers] prevails, e. g. **181** obtained from 4-methyl-2-pentanone.[294]

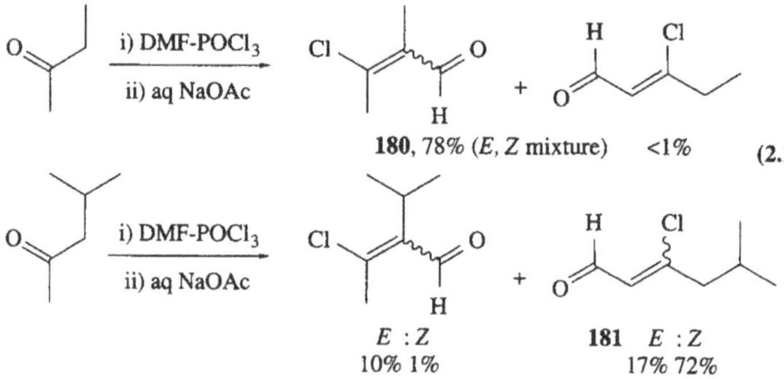

(2.121)

Cycloalkanones, from cyclobutanone to cyclooctanone, have all been converted into the corresponding β-chlorovinylaldehydes; the yields are generally good, and are improved by using a solvent such as trichloroethene in the higher range of reaction temperatures, typically 55-60°C.

Aryl ketones give good yields of relatively stable β-chlorovinylaldehydes. A variety of substituents at the 3- and/or 4-position of the aromatic ring are tolerated. Aryl methyl ketones react in good yield; thus acetophenone reacts with DMF-POCl$_3$ in dichloroethane (60-70°C, 3 hours) to give the corresponding β-chlorovinylaldehyde in 98% yield.

A 1:1 mixture of (*E*)- and (*Z*)-isomers of β-chloro-3-phenyl-2-(trifluoromethyl)acrylaldehyde results by treating the corresponding fluorinated aryl ketone with DMF-$POCl_3$.[301]

Phenylacetaldehydes usually afford predominantly the (*E*)-β-chlorovinylaldehydes.[302]

Acetylferrocenes react with Vilsmeier reagents at 0°C to give high yields of 2-formyl-1-chlorovinylferrocenes.[303]

Terminal alkynes, RC≡CH, are generally converted into β-chlorovinylaldehydes, RCCl=CHCHO, by Vilsmeier reagents.[304]

β-Chlorovinylaldehydes can be generated in the presence of an ester group.[305]

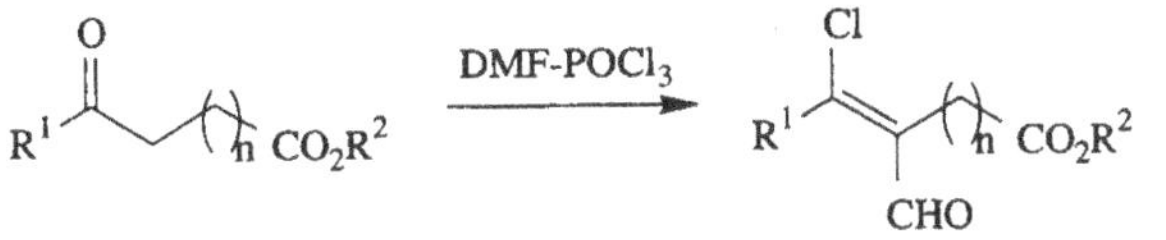

(2.122)

R^1=alkyl, aryl, thienyl; R^2=methyl, ethyl; n=1,2,3

2.11.3.2 β-Chlorovinylaldehydes from Alkyl Aryl Ketones

Simple alkyl aryl ketones usually give the chloroformylalkene in good yield: acetophenone and propiophenone afford the β-chlorocinnamaldehydes (47% and 90% respectively).[152,306] However, yields from *p*-substituted acetophenones are typically below 30%.[298] A procedure for the preparation of substituted dihydrocinnamaldehydes from aryl ketones involves addition at 70-80°C to a mixture of DMF-$POCl_3$, followed by cooling and treatment with aqueous NaOH.[307] Desoxybenzoin affords a mixture of diastereoisomeric β-chlorovinylaldehydes upon treatment with DMF-$POCl_3$ (scheme 2.123).[298]

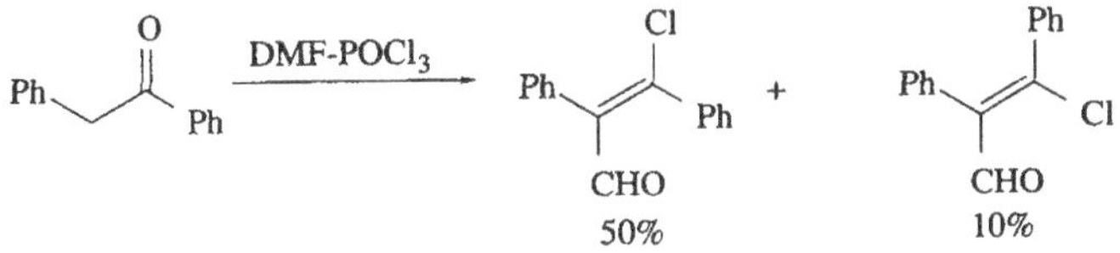

(2.123)

Reaction of the arylketones **182** with Vilsmeier reagents affords a mixture of the corresponding β-chlorovinylaldehyde and the iminium salts.[141]

RCH = CHCOMe —[i) DMF-$POCl_3$; ii) $HClO_4$; iii) $NaHCO_3$]→ RCH = CH − CCl = CH − CHO + RCH = CH − $(Me_2N)C$ = CH − CH = $\overset{+}{N}Me_2$ ClO_4^- (2.124)

182 R= C_6F_5, C_6H_5

Diacetylbenzenes undergo diformylation: when DMF-$COCl_2$ is used as the Vilsmeier reagent, a mixture of $CHCl_3$ and CH_2Cl_2 as solvent gave

excellent yields of β,β′-dichloro-*m*-benzenediacrylaldehyde.[308] Bis- and tris-
(β-chlorovinylaldehydes) were formed from 1,4-diacetylbenzene and 1,3,5-
triacetylbenzene respectively.[309] Diacetyl- and triacetyl-benzenes react with
Vilsmeier reagents to give the corresponding bis- and tris-(β-chlorovinylaldehydes), respectively (figure 2.125).[310]

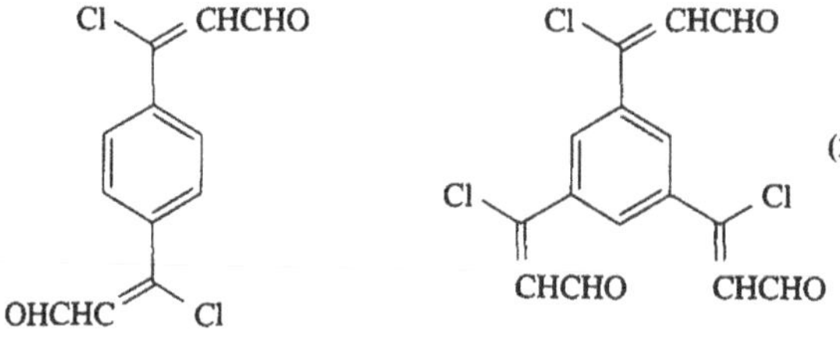

(2.125)

3,4-Dimethoxyacetophenone can undergo both mono- and di-
formylation,[135] whereas an indene is formed from propioveratrone[311] (section
4.2.2).

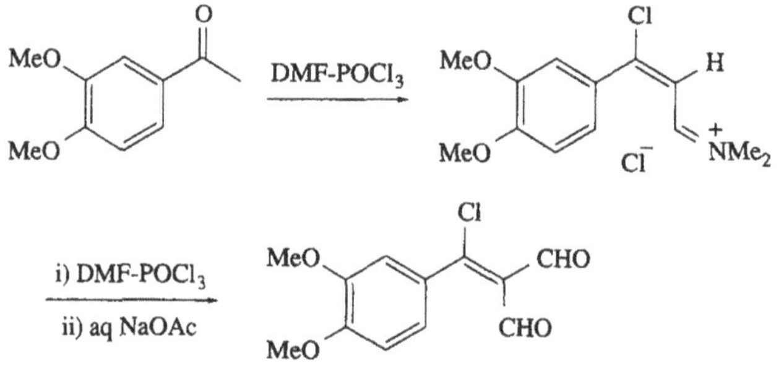

(2.126)

During the chloroformylation of 1-acetonaphthone, migration of the
acetyl group occurs, the same aldehyde being formed as that obtained from
2-acetonaphthone.[43,44]

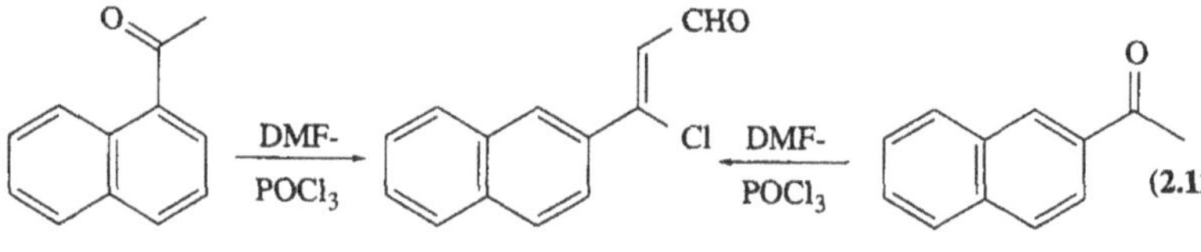

(2.127)

2.11.3.3 *β-Chlorovinylaldehydes from Saturated Monocarbocyclic Ketones*

Unsubstituted Cycloalkanones

β-Chlorovinylaldehydes are the most abundant of the β-halovinylaldehydes prepared using Vilsmeier reagents, although Arnold and Holy[159]
showed that cycloalkanones of five- to eight-membered rings are converted
into the corresponding β-bromoacrylaldehydes by DMF-PBr$_3$ complexes.

The corresponding β-chloroacrylaldehydes were reported by Ziegenbein and Lang.[296]

Cyclobutanone affords the corresponding β-chlorovinylaldehyde (scheme 2.128).[125] No other product was reported, even when an excess of formylating agent was used. With DMF-POCl₃, cyclopentanone was converted into 2-chlorocyclopentene-1-carboxaldehyde (54%),[312] the reaction for cyclohexanone proceeding analogously.[293,296]

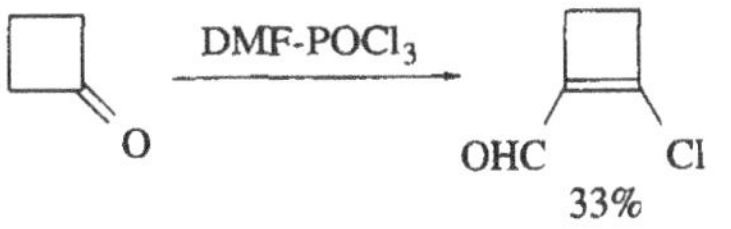

(2.128)

If a large excess of the Vilsmeier reagent is used, cyclopentanone and cyclohexanone afford the corresponding 3-chloropentamethinium salts (section 2.29.4).

The general pathways depicted in scheme 2.134 are consistent with the products formed by the action of DMF-POCl₃ on dimedone **183** (scheme 2.129).[313] The dialdehyde **184** could be derived from the *gem*-dimethyl analogue of the dication **203** by nucleophilic attack leading to dealkylation; alternatively, the reaction might not proceed beyond the formation of the *gem*-dimethyl analogue of the cation **200**.

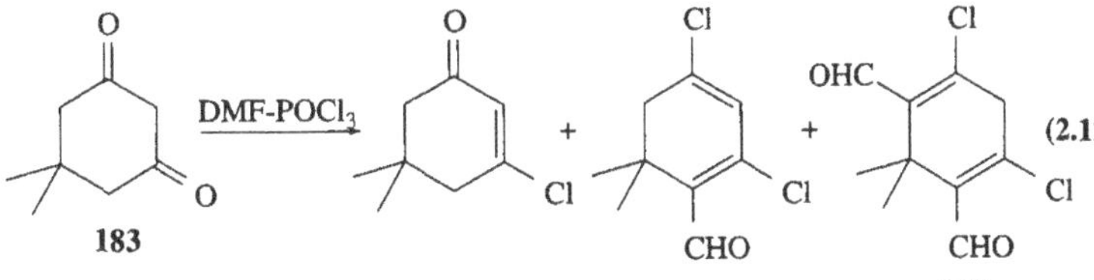

(2.129)

Chloroformylation of cyclooctane-1,5-dione under mild conditions gave the keto-aldehyde **185**. The dialdehyde **186** is formed from cyclooctane-1,5-dione and DMF-POCl₃.[314]

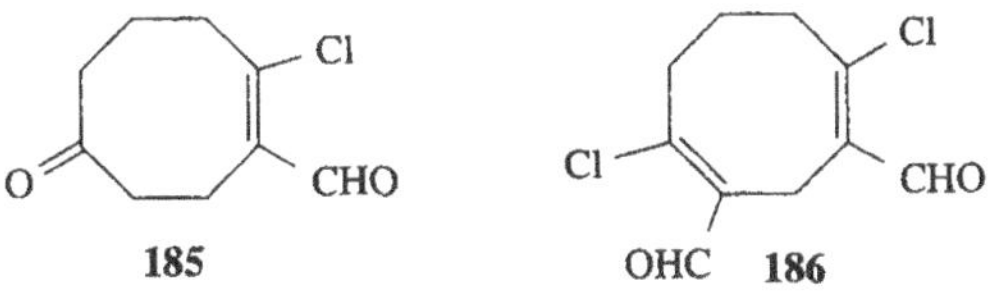

(2.130)

Monosubstituted Cycloalkanones

The regiochemistry in the reaction of unsymmetrical monosubstituted cycloalkanones has been little explored. Karlsson and Frejd[315] showed that a relatively small group, such as a 3-methyl substituent has a large steric influence on the attack of the Vilsmeier reagent, in the cases of six-, seven-, and eight-membered rings, although not for the five-membered case (scheme

2.131). An equilibrium mixture of enols was proposed as being present under Vilsmeier-Haack conditions. Interestingly, the larger rings, despite possessing more conformational mobility, exhibit greater regioselectivity. However, 4-methylcycloheptanone gave a 1:1 mixture of the regioisomers **187** and **188**, showing that the 4-methyl group has little influence on the regiochemistry of the reaction.

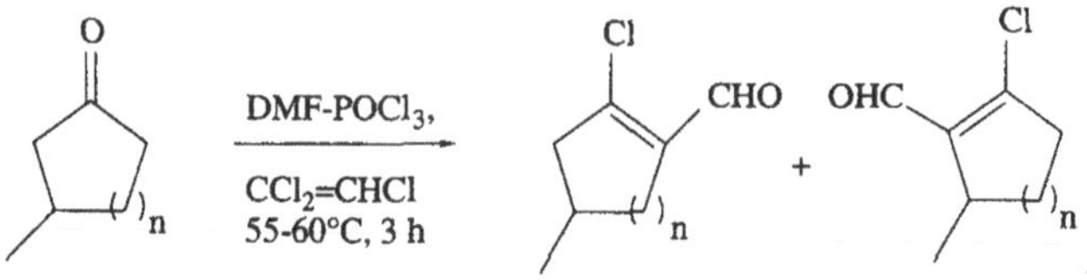

			total yield, %
60	n=1	40	43
90	n=2	10	52
95	n=3	5	42
100	n=4	0	56 (2.131)

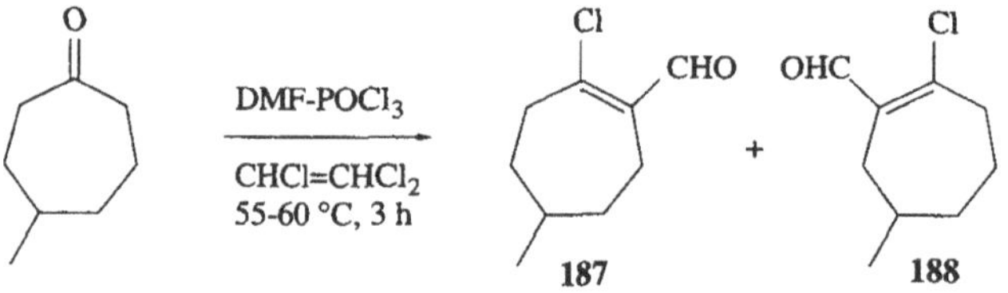

187   **188**   48%

Steroidal Cycloalkanones

3-Keto-5α-steroids afford the 3-chloro-2-formyl-2-ene derivatives with DMF-POCl$_3$; thus 17-acetoxy-5α-androstan-3-one,[316] 17β-acetoxy-17α-methyl-5α-androstan-3-one,[317] and 5α-pregnan-3,20-dione[122] give the corresponding 3-chloro-2-formyl derivatives **189** (22-27%) (scheme 2.132). The reaction proceeds with 4,4-disubstituted precursors: 17β-acetoxy-4,4-dimethylandrost-5-en-3-one affords 17β-acetoxy-3-chloro-2-formyl-4,4-dimethylandrosta-2,5-diene (62%) from a reaction with DMF-POCl$_3$ at 50-60°C for 4 hours.

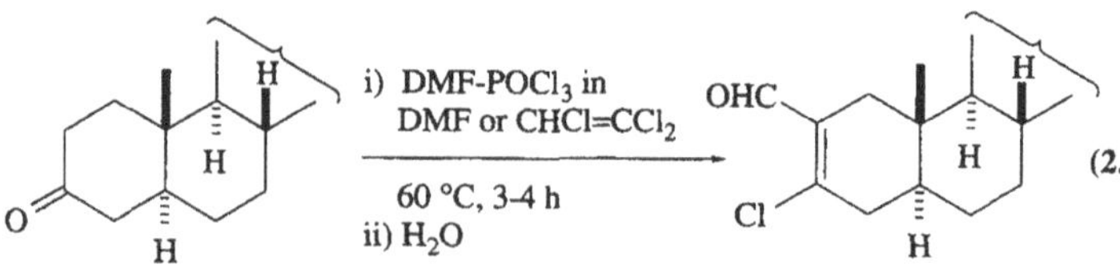

189

The regioselectivity of formylation in ring A is markedly influenced by the relative configuration of the A-B ring junction. Thus, the 5β-androst-3-ene **190** is converted into the 3-chloro-4-formyl derivative **191**;[289,316] the use of acetyl chloride as the solvent is unusual. Chloroformylation of a steroidal ketone having the carbonyl group at a position other than C-3 usually affords the corresponding β-chlorovinylaldehyde. Thus, 3β-hydroxyandrost-5-en-17-one 3-acetate **192** is chiefly converted into the aldehyde **193a**, with the chloroalkene **193b** being formed as a by-product (scheme 2.133). A *D*-homosteroidal ketone has been converted by DMF-$POCl_3$ into the chlorovinylaldehyde **194**.[67]

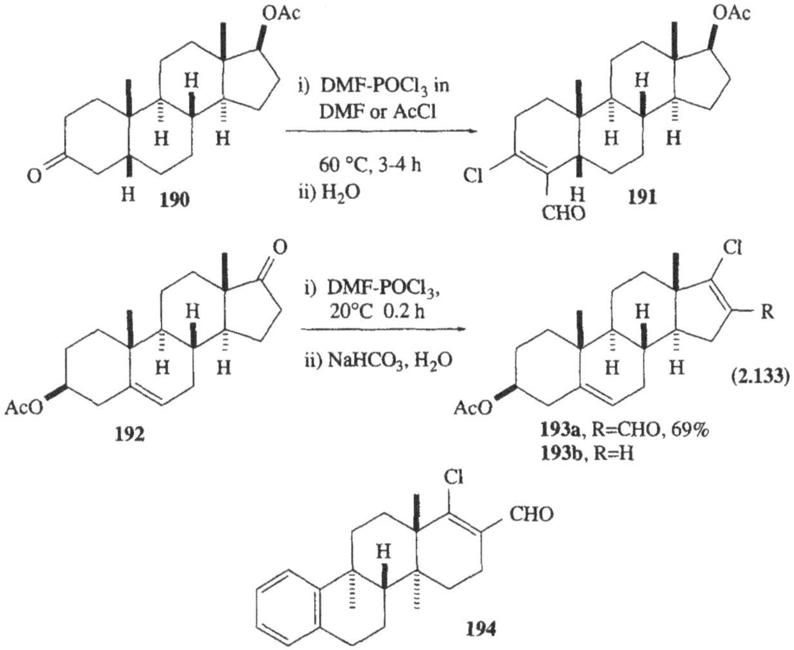

190 **191** **192**

193a, R=CHO, 69%
193b, R=H

194

(2.133)

2.11.3.4 *β-Chlorovinylaldehydes from α,β-Unsaturated Carbocyclic Ketones*

2-Cycloalken-1-ones

Of the 2-cycloalken-1-ones reacted with Vilsmeier reagents, 2-cyclohexen-1-ones are the most common. An interesting variety of chloroformyl compounds can be obtained from substituted cyclohexenones and their derivatives. The exothermic reaction of 2-cyclohexen-1-one with *N*-formylmorpholine (NFM)-$POCl_3$ in trichloroethene at 20°C afforded a deep red mixture which upon hydrolysis gave the somewhat unstable dialdehyde **205**;[318] upon keeping at 25°C for several weeks, aerial oxidation afforded the trialdehyde **206**.

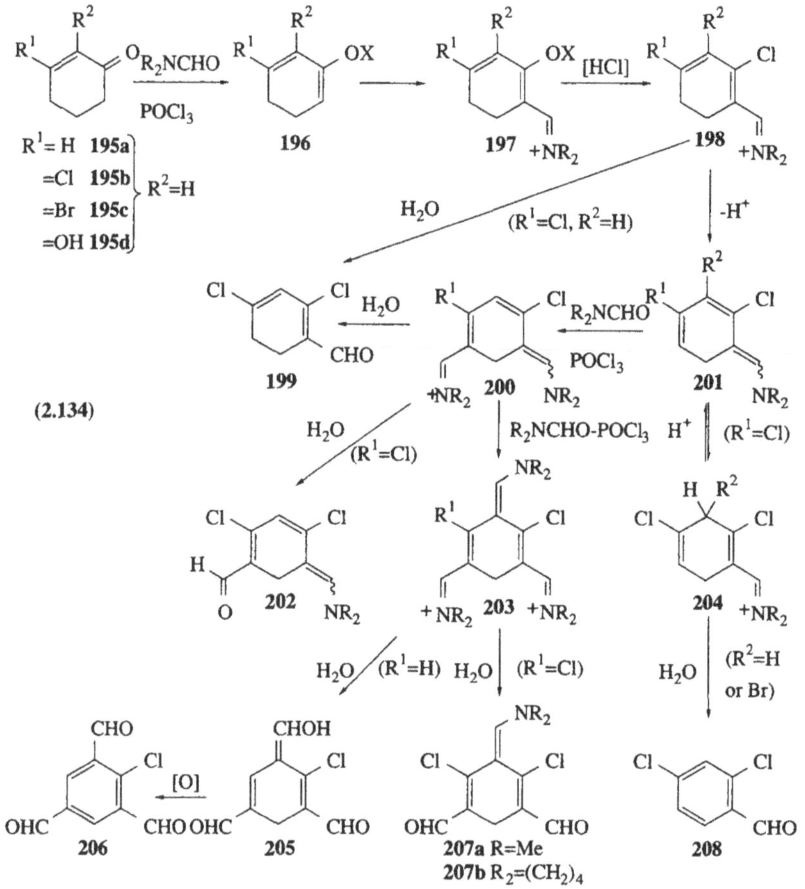

(2.134)

Scheme 2.134 provides a pathway which unifies the experimental observations on the reaction of the 3-substituted-2-cycloalken-1-ones **195a-195c** with Vilsmeier reagents; cyclohexane-1,3-dione **195d** is discussed in section 3.1.1. There is substantial evidence for the involvement of an enolic intermediate **196** (in which X may be CH(Cl)NR$_2$ rather than the enol, X=H, or a phosphate derivative). For ketones **195a-c**, the early stages of the reaction apparently do not involve the C=C double bond. A common intermediate of the form **203**, stabilized by extensive delocalization, is considered to be present prior to hydrolytic work-up. For the ketones **195b** and **195d**, partial hydrolysis affords the dialdehydes **207**, whereas for 2-cyclohexen-1-one, the intermediate **203** (R^1=H) is more susceptible to nucleophilic attack, and the enolic dialdehyde **205** is formed, which is remarkable for its stability over the benzenoid tautomer.

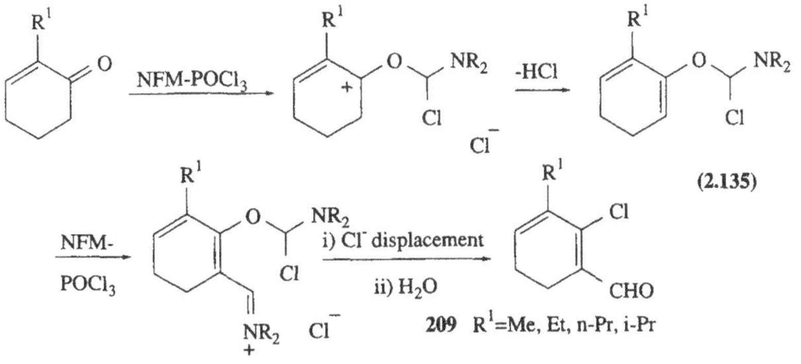

(2.135)

For 2-alkyl-2-cyclohexen-1-ones, intermediates of the form **198** (R^1=H) are evidently involved because hydrolytic work-up affords the aldehydes **209**. Such mechanistic features are compatible with the formation of the dialdehyde **210** from 4,4-dimethyl-2-cyclohexen-1-one and NFM-POCl$_3$ (scheme 2.136).[32,319]

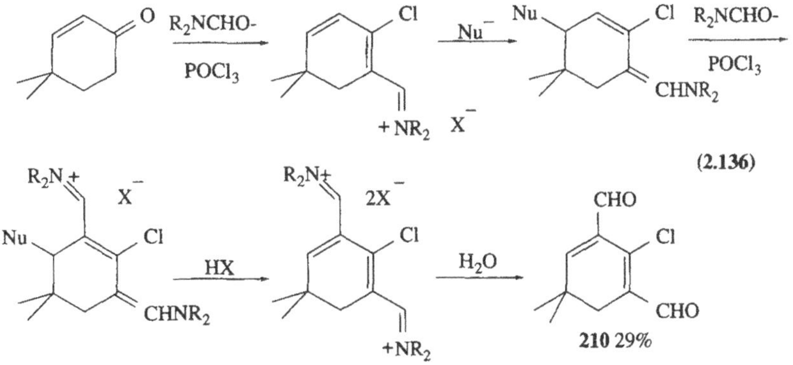

(2.136)

A unified pathway[32] illustrating typical processes involving 2-cyclohexen-1-ones lacking a methyl group at C-3 is given in scheme 3.4. For 3-methyl-2-cyclohexen-1-one and its 5-substituted derivatives **211** an entirely different course is followed (scheme 2.137).[32,318,319] Although it is uncertain whether Y=Cl or an oxygenated moiety in the exocyclic alkene **212**, several of the analogous endocyclic alkenes are known to be thermodynamically favoured over their endocyclic isomers.

Isophorone affords **212c**[320] with 2 equiv. of the Vilsmeier reagent, but with a large excess of the reagent undergoes iminoalkylation to give the polymethinium species **213**, which is converted, during hydrolysis into the 4*H*-pyran **214** (scheme 2.137).[34,319] If the reaction of isophorone with DMF-POCl$_3$ (2 equiv.) is quenched after 15 minutes an 80% yield of a mixture of three chloro-dienes is obtained.[319] From the aldehydes 3-methyl-2-

cyclohexen-1-one, 3,5-dimethyl-2-cyclohexen-1-one and isophorone, a mixture of *(E)-* and *(Z)*-isomers was obtained. *(1S)*-(-)-Verbenone afforded the chlorinated aldehyde **215**.

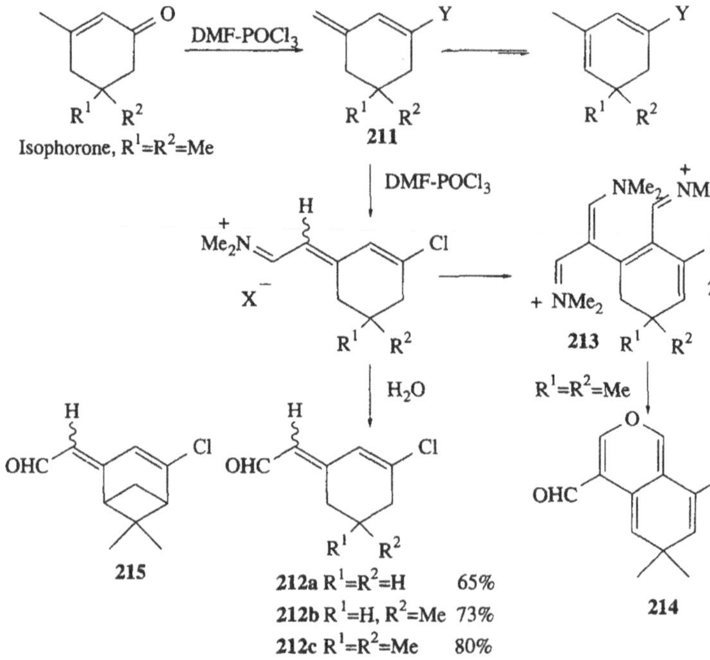

α,β-Unsaturated Steroidal Ketones

Reviews in this area are available.[3,152] α,β-Unsaturated steroidal ketones usually afford a mixture of products of which the halo-diene is the major component. Thus, 3-oxo-4-ene steroids when heated with DMF-POCl$_3$ in an inert solvent afforded the corresponding 3-chloro-3,5-dienes; 3-oxo-1,4,6-trienes gave the 3-chloro-1,3,5,7-tetraenes.[321] However, 19-nortestosterone acetate **216** gave the aldehydes **217b** and **217c** in equal amounts, in addition to the chlorodiene **217a** (scheme 2.138).[322] Whereas enolization of the 3-oxo group was postulated as the first step,[322] both for the 19-methyl-3-oxo-enes and for the 19-nor-compounds, only in the latter series was the Vilsmeier reagent able to attack the C-4 and C-6 positions, prior to subsequent displacement of the 3-oxygenated function by chloride. The steric hindrance of the 19-methyl group prevents further reaction of the 3-halo-3,5-dienes with the Vilsmeier reagent.

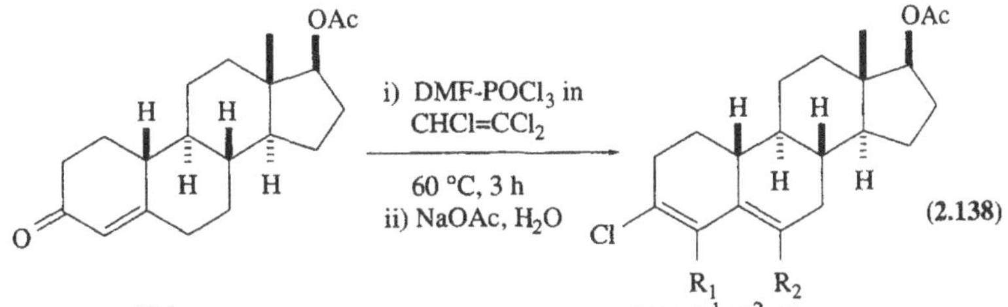

(2.138)

216

217a R^1=R^2=H
217b R^1=H, R^2=CHO
217c R^1=CHO, R^2=H

A series of steroidal 4,6-dien-3-ones afforded with DMF-POCl$_3$ a mixture of 3-chloro-2,4,6-trienes **219a**, 3-chloro-2-formyl-2,4,6-trienes **219b**, and 3-chloro-3,5,7-trienes **220** (scheme 2.139). The steroid **219a** is implicated as an intermediate since it was converted by DMF-POCl$_3$ into the aldehyde **219b**.[323]   In the cases of 17β-acetoxyestra-4,6-dien-3-one **221a** and 17-acetoxy-19-norpregna-4,6-dien-3,20-dione **212b**, the expected aldehydes **222a** and **222b**, respectively, are accompanied by the aromatic dialdehydes **223a** and **223b**, formed by oxidation.[323] 3β-Acetoxy-5-ene-7-ketosteroids **224** eliminate acetic acid under Vilsmeier conditions giving the 3,5-dien-7-ones **225** which are then converted into a mixture of chlorinated steroids **226a** and **226b**. Formylation of triene **226a** gives the aldehyde **226b**.[324] 3-Alkoxy-3,5-dienes undergo attack by Vilsmeier reagents at C-6 only.[120]

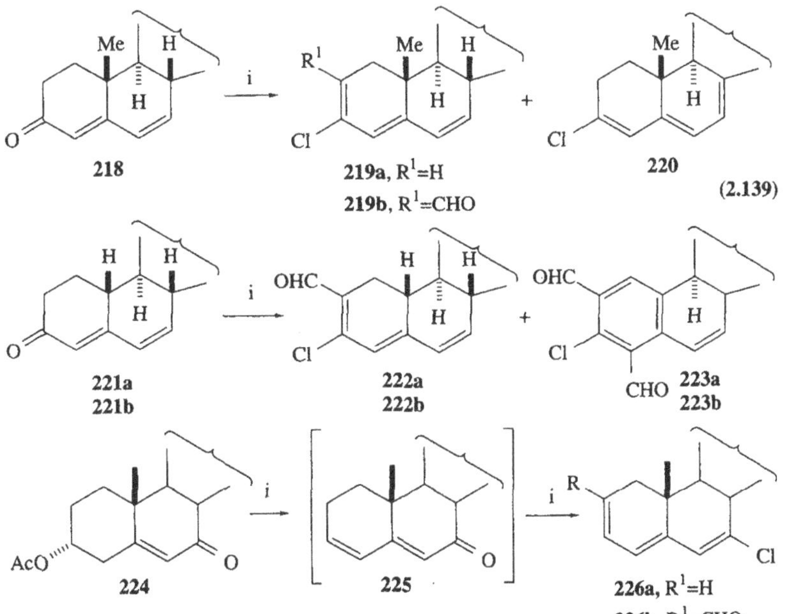

218

219a, R^1=H
219b, R^1=CHO

220

(2.139)

221a
221b

222a
222b

223a
223b

224

225

226a, R^1=H
226b, R^1=CHO

i, DMF-POCl$_3$

2.11.3.5 β-Chlorovinylaldehydes from Benzo-Fused Cycloalkanones

Benzo-fused cycloalkanones are usually converted by Vilsmeier reagents into the corresponding β-chlorovinylaldehydes in good yield and under mild conditions (figure 2.140). Under normal conditions, formylation of the aromatic ring does not occur. The resulting β-chlorovinylaldehydes have been used in the synthesis of a wide variety of polycondensed heterocycles (chapter 4).[325-328]

1-Indanone, α-tetralone and benzosuberone afford the corresponding β-chlorovinylaldehydes **227a**, **227b** (77%)[159] and **227c** (75%)[297] respectively. Derivatives of α-tetralone with alkyl groups on either ring afford the expected β-chlorovinylaldehydes **228a-228c**.[327] No aromatic formylation is observed even when 6- or 7-methoxy groups are present. β-Tetralones usually give the 2-chloro-3,4-dihydro-1-naphthalenecarboxaldehyde derivatives [*e.g.* **229** is formed[329] (44%)], although these decompose much more readily than the isomers such as **228**.[330] Some β-tetralones afford the corresponding 2-chloro-1,3-naphthalenedicarboxaldehydes. The reaction of β-tetralone with $HCONH_2$-$POCl_3$ afforded 5,6-dihydrobenzo[*f*]quinazoline in very low yield.[331]

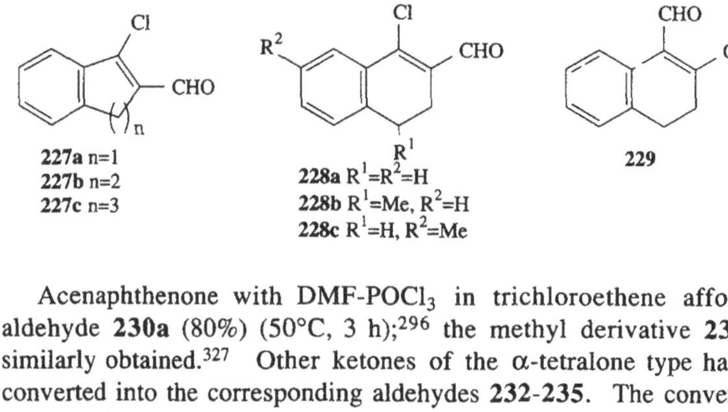

227a n=1
227b n=2
227c n=3

228a R^1=R^2=H
228b R^1=Me, R^2=H
228c R^1=H, R^2=Me

229

(2.140)

Acenaphthenone with DMF-$POCl_3$ in trichloroethene affords the aldehyde **230a** (80%) (50°C, 3 h);[296] the methyl derivative **230b** was similarly obtained.[327] Other ketones of the α-tetralone type have been converted into the corresponding aldehydes **232-235**. The conversion of anthrone into 10-chloro-9-anthracenecarboxaldehyde **231** is probably the earliest reported example of a 'chloroformylation' of a methylene ketone.[332]

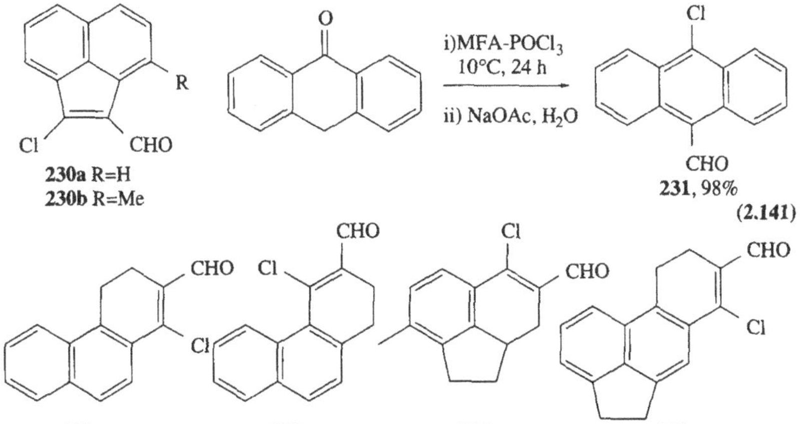

230a R=H
230b R=Me

i)MFA-POCl$_3$
10°C, 24 h
ii) NaOAc, H$_2$O

231, 98%

(2.141)

232   **233**   **234**   **235**

2.11.3.6 β-Chlorovinylaldehydes from Cycloalkanones Fused to Heteroaromatic Rings

The aldehyde **236** was prepared from the corresponding ketone.[327] The indazolecarboxaldehyde **237** was also directly prepared by the action of a Vilsmeier reagent on the corresponding ketone.[333]

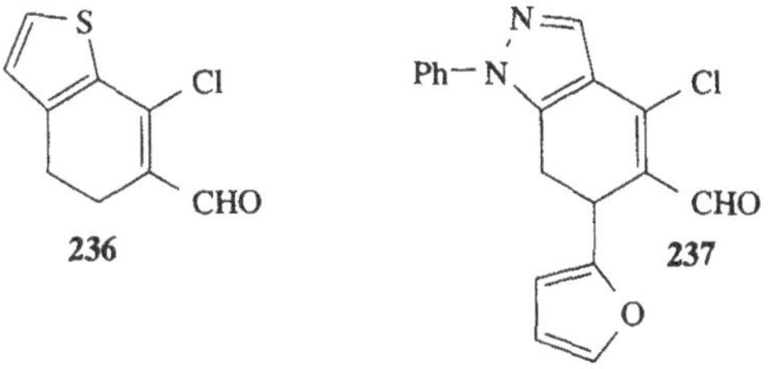

(2.142)

236   **237**

The Vilsmeier formylation of 1-ketotetrahydrocarbazoles **238** gave appreciable quantities of aromatized products **239**, in addition to the chlorovinylaldehydes **240** as the major products (scheme 2.143).[49]

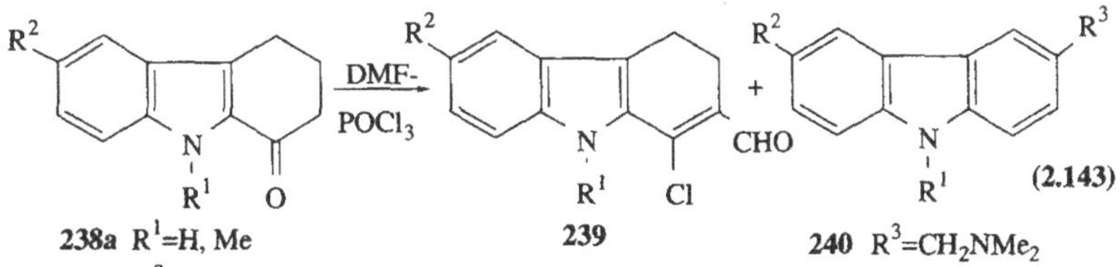

DMF-
POCl$_3$

(2.143)

238a R^1=H, Me
238b R^2=H, Cl, Br, CO$_2$Et
239
240 R^3=CH$_2$NMe$_2$

The benzo[*b*]thiophen-4-one **241** reacts with the Vilsmeier-Haack reagent giving four products, a reaction which demonstrates competition between formylation of the carbonyl group and the thiophene ring (scheme 2.144).[334]

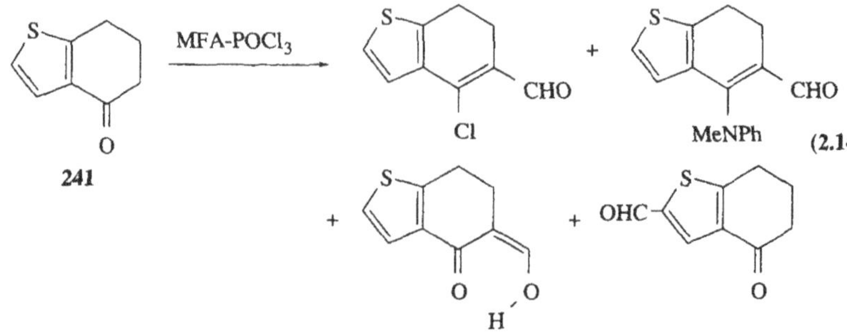

(2.144)

2.11.3.7 β-Chlorovinylaldehydes from Cyclic Ketones with One or More Heteroatoms in the Ketone Ring:

The principal reactions are of five- and six-membered heterocyclic ketones containing either a sulfur or an oxygen atom in the ring. Aromatization to formylthiophenes is typical.

Tetrahydro-4H-thiopyran-4-one and tetrahydro-4H-pyran-4-one are converted at ambient temperatures by DMF-POCl$_3$ into the respective chlorovinylaldehydes[335] which are suitable for further functionalization (scheme 2.145). Two 2,3-dihydro-4H-thiopyran-4-ones afforded the 4-chloro-2H-thiopyran-3-carboxaldehydes 242[335] and 243[336] (scheme 2.145). The chloroformylation of a piperidin-4-one using one equivalent of DMF-POCl$_3$ afforded the aldehyde 244 (24%).[33]

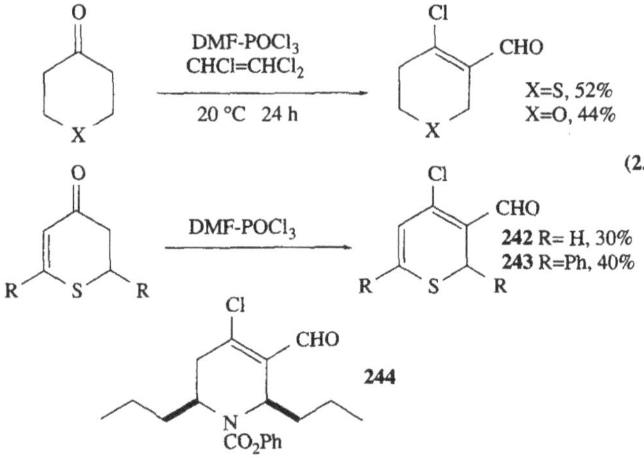

(2.145)

The action of DMF-POCl$_3$ on thiochroman-4-one 245 was shown[297,335] to afford the β-chlorovinylaldehyde 247 exclusively at temperatures below 50°C, but at 100°C, 3-formylthiachromone 246 was formed in appreciable quantity (scheme 2.146).[335] A mechanism involving oxidation by the Vilsmeier reagent was tentatively proposed.[335b]

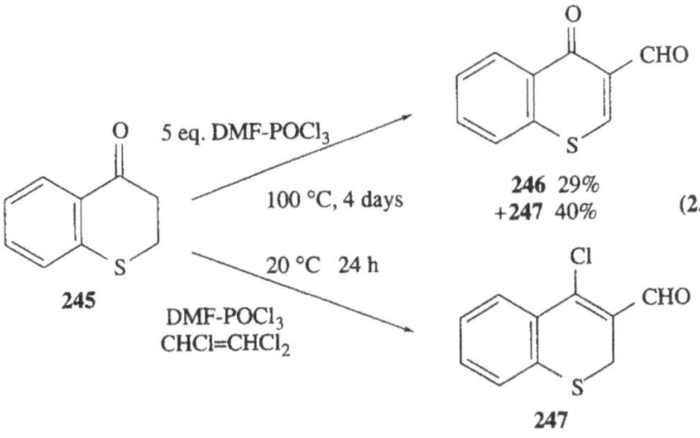

$$\text{(2.146)}$$

The reaction of chroman-4-one **248** is also temperature-dependent,[335] the expected aldehyde **250**[297,335] being obtained at 35°C (scheme 2.147). The formation[335] of 3-(chloromethyl)chromone **249** at 100°C is thought to proceed by isomerization to a 3-(chloromethyl)benzo[*b*]pyrylium cation **251** (scheme 2.148).

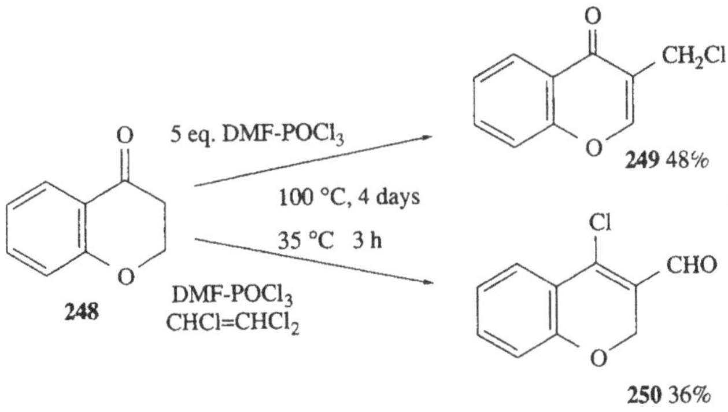

$$\text{(2.147)}$$

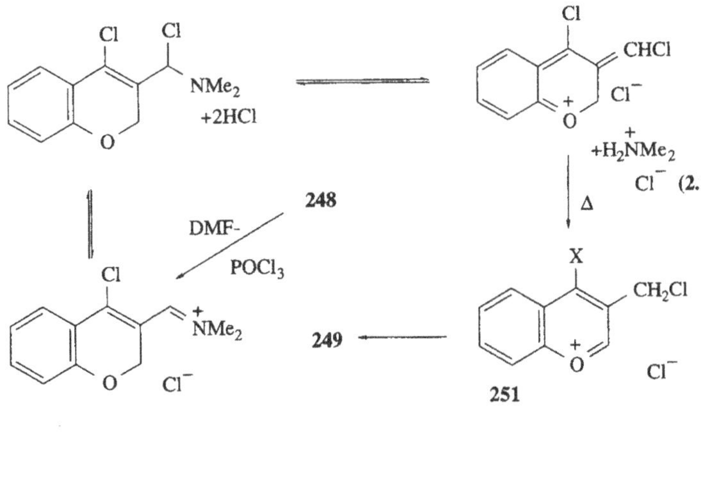

$$\text{(2.148)}$$

The reaction of substituted chroman-4-ones with Vilsmeier reagents illustrates the importance of intricate steric effects in Vilsmeier reactions with ketones. 4-Chloro-3-carboxaldehydes can be obtained from many C-2 unsubstituted chroman-4-ones[298,337] but a single 2-methyl group is sufficient to block 3-formylation. 7-Methoxy-2-methylchroman-4-one[338] affords the 4-chlorochromene in high yield.[339] The Vilsmeier formylation of chromenes is also usually prevented by substitution at C-2,[105,340] although the electronic effect of a 7-methoxy group, if introduced into 2,2-dimethyl-2*H*-chromene is sufficient to promote 6-formylation.[340] The interplay of steric and electronic effects operating during the Vilsmeier formylation of 2,2-dimethylchroman-4-one derivatives has been studied.[339] The high degree of solvation of the Vilsmeier reagent in the mildly polar solvents normally employed[3] renders the reaction particularly susceptible to steric effects.

2,2-Dimethylchroman-4-one derivatives **252** gave high yields of the corresponding 4-chloro-2*H*-chromenes **253** and only a small quantity (or none) of the chlorocarboxaldehydes **254** (scheme 2.149). Prolonged reaction times did not increase the yield of the latter; instead, formylation at C-6 of **253** occurred.[339,341] The results indicate that the chlorocarboxaldehydes such as **254** arise not by formylation of **253**, but presumably by formylation of an enolic precursor.

(2.149)

Flavanone affords the aldehyde **255** when treated with DMF-POCl$_3$.[342,343] However, at elevated temperatures and prolonged reaction times, a mixture of 3-chloromethyleneflavanone **256** (25%) and 3-formyl-4-hydroxyflav-3-ene **257** (30%) were obtained.[343] No trace of the 3-chloromethyl-2-phenyl compound analgous to **249** was detected. 4-Chloro-3-formylcoumarin **258** has also been prepared by a Vilsmeier reaction.[344] The heterocyclic aldehydes **259a** and **259b** have been prepared,[297] as has aldehyde **261**.[345] Reaction of heterocycles containing a carbonyl group and a nitrogen atom in the same ring with Vilsmeier reagents are rare, although Comins reported the synthesis of aldehyde **244**, albeit in low yields.[33] Hydroxytriazolopyrimidines have been converted by Vilsmeier-Haack reagents into chlorovinylaldehydes such as **260** (3 h; 70-80°C; 85%).[346]

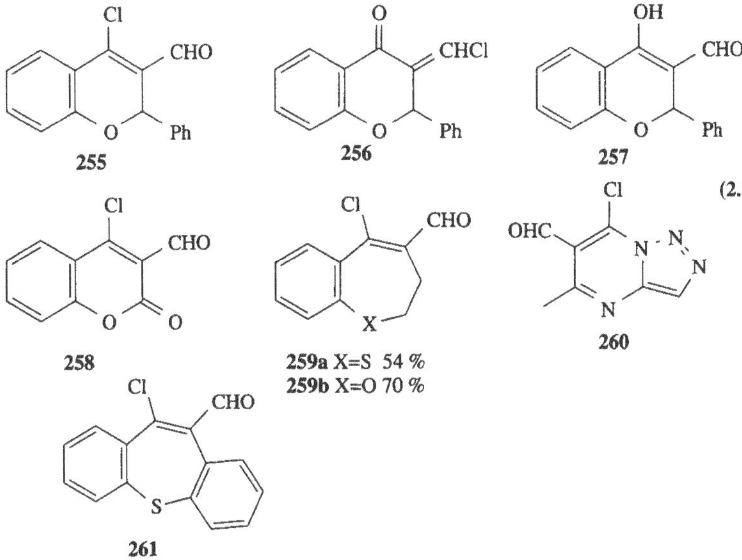

255   256   257   (2.150)

258   259a X=S 54 %   260
       259b X=O 70 %

261

Some unusual substitution reactions under Vilsmeier conditions have been reported with coumarin derivatives (scheme 2.151).[347]

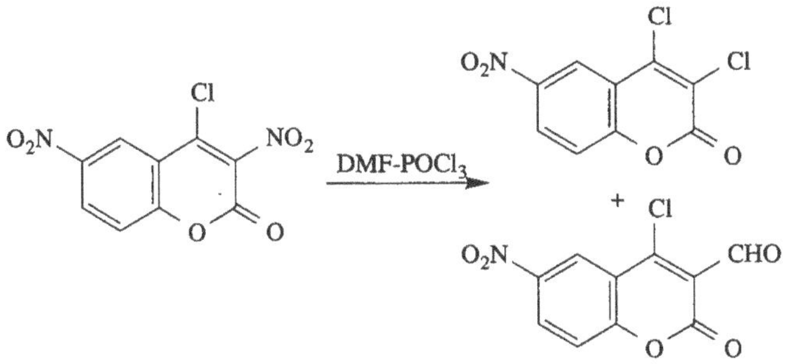

(2.151)

Several thiazinones have been reacted with Vilsmeier reagents. A variety of 2-arylthiazine derivatives **262** afforded mixtures of the chlorinated thiazinones **263a** and **263b** in ratios depending on the electronegativity of the *p*-substituents in the aryl ring; electron-donating substituents (*e.g.* NMe$_2$) favored **263a**, whereas electron-withdrawing ones (*e.g. p*-NO$_2$) favoured **263a**.[348] Vilsmeier reactions of 2-aryl-4-hydroxy-1,3-thiazin-6-ones **262** were analyzed by CNDO/2 and MNDO/3 MO methods, and by photo-electron spectroscopy; the product ratios (R=H or CHO in **263**) are chiefly determined by the energetic and structural parameters of the HOMO.[349]

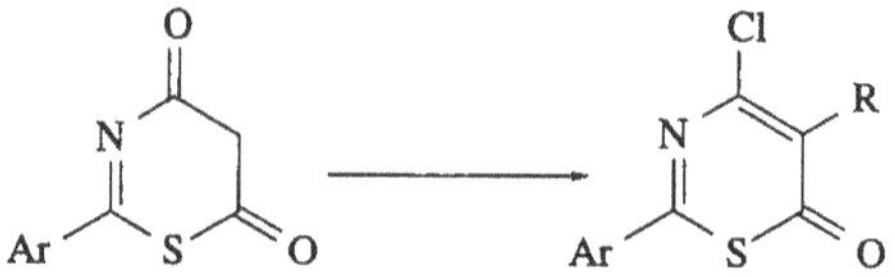

(2.152)

262   **263a, R=H**
263b, R=CHO

The chloroformyl derivatives **264** and **265** were formed by the action of DMF-POCl$_3$ on the corresponding lactams.[350] 2*H*-1,4-Benzoxazin-3-one and its derivatives react with DMF-POCl$_3$ to give iminium salts **266** which form a variety of products with alkali, some lacking chloro-substituents (scheme 2.153).[51]

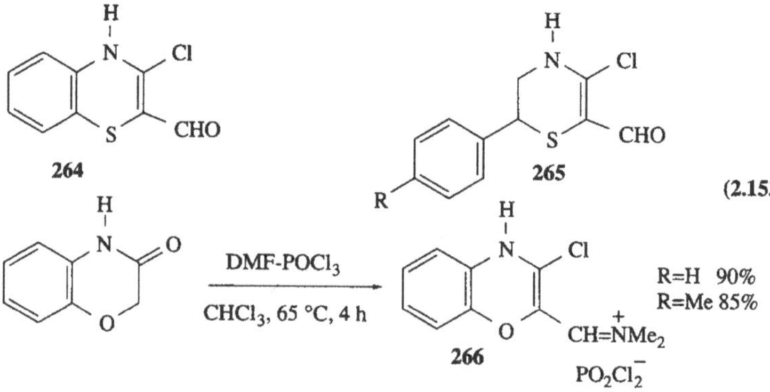

264   **265**

(2.153)

DMF-POCl$_3$

CHCl$_3$, 65 °C, 4 h

266

R=H 90%
R=Me 85%

CH=NMe$_2$

PO$_2$Cl$_2^-$

The chloroformylated steroid derivatives **267** and **268** are formed by treating the appropriate azacholestanone and azahomoandrostenone derivatives with DMF-POCl$_3$ in refluxing chloroform.[351]

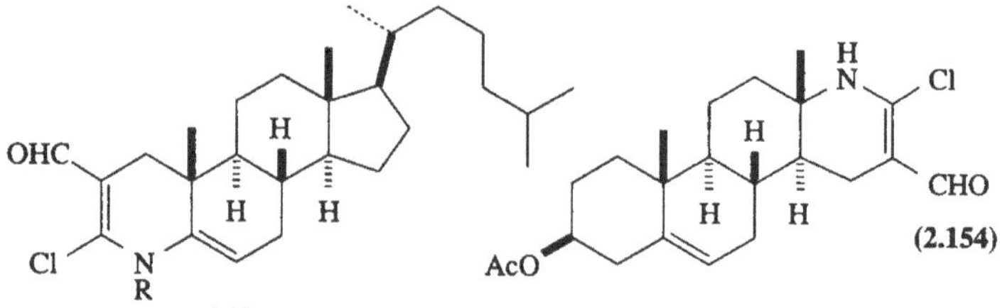

(2.154)

267   **268**

Amides and Imides

Glutarimides are converted by DMF-POCl$_3$ into the lactams **269**;[352] the diformyldihydropyridines **270** can also be obtained.[353] Imides of five-, six-, and seven-membered rings react with excess DMF-POCl$_3$ to give the useful di-β-chlorovinylaldehydes **272** (scheme 2.156).[354]

(2.155)

269 R=Me, Et, Pr, Ph   **270** R=H, Ph

The azepinones **271** are converted by excess DMF-POCl₃ into the aldehydes **272**; 4 equiv. of the Vilsmeier reagent diluted with CHCl₃ afforded a mixture of azepines **272** (10%) and **273** (35%).

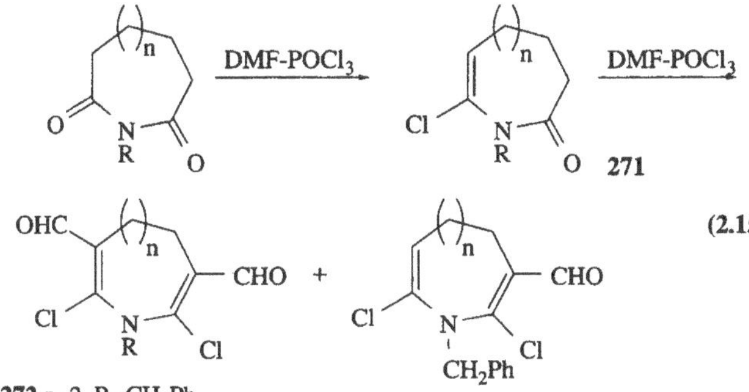

(2.156)

272 n=2, R=CH₂Ph
n=0, 1, 2; R=CH₂Ph, CH₂CO₂Et, Ph   **273**, n=2

2.11.3.8 β-Chlorovinyllaldehydes from Polycyclic Systems

2,3-Dihydro-1*H*-pyrrolizin-1-one **274** is converted by DMF-POCl₃ into **275** and **276**, depending on the reaction conditions (scheme 2.157).[355] The chloroalkene **277** was detected, and it is believed that enolizable C=O groups of acylated pyrroles are first converted into chloroalkenes, formylation occurring subsequently.

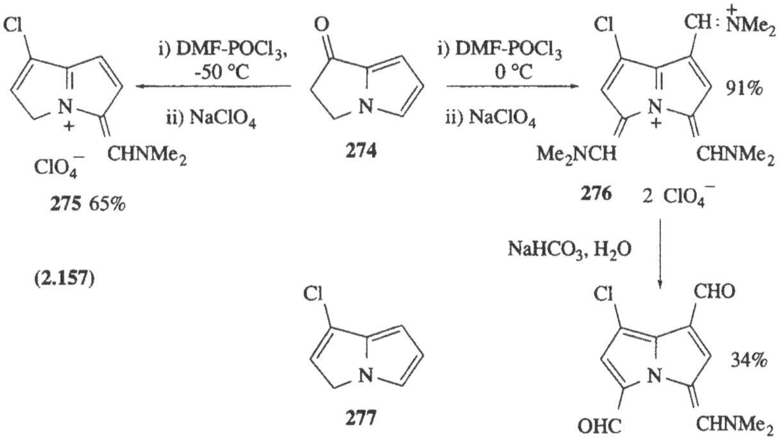

(2.157)

Angularly annellated derivatives of chroman-4-one are also converted into the corresponding β-chlorovinylaldehydes **278**[356,357] and **279**.[356] The formylated naphtho[2,3-*b*]furan **280** was similarly prepared from the corresponding naphtho[*b*]furanone. Vilsmeier reaction of 10,11-dihydro-benzo[*b,f*]thiepin-10-one with DMF-$POCl_3$ in trichloroethene at 80°C gave the carboxaldehyde **281** (52%).[345] The chloroformyl derivatives **282** and **283** are obtained from the corresponding benzazepin-2-one and Vilsmeier reagents.[297,350] Analogous reactions are known for related benzazepine-2-thiones.[350]

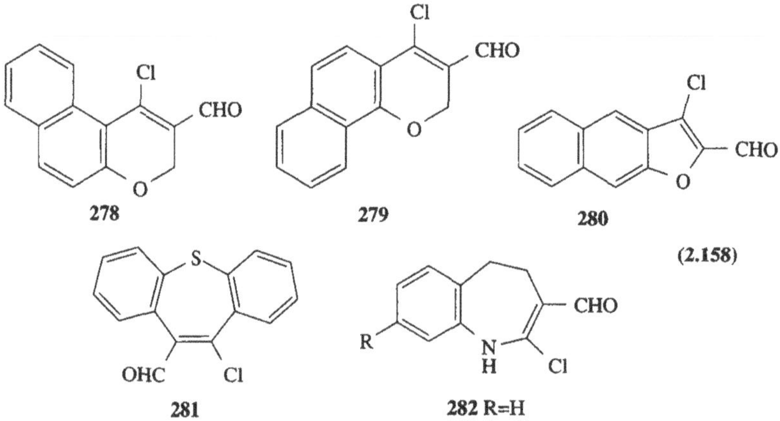

278 **279** **280**

(2.158)

281

282 R=H
283 R=Cl

Treatment of 2-coumarinone with a Vilsmier reagent afforded three products, two of them as mixtures of (*E*)- and (*Z*)-isomers (scheme 2.159).[358]

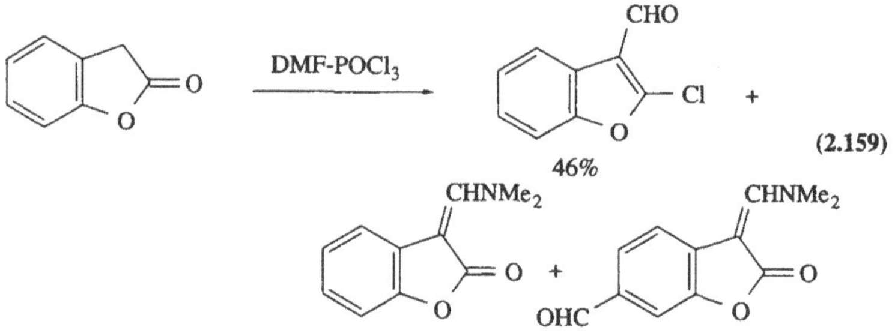

(2.159)

2.11.3.9 β-Chlorovinylaldehydes in Synthesis

β-Chlorovinylaldehydes have been hydrodehalogenated, thus allowing the overall conversion of an α-methylene ketone into an α-aldehyde.[359] Dehalogenation has been achieved using zinc dust in ethanol[70,360] or by hydrogenolysis using a Pd-C catalyst,[71] giving α,β-unsaturated aldehydes (scheme 2.160).

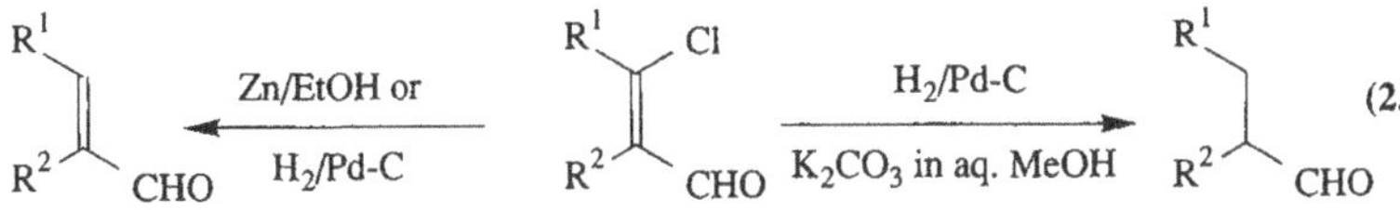

$$(2.160)$$

Simple β-chlorovinylaldehydes have been used to provide key synthetic intermediates in the synthesis of complex natural products. One such reaction is the synthesis of tabersonine and catharathine from **284**. After a Grignard reaction and oxidation, the intermediate **285** was reacted with the indole **286** to afford 15-oxo-$\Delta^{20(21)}$ secodine **287**, which was further elaborated to the natural products.[361]

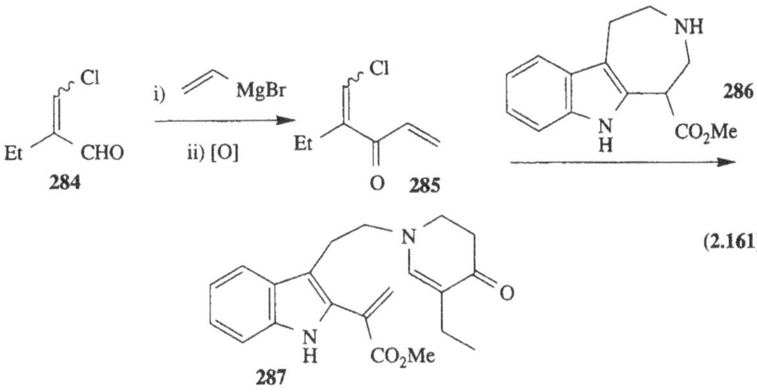

$$(2.161)$$

β-Chlorovinylaldehydes have been little used as dienophiles in Diels-Alder reactions; however, Willard and de Laszlo[362] reported the use of the β-chlorovinylaldehydes **288a** and **288b** in Diels-Alder reactions with butadiene, allowing the synthesis of the monoterpene natural products of the *Plocamium* series from the Diels-Alder adduct **289a**. The Diels-Alder reaction proceeded in higher yields with methyl butadiene and cyclopentadiene. Aldehyde **288a** also gives poor yields with those dienes, probably because of facile isomerization of **288a** to **288b** by traces of acid.

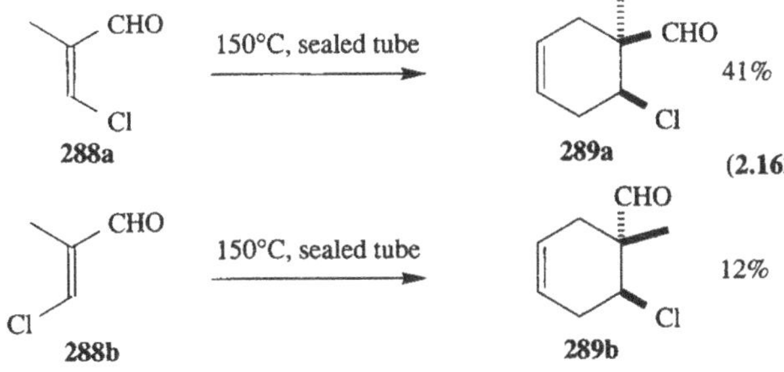

$$(2.162)$$

β-Chlorovinylaldehydes are available by other, more troublesome routes than by Vilsmeier methodology. Willard[362a] had to prepare the (Z)-β-chloro-vinylaldehyde **290** by an alternative route for the synthesis of costatolide

291, since only the (*E*)-β-chlorovinylaldehyde is obtained by the reaction of propanal with a Vilsmeier reagent.

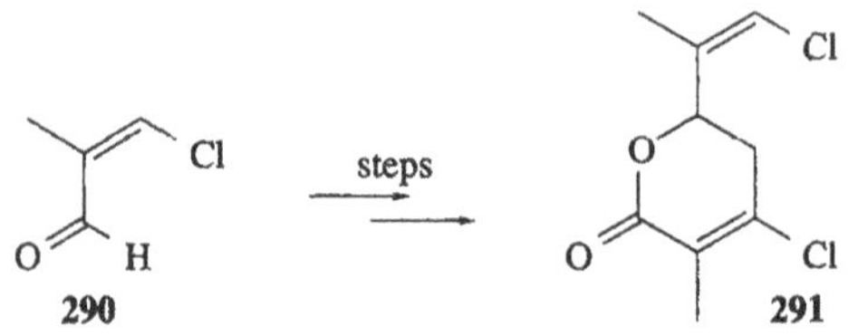

(2.163)

2.11.4 β-Iodovinylaldehydes

Since carbonyl iodide is not known, iodomethylenedimethyliminium iodide **292** is usually prepared by reacting chloromethylenedimethyliminium chloride with HI. (scheme 2.164).[288,289] β-Iodovinylaldehydes are usually unstable; 3-iodoacrylaldehyde has been prepared and derivatized.

(2.164)

2.12 β-Halovinylketones

The chlorovinylketone **294** was formed from the 6,6-*gem*-dimethyl ketone **293**.[363] The reaction took a different course when C-6 was not disubstituted; 5-formylated products were obtained.

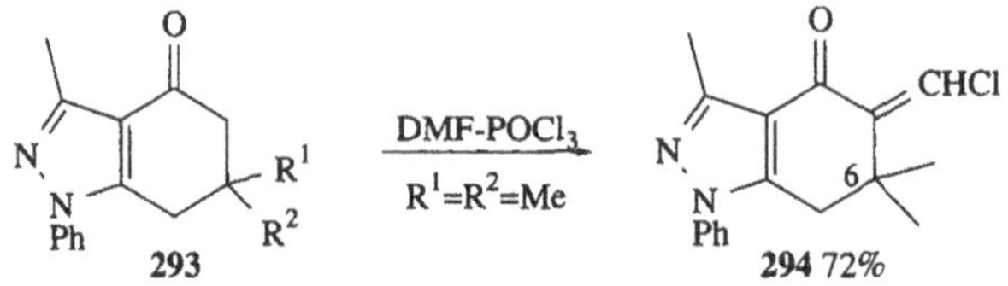

(2.165)

3-Halo-2-cycloalken-1-ones **295** can be prepared[364] in excellent yield by the reaction of cycloalkane-1,3-diones with Vilsmeier reagents prepared from DMF and (COCl)$_2$ or (COBr)$_2$. A mechanism involving initial attack at oxygen is proposed (scheme 2.166). The absence of formylated products is a notable feature since β-chlorovinylaldehyde moieties are usually introduced (section 2.11.3).

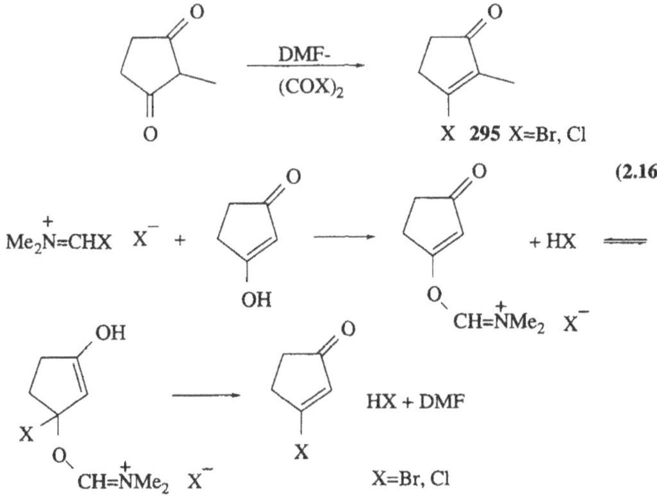

The reaction of dimedone (5,5-dimethylcyclohexane-1,3-dione) with DMF-POCl$_3$ affords some 3-chloro-5,5-dimethylcyclohex-2-en-1-one, but two other major products, β-chlorovinylaldehydes, are also obtained (section 2.11.3).[313]

2.13 Carboxylic Acids

A number of cyclic β-chlorovinylaldehydes has been converted into α,ω-dicarboxylic acids by silver nitrate oxidation, followed by oxidative cleavage using alkaline hydrogen peroxide.[296]

2.14 Carboxylic Acid Halides and Sulfonyl Chlorides

2.14.1 Carboxylic Acid Bromides
The action of the DMF-SOBr$_2$ complex on carboxylic acids affords the corresponding acid bromides[365] in the same way as DMF-SOCl$_2$ affords the corresponding acid chlorides.

2.14.2 Carboxylic Acid Chlorides
N,N-Disubstituted formamides catalyze the formation of carboxylic acid chlorides (and sulfonyl chlorides) form the corresponding carboxylic acids and (sulfonic acids) and SOCl$_2$ or COCl$_2$.[366-371] The yields are usually greater than 90%.

The salt [Me$_2$N=CHCl]$^+$ Cl$^-$, formed from DMF-SOCl$_2$ with loss of SO$_2$, has been established[367] as the intermediate that reacts with the carboxylic acid to give the acid chloride. The intermediates depicted in brackets in

Acid	Acid Chloride	Reference
RCO$_2$H	RCOCl	16
R=alkyl, Ph, CCl$_3$		
CNCH$_2$CO$_2$K	NCCH$_2$COCl	372
HO$_2$C-(CH$_2$)$_4$-CO$_2$H	ClO$_2$C-(CH$_2$)$_4$-CO$_2$Cl	16, 366, 367
Ph-SO$_3$H	Ph-SO$_2$Cl	367
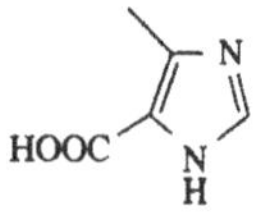	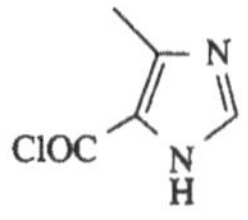	369
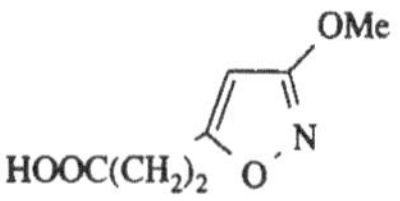	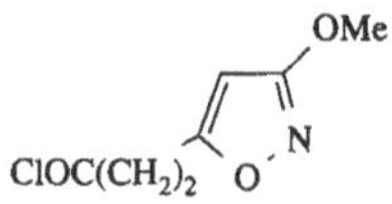	373
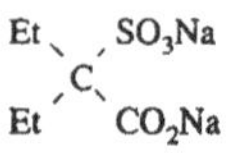	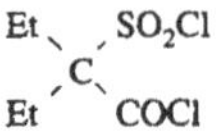	374
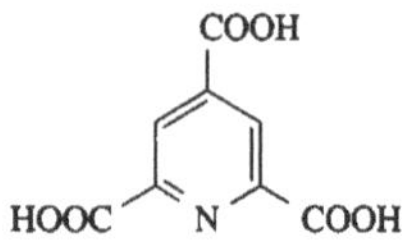	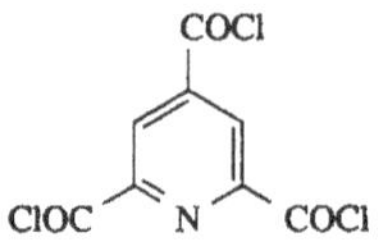	375
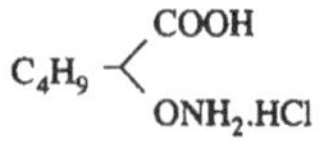	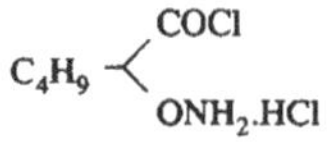	376

Table 2.5. Preparation of Acid chlorides from Acids or their Salts
and DMF-SOCl$_2$

scheme 2.167 are presumed. Kinetic studies on the reaction of isomeric
naphthalene carboxylic acids with SOCl$_2$ in the presence of DMF have been
undertaken.

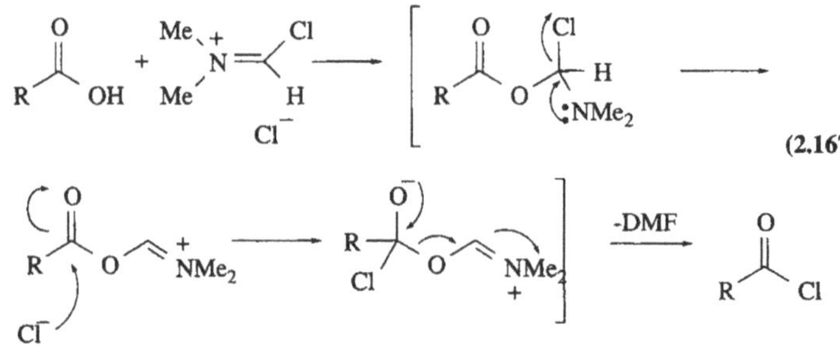

(2.167)

A wide variety of carboxylic acids have been converted into acid chlorides in the presence of DMF; $COCl_2$ is often used in place of $SOCl_2$. Acid chlorides containing a keto group and an ester group have been so prepared.[377] The use of DMF-$SOCl_2$ is illustrated in Table 2.5.

An interesting example of the tolerance of functionality is the preparation of α-isocyanoacetyl chloride,[372] which on treatment with triethylamine affords isocyanoketene, CN-CH=C=O.

α,α-Dihalocarboxylic acids,[378] sulfonic acids[367] and their salts,[379] and diarylphosphonic acids[380] and heterocyclic carboxylic acids have all been converted into the corresponding acid chlorides, as have a variety of dicarboxylic acids.[367, 381]

Limitations on the reaction have been found using α,β-unsaturated carboxylic acids; the corresponding β-chloro saturated acid chlorides are formed, although alkyl or aryl substituents at the β-position suppresses the addition of HCl.[373a] But-2-yndioic acid reacts with DMF-$SOCl_2$ to give dichloromaleic anhydride.[382] Carboxylic acids that contain strongly acidic phenolic or heterocyclic OH groups[383,384] or SH groups[385] are converted into the chlorinated acid chlorides such as **296**. Thiol groups may dimerize giving disulfides.[385]

$$\text{(2.168)}$$

296

2.14.3 Carboxylic Acid Fluorides

Reaction of carboxylic acids with (difluoromethyl)dimethylamine, Me_2NCHF_2 at 0°C, affords the corresponding acid fluorides.[386]

2.14.4 Sulfonyl Chlorides

Moderately activated aromatic compounds such as anisole undergo chlorosulfonylation (rather than chloroformylation) upon reaction with acid amide-sulfuryl chloride adducts.

$$\text{(2.169)}$$

Conversion of the intermediate sulfonic acid into the acid chloride is explained by Kojtscheff's proposal[387] that the acid amide-sulfuryl chloride adduct **297** undergoes thermal degradation (with loss of SO_3) to give dimethylchloromethyleneiminium chloride which converts the sulfonic acid into the sulfonyl chloride.

2.15 Anhydrides

p-Nitrobenzyl chloride is converted into p-nitrobenzoic anhydride by the DMF and sodium perchlorate in acetonitrile.[388]

2.16 Carbamates, Carbamoyl Halides and Isocyanates

Aryl chloroformates react with DMF initially to give the primary products of acylation 298 (which can then be decarboxylated); reaction of the salts 298 with primary arylamines affords carbamates (scheme 2.170).[389]

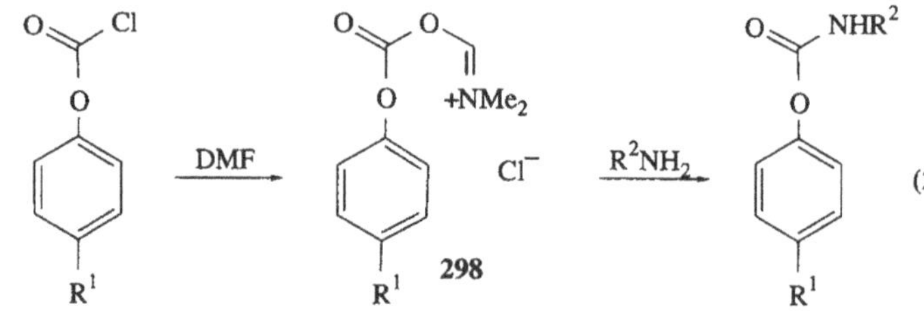

Carbamoyl chlorides are formed *en route* to isocyanates by the action of SO_2Cl_2 or $SOCl_2$ on N-monoalkylated formamides (section 2.24).[390] Where R is methyl or ethyl in RNHCOCl, such compounds have been isolated.[390]

N-Monoalkylated urethanes 299 are converted into isocyanates 300 by phosgene in the presence of catalytic amounts of DMF.[391] An explanation proposed was the iminoalkylation of the intial tautomer, resulting in a chloroethoxyimine that undergoes α-elimination (scheme 2.171).

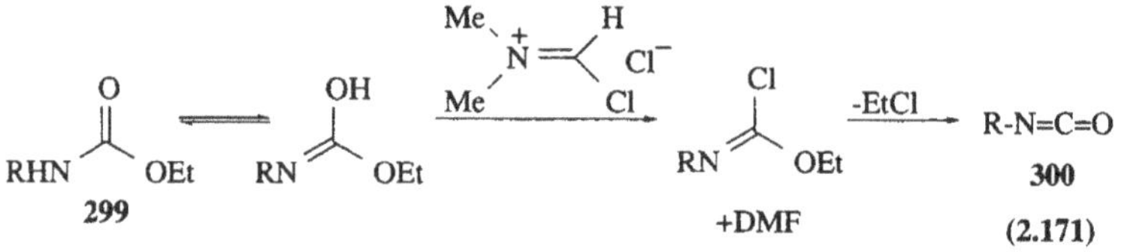

2.17 Carbodiimides

p-Toluenesulfonyl chloride effects dehydration of N,N-disubstituted ureas, giving carbodimides (scheme 2.172).[392]

$$\text{RHN} \overset{O}{\underset{}{\bigwedge}} \text{NHR} \xrightarrow{\ p\text{-TsCl}\ } \text{R-N=C=NR} \qquad (2.172)$$

2.18 Thiocyanates

Treatment of formamide with thionyl chloride affords the salt **301** which reacts with sulfenyl chlorides by elimination of HCl to give thiocyanates.[393]

$$\text{H} \overset{O}{\underset{}{\bigwedge}} \text{NH}_2 \xrightarrow[-\text{SO}_2]{\text{SOCl}_2} \underset{\text{H}}{\overset{\text{H}}{\underset{\mathbf{301}}{\overset{+}{\text{N}}}}} \underset{\text{Cl}}{\overset{\text{H}}{{=}\!\!\!<}} \text{Cl}^- \xrightarrow[-2\text{HCl}]{\text{RSCl}} \text{HN}{=}\!\!\!< \underset{\text{Cl}}{\overset{\text{SR}}{}} \xrightarrow{-\text{HCl}} \text{RSCN}$$

$$(2.173)$$

Arylthiocyanates can be prepared, *via* a rearrangement, by reaction of *N*-formylsulfenamide with phosgene in the presence of triethylamine (scheme 2.174).[394]

$$\text{ArSNCHO} \xrightarrow[\text{Et}_3\text{N}]{\text{COCl}_2} \text{ArS}{-}\overset{+}{\text{N}}{\equiv}\overset{-}{\text{C}} \longrightarrow \text{ArS}{-}\text{C}{\equiv}\text{N} \qquad (2.174)$$

2.19 Esters and Derivatives

2.19.1 Formates and Higher Esters

Simple alcohols, ROH, are converted into ROCHO by DMF-PhCOCl.[395,396] Many Vilsmeier reagents convert alcohols into halides. However, reaction of benzoyl chloride with DMF affords the reagent **302** which reacts with alcohols to give imidate ester chlorides **303** which can be hydrolyzed (very dilute H_2SO_4 or a 1M chloroacetate buffer) to give the corresponding formates.

$$\underset{\text{Me}}{\overset{\text{Me}}{\underset{\mathbf{302}}{\overset{+}{\text{N}}}}}{=}\!\!\!< \underset{\text{Cl}^-}{\overset{\text{OCOPh}}{}} + \text{ROH} \longrightarrow \underset{\text{Me}}{\overset{\text{Me}}{\underset{\mathbf{303}}{\overset{+}{\text{N}}}}}{=}\!\!\!< \underset{\text{Cl}^-}{\overset{\text{OR}}{}} \longrightarrow \text{ROCHO} \quad (2.175)$$

The reaction succeeds for a variety of primary and secondary alcohols.[395] The formates $O_2NCFClCH(R)O_2CH$ have also been prepared by the action of DMF-SOCl$_2$ on $O_2NCFClCH(R)OH$.[397] Interestingly, aryl formates[398] can be similarly prepared in good yield (scheme 2.176). Under these conditions (1.5 equiv. Vilsmeier reagent, 75-80°C) phenol itself is converted into phenyl formate (59%). Contrary to Buu Hoi's report,[200] no *p*-hydroxybenzaldehyde was detected.

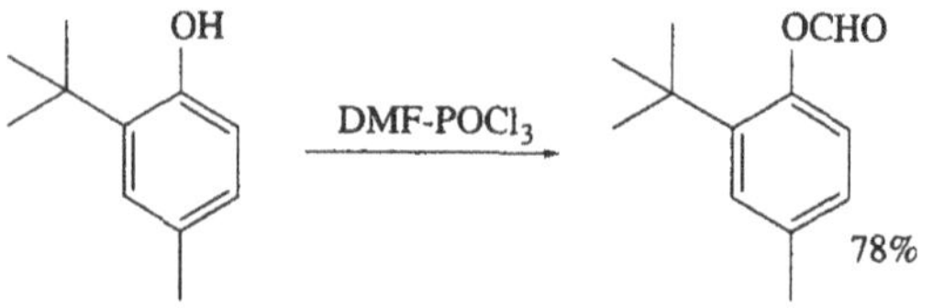

$$(2.176)$$

The polyfunctional alcohols erythritol and mannitol are converted into the corresponding formates by chloromethyleneiminium chloride.[399]

In a careful study, a mixture of DMF and Ph_3PBr_2 below 0°C was shown to convert primary alcohols into the bromides, but *secondary* alcohols into the corresponding formates.[400] Primary and secondary hydroxy groups in the same compound are converted respectively into bromide and formate; for example the bromo ester **304** (64%) is obtained from the corresponding diol. A useful feature is the inertness of C=C double bonds to the reagent.

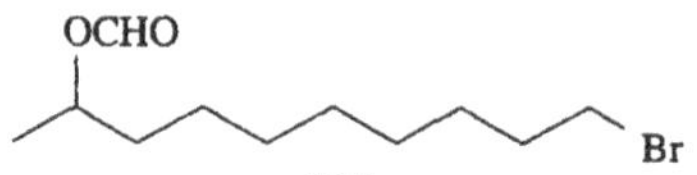

$$(2.177)$$

A general procedure for the esterification of primary, secondary, and tertiary alcohols, and of phenols, mediated by a Vilsmeier reagent has been described.[401] To the adduct of DMF-$(COCl)_2$ in a suitable solvent is added the carboxylic acid. Pyridine and the alcohol (or phenol) is then added. The reaction is usually conducted between 0 and 20°C.

2.19.2 1-Acyloxy-2-chloroalkanes

The reaction of certain epoxides with DMF-POCl₃ or *N,N*-dimethylacetamide-POCl₃, followed by hydrolysis of the intermediates, affords 1-acyloxy-2-chloroalkanes.[20]

2.19.3 Thioesters and Dithioesters

Chloromethyleneiminium chlorides **305** react with thiols to give *N,N*-dialkylmercaptoalkylmethyleneiminium chlorides **306**; these are not isolated, but may be converted into thio esters by hydrolysis, or dithioesters by thiolysis (scheme 2.178).[17]

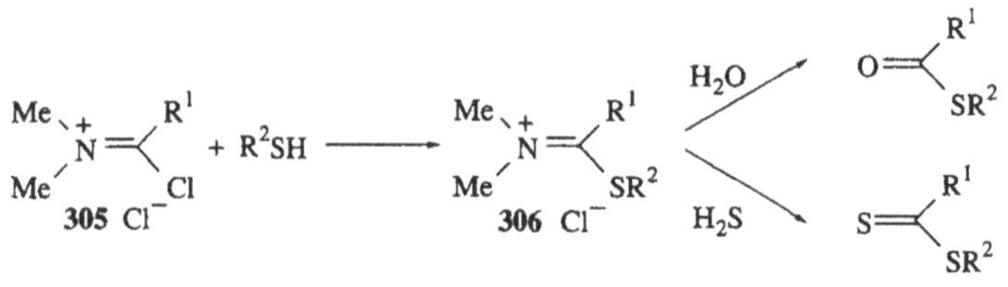

$$(2.178)$$

An example is the reaction of the adduct **305** (R^1=Ph), obtained from *N,N*-dimethylbenzamide and $POCl_3$, with thiophenol to give, after hydrolysis, phenyl thiobenzoate.[402]

2.19.4 β-Amino-α,β-Unsaturated Esters

Vilsmeier reactions of benzazoles **307** generated the enamines **308** (X = NH, S, and O).[403] Ethyl-2-pyridineacetate reacted analogously.

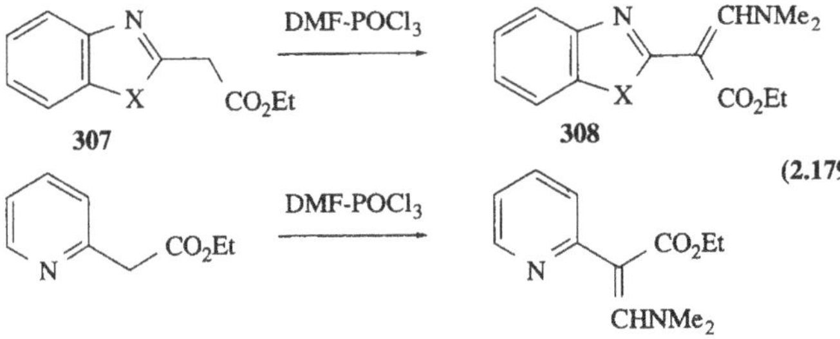

(2.179)

2.19.5 Phosphate Esters

Monoesters of phosphoric acid can be converted into diesters by reaction with Vilsmeier reagents, followed by alcoholysis of the adducts **309** (scheme 2.180).[404]

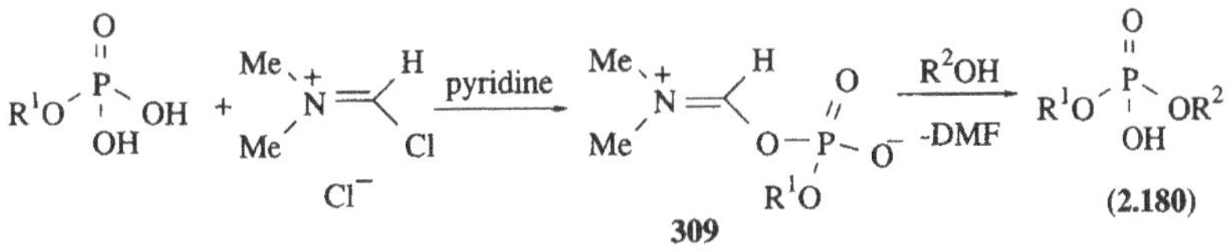

(2.180)

2.19.6 Orthoesters

Two orthoesters, ethyl orthopropionate and ethyl orthobenzoate, have been prepared by allowing the appropriate chloromethyleneiminium chloride **310** to stand in the presence of ethanol and a tertiary amine (scheme 2.181).[16,17] The reaction has been shown to proceed through amide acetals.

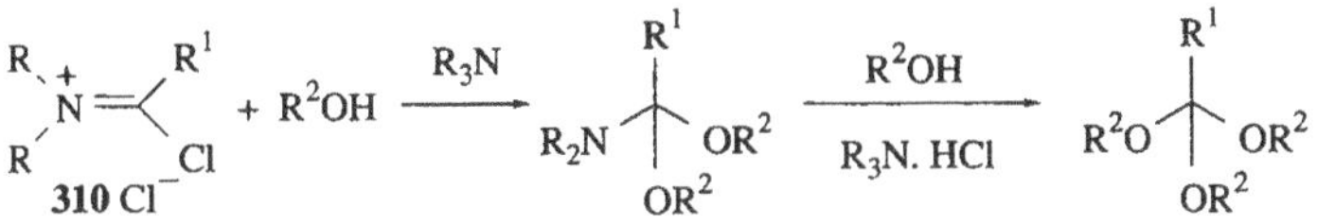

(2.181)

2.20 Amides and Derivatives

2.20.1 Formamides

Bredereck prepared triformamidomethane, HC(NHCHO)$_3$, by reaction of formamide with sulfuryl chloride.[405]

2-Aroyl-*N*-(dimethylamino)formanilides are formed by the ring cleavage of 2,1-benzisoxazoles introduced by DMF-POCl$_3$; initial attack by the Vilsmeier reagent is at the nitrogen atom of the heterocycle.[406]

2-Aminobenzothiazoles undergo *N*-formylation with Vilsmeier reagents generated using aryl sulfonyl chlorides.[407]

The subsidiary product of the treatment of 1,2,3,4-tetrahydrocarbazoles with DMF-POCl$_3$ is the unusual dialdehyde **311** that lacks enolizable hydrogen atoms; the major product (36%) is simply the *N*-formyl derivative of tetrahydrocarbazole.[263]

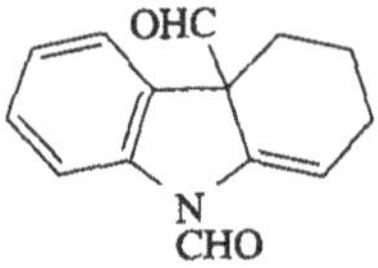

(2.182)

311

2.20.2 Thioamides

The reaction of halomethyleneiminium halides in dichloromethane or chloroform with hydrogen sulfide affords a general route to thioamides **312** in good yields (scheme 2.183).[16,17,408-412] In this way, aliphatic and aromatic *N,N*-disubstituted thioamides and *N*-alkylated thiolactams have been prepared, and also thioamides of squaric acid.[413] For example, Me$_2$NCHS was obtained in 85% yield. A useful alternative procedure involves the reaction of a secondary amine with chloroform; the intermediate chloromethyleneiminium salt need not be isolated, and reaction with hydrogen sulfide affords the corresponding thioamide **313**.[414]

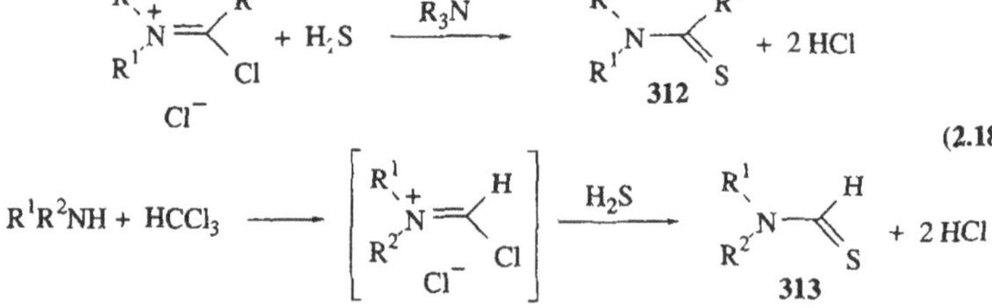

(2.183)

2.20.3 Amide Acetals

Chloromethyleneiminium chlorides **314** react with alkoxides to give the highly reactive *N,N*-dialkylalkoxymethyleneiminium salts **315** which react further with alkoxide to give the very reactive amide acetals **316** (scheme 2.184).

$$R^1, R \quad \text{NaOR}^3 \quad R^1, R \quad \text{NaOR}^3 \quad R^1 \quad R \quad (2.184)$$

314 Cl⁻   **315** Cl⁻   **316**

The amide can be added to an alcoholic solution of the alkoxide,[16,17] or to a suspension of the alkoxide in dichloromethane,[415] or an alcoholic solution of a tertiary amine.[416] The substituents in **316** may be a combination of alkyl, or R may be H or alkyl.[417,418] In all cases the temperature should be kept below 0°C, in order to avoid dealkylation of the intermediate **315** by alkoxide. The iminium salts **314** are generated when secondary amines react with dichlorocarbene, and in the presence of alkoxide this provides another route to amide acetals **316**, based on iminium salts **314**.[16]

2.20.4 Amide Mercaptals

The adduct of DMF and POCl₃ reacts with thiols providing a useful method of preparing amide mercaptals (scheme 2.185).[402]

$$\underset{\text{Me}_2\text{N}}{\overset{H}{\diagdown}} \overset{\text{SEt}}{\underset{\text{SEt}}{}} \quad \xleftarrow{\text{EtSH}} \quad \text{DMF + POCl}_3 \quad \xrightarrow{\text{HSCH}_2\text{CH}_2\text{SH}} \quad \underset{\text{Me}_2\text{N}}{\overset{H}{\diagdown}} \overset{S}{\underset{S}{}} \quad (2.185)$$

2.20.5 α-Chloroamides and α,α-Dichloroamides

One of the simplest preparations of α,α-dichloroamides **320**, and also of α-chloroamides, is the α-chlorination of substituted chloromethyleneiminium salts **317** ultimately to give the salts **319** which may be hydrolyzed to give the dichloroamides **320**.[16,419]

$$\underset{\text{Cl}^-}{\overset{R, +}{\underset{R}{N}}} \overset{R^1}{=\!\!\!<} \overset{}{\underset{\text{Cl}}{}} \quad \xrightarrow[\text{-HCl}]{\text{Cl}_2} \quad \underset{\text{Cl}^-}{\overset{R, +}{\underset{R}{N}}} \overset{\text{Cl}}{=\!\!\!<} \overset{R^1}{\underset{\text{Cl}}{}} \quad \xrightarrow[\text{-HCl}]{\text{Cl}_2}$$

317   **318**   (2.186)

$$\underset{\text{Cl}^-}{\overset{R, +}{\underset{R}{N}}} \overset{\text{Cl}\quad \text{Cl}}{=\!\!\!<} \overset{R^1}{\underset{\text{Cl}}{}} \quad \xrightarrow{\text{H}_2\text{O}} \quad \underset{R_2N}{\overset{\text{Cl}\quad \text{Cl}}{}} \overset{R^1}{\underset{O}{}}$$

319   **320**

The reaction can be stopped after the first chlorination and the salts **318** isolated or hydrolyzed to give the corresponding α-chloroamides.[420,421] Bromination gives side reactions, chiefly the monobromo derivatives contaminated with chlorine.[419]

2.20.6 β-Ketocarboxylic Acid Amides

N,N-disubstituted acid amides and *N*-monosubstituted lactams undergo self-condensation in the presence of POCl₃. Hydrolysis affords β-keto-carboxylic acid amides.[16,416,419,422] *N*-Unsubstituted amides and lactams undergo condensation with POCl₃ to give pyrimidines and condensed pyrimidines (section 4.4.2.9).

$$\begin{array}{ccc} & \text{i) POCl}_3 & \\ R_2N \diagdown O & \xrightarrow{\hspace{1cm}} & \\ & \text{ii) H}_2O & \end{array} \qquad (2.187)$$

2.20.7 α,β-Unsaturated Amides

At low temperatures, acid amides can react with Vilsmeier reagents to give salts such as **85** (scheme 1.38), in which the oxygen moiety is not replaced by halide. Consequently hydrolysis liberates the unsaturated amides **87** (scheme 1.38). However, even if chlorine does become bonded to carbon, as in **86** (scheme 1.38), hydrolysis may also lead to the α,β-unsaturated amides **87**, particularly if cyclic amides are used.

N,N-Dimethylacetamide reacts with DMF-POCl₃ below 0°C to give [Me₂NCH=CHC(Cl)=NMe₂]⁺ PO₂Cl₂⁻, but at higher temperatures further iminoalkylation in an exothermic reaction leads to the dication **321** which can be hydrolyzed to give the α,β-unsaturated amide **322** (scheme 2.188).[159]

$$\begin{array}{cccc} & \text{DMF-POCl}_3 & & \text{K}_2\text{CO}_3 \\ \text{Me}_2\text{N} \diagdown O & \xrightarrow{70°\text{C, 3 h}} & \mathbf{321} \; 2\,\text{X}^- & \xrightarrow{\text{H}_2\text{O}} & \mathbf{322} \; 76\% \end{array} \qquad (2.188)$$

2.20.8 Ureas

Reaction of hydroxylamine with DMF-POCl₃ affords *N,N*-dimethylurea, presumably by Beckmann rearrangment of dimethylformamidoxime[416] brought about by the dichlorophosphoric acid liberated during the reaction (scheme 2.189). An oxime is the first product when hydroxylamine reacts with dimethylchloromethyleneiminium chloride.

$$H_2NOH \xrightarrow{\text{DMF-POCl}_3} \underset{\substack{H \quad NMe_2}}{\overset{HON}{\|}} \longrightarrow \underset{\substack{H_2N \quad NMe_2}}{\overset{O}{\|}} \qquad \textbf{(2.189)}$$

2.20.9 Peptides and Nucleotides

DMF-SOCl$_2$ has been used to make peptide bonds,[370,423] and oligo- and poly-nucleotides.[404,424,425]

2.21 Enamines

2.21.1 α-Chloroenamines

α-Chloroenamines are valuable precursors of keteniminium salts. Alkyl- and aryl- substituted α-chloroenamines can be prepared by deprotonation of chloromethyleneiminium chlorides with a tertiary amine.[420,421,426-431] By operating between -20 and +20°C, yields of between 60 and 70% are typically obtained. One of the substitutents R^3 or R^4 may be hydrogen.

$$\underset{\substack{Cl}}{\overset{R^3}{\underset{R^1R^2\overset{+}{N}=}{\bigvee}}}\kern-1em{\overset{R^4}{}} \quad Cl^- \xrightarrow[\text{pyridine}]{\text{Et}_3N} R^1R^2N-\underset{\substack{Cl}}{\overset{CR^3R^4}{\bigvee}} \qquad \textbf{(2.190)}$$

The formation of α-chloroenamines provides an explanation of the self-condensation of halomethyleneiminium halides, and also their facile α-halogenation.[16,419]

2.21.2 α-Cyanoenamines

Some α-cyanoenamines **324** can be prepared from the iminium salts **323** by treatment with potassium cyanide in acetonitrile or zinc cyanide and triethylamine in chloroform.[426]

$$\underset{\substack{323 \quad Cl}}{\overset{R^3}{\underset{Me_2\overset{+}{N}=}{\bigvee}}}\kern-1em{\overset{R^4}{}} \quad Cl^- \xrightarrow[\text{Et}_3N]{\text{KCN or Zn(CN)}_2} \underset{\substack{324 \quad CN}}{Me_2N-\overset{CR^3R^4}{\bigvee}} \qquad \textbf{(2.191)}$$

2.21.3 N-Formylenamines

By analogy with the inital attack of Vilsmeier reagents on the carbonyl oxygen atom of ketones, an acetophenone anil reacts at nitrogen with DMF-POCl$_3$ to give an isolable salt (scheme 2.192). The latter can be hydrolyzed with aqueous sodium hydroxide, giving a stable *N*-formylenamine.[432]

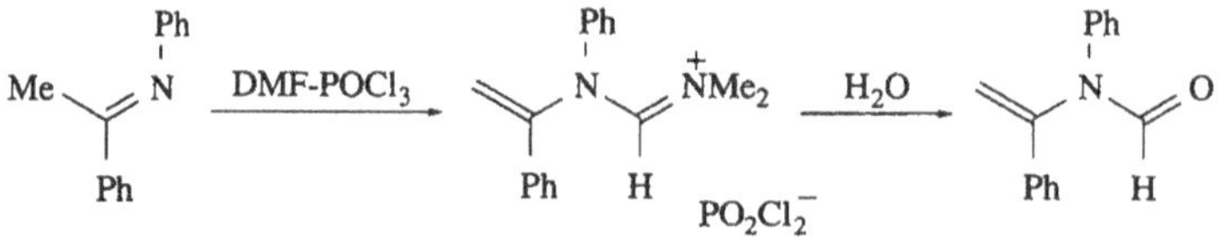

(2.192)

2.21.4 Cyclic enamines

The Vilsmeier reagent prepared from pyrrolidin-2-one and POCl$_3$ reacts with 4,6-dimethoxy-2,3-diphenylindole (and some other activated indoles) to give the enamine **325**.[433] The corresponding reactions with acetanilide and its derivatives are similiar, although the Schiff bases are the products isolated.[433]

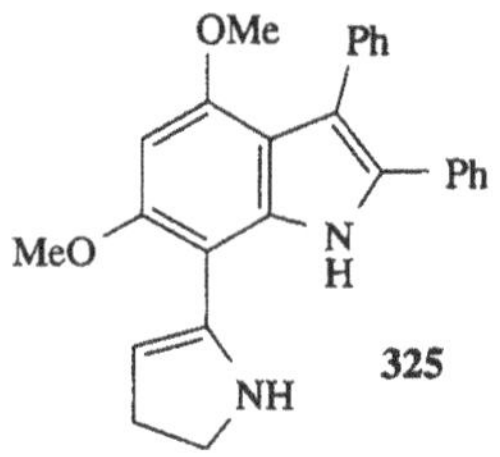

(2.193)

325

2.22 Ynamines

A useful preparation of ynamines is the reaction of two mole equivalents of a lithium dialkylamide with a chloromethyleneiminium salt.[420] The yield much depends upon the dialkylamine employed; bulky amides such as dicyclohexylamide give the highest yields.

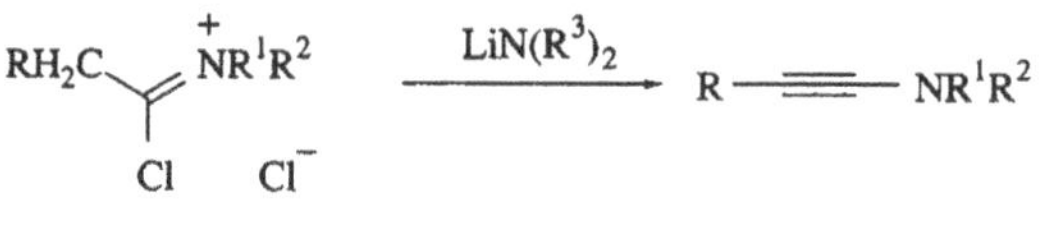

(2.194)

2.23 Nitriles

The von Braun thermal degradation of halomethyleneiminium halides has been extensively investigated; decomposition to the nitriles proceeds *via* the haloimines **326**.[434-436]

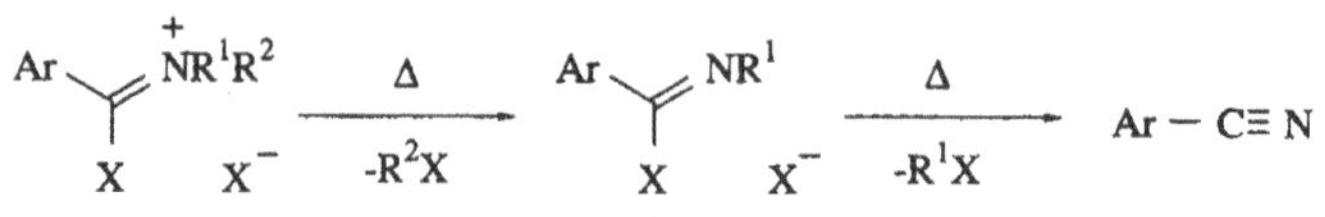

(2.195)

326

For cyclic and benzoylated compounds such as **327**, the pure iminium bromides degrade much better than the chlorides.[437-440]

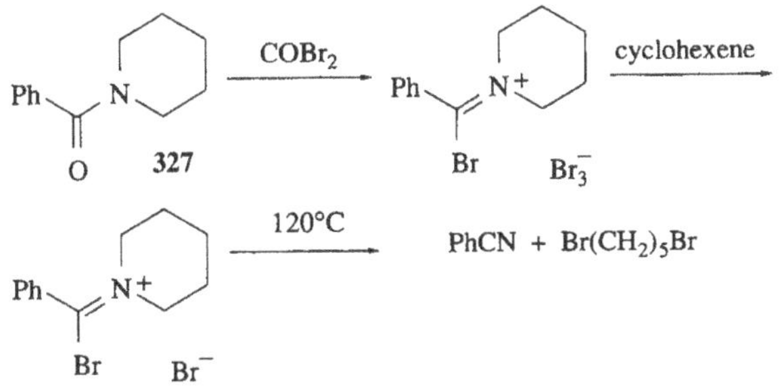

$$(2.196)$$

Primary carboxylic acid amides are dehydrated to the corresponding nitriles by a wide variety of halogenating agents, including PCl_5, $POCl_3$, $SOCl_2$, $COCl_2$, $ClCO_2CCl_3$ (dimeric phosgene), and $(COCl)_2$, in the presence of a tertiary amine. The reaction most likely proceeds through the chloromethyleneiminium chlorides. There are reviews of this reaction of wide scope.[441]

Primary amides are also converted into nitriles by chloromethylene-iminium chlorides (scheme 2.197). The reaction takes a few minutes at room temperature, and the yields usually exceed 80%.[16] DMF is regenerated, and 5-10 mol% is sufficient to dehydrate amides with thionyl chloride or phosgene. α,ω-Dinitriles[16] and even tetracyanoarenes have been so prepared.[442-444]

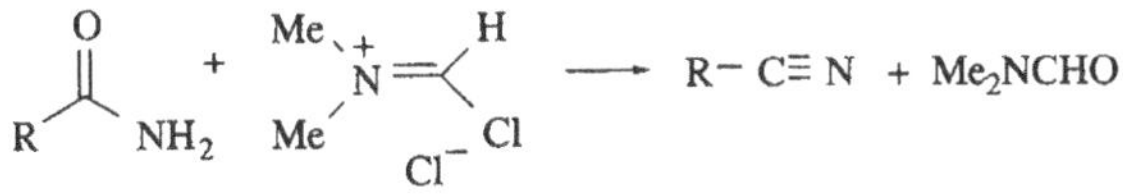

Aldoximes are generally converted into nitriles in excellent yields by $[Me_2N=CHCl]^+$ Cl^-.[445] Some oximes have been dehydrated to nitriles using DMF-$SOCl_2$.[446] The nitrile **328** has been obtained by reacting the corresponding unsaturated amide with DMF-$POCl_3$; related conversions are also known.[447] Dialkylaminomalonodinitriles **329** can be be prepared by treating chloromethyleneiminium chloride with HCN and NaCN, or simply with CuCN.[448]

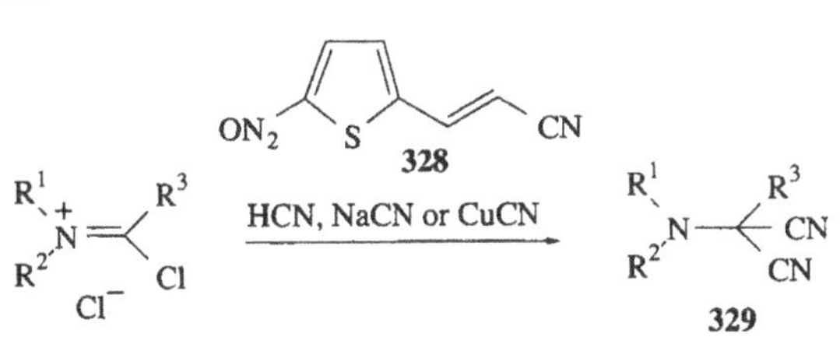

$$(2.198)$$

2-Cyanopyrrole (64%) can be prepared in a modified Vilsmeier-Haack reaction by treatment of pyrrole with DMF-(COCl)$_2$ and then with hydroxylamine hydrochloride-pyridine.[449]

Reaction of Vilsmeier products with hydroxylamines can lead to nitriles (scheme 2.199).[450]

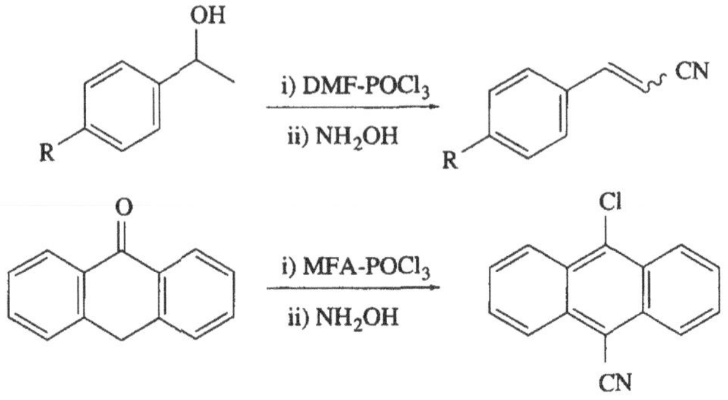

(2.199)

2.24 Isocyanides

Ugi discovered that isocyanides are formed by reaction of *N*-monosubstituted formamides with phosphorus pentachloride, thionyl chloride or phosgene in the presence of a tertiary amine such as pyridine.[451] PBr$_3$, PCl$_3$ and BCl$_3$ have also been used as halogenating agents,[452-454] as has POCl$_3$.[455,456]

The reaction proceeds *via* the chloromethyleneiminium chloride **330** that undergoes α-elimination. The reaction is widely applicable, and any isonitrile can be so prepared, provided that the formamide does not contain other functionality which is incompatible with the halogenating reagent. Even β- and α-hydroxyalkyl isocyanides can be prepared by the Ugi reaction.[457]

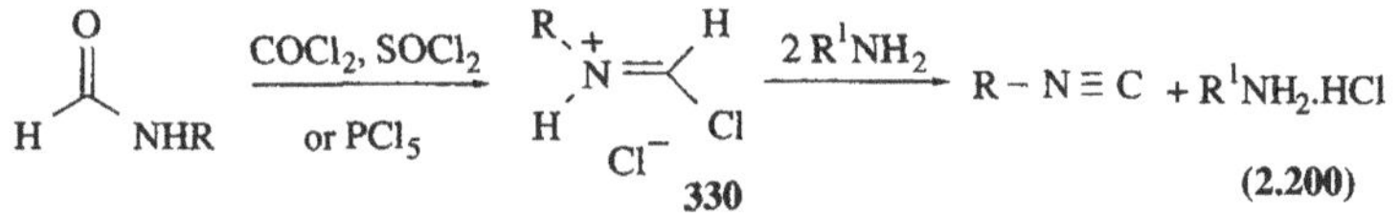

(2.200)

Using tertiary amines, the yield of arylamines is usually only moderate, but can be much improved by using potassium ^{t}butoxide as the base,[451c] which is considered to generate the amide anion that reacts rapidly with POCl$_3$.

A recent improvement is the replacement of phosgene by the safer and more convenient diphosgene, ClCO$_2$CCl$_3$.[458] Organometallic isonitrilium salts have also been prepared by a reaction related to the Ugi method.[459,460]

Treatment of α-amino- and α-thio-formamides with POCl$_3$, followed by aqueous sodium hydrogen carbonate has been reported to give isocyanides (scheme 2.201).[461]

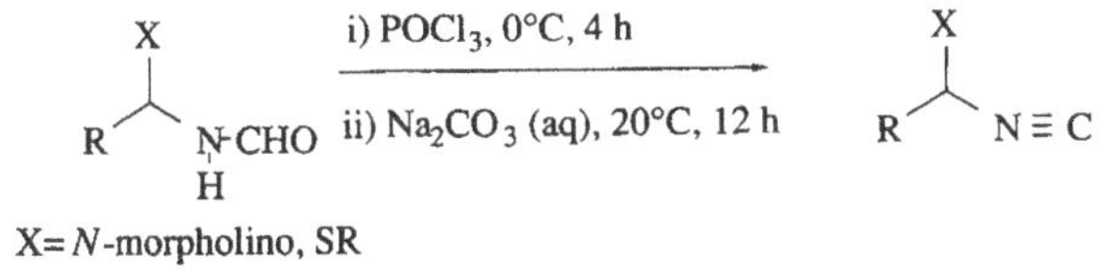

X= *N*-morpholino, SR

2.25 Isocyanide Dichlorides

The final product of treating a formanilide with sulfuryl chloride or thionyl chloride and chlorine is an isocyanide dichloride **331**. A mixture of thionyl chloride and oxygen (which presumably acts as sulfuryl chloride) may also be used.[393]

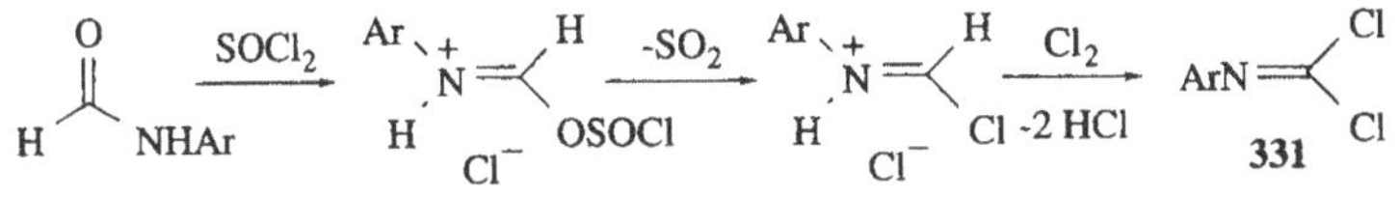

2.26 Vinylogous Amides

2-Acyl-6-aminofulvenes **332** result from the reaction of acetylcyclo-pentadienes with DMF-POCl$_3$; it is notable that no formyl group is introduced at the methyl carbon atom.[462]

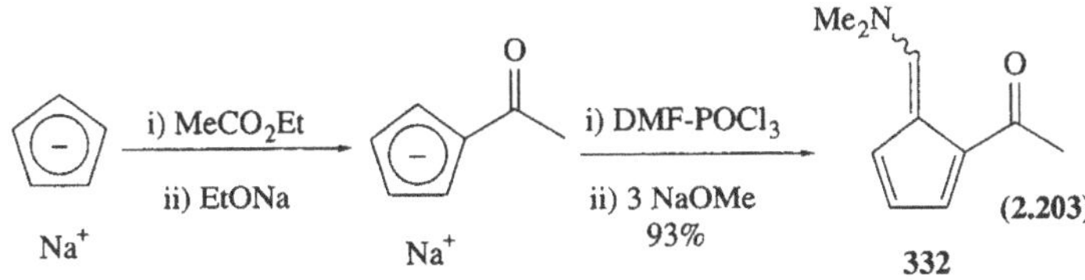

2.27 Amidines and Derivatives

2.27.1 Amidines from Acyclic Amides

A general route to amidines **335** involves the reaction of chloro-methyleneiminium salts **333** with amines to give amidinium salts **334** (section 2.29.2) from which amidines are obtained by treatment with alkali.[416,463-466]

$$\underset{\substack{\text{333} \quad PO_2Cl_2^-}}{\overset{R^1}{\underset{R^2}{\diagdown}}\overset{+}{N}=\overset{R}{\underset{Cl}{\diagup}}} + R^3NH_2 \longrightarrow \underset{\substack{\text{334} \quad Cl^-}}{\overset{R^1}{\underset{R^2}{\diagdown}}\overset{+}{N}=\overset{R}{\underset{NHR^3}{\diagup}}} \overset{NaOH}{\longrightarrow} \underset{\text{335}}{\overset{R^1}{\underset{R^2}{\diagdown}}N-\overset{R}{\underset{NR^3}{\diagup}}} \quad (2.204)$$

The reaction tolerates a variety of substituents for R, or simply R=H in **333**. Frequently R^3 is aromatic, but alkylamines have been used. The synthesis of N-mono- and N,N-disubstituted amidines by this method has been reviewed.[467] In early work,[468,469] alkylamines were shown to react with complexes of N-alkylformamides and POCl$_3$. In this way, Davis and Yelland isolated the picrate of N,N'-dibutylformamidine from the reaction of butylamine with BuNHCHO-POCl$_3$.[469]

Although N,N,N',N'-tetraalkylformamidinium salts can be prepared by reacting chloromethyleneiminium chlorides with dialkylamines, the dialkyl-ammonium chloride is often inseparable from the formamidinium salt. This was so for the amidinium salt **336** which could not be obtained pure by recrystallization.[389]

$$\underset{\substack{\\ Cl^-}}{\overset{Me}{\underset{Me}{\diagdown}}\overset{+}{N}=\overset{H}{\underset{Cl}{\diagup}}} + Me_2NH \longrightarrow \underset{\text{336} \quad Cl^-}{\overset{Me}{\underset{Me}{\diagdown}}\overset{+}{N}=\overset{H}{\underset{NMe_2}{\diagup}}} + Me_2NH.Cl \quad (2.205)$$

The sequence above was shown to succeed with aromatic chloro-methyleneiminium salts and aromatic amines, giving N,N,N'-trisubstituted amidines, as described in very early work.[409,470-474] [Me$_2$N=CHCl]$^+$ Cl$^-$ has been used to convert aminoanthraquinones into the corresponding amidines.[16,475]

With DMF-SOCl$_2$, aniline forms the two amidines (scheme 2.206), whereas diphenylamine simply undergoes N-formylation.[476]

$$PhNH_2 \xrightarrow[\text{ii) HO}^-]{\text{i) DMF-SOCl}_2} PhN=\overset{H}{\underset{NMe_2}{\diagup}} + PhHN-\overset{H}{\underset{NPh}{\diagup}}$$

$$Ph_2NH \xrightarrow[\text{ii) HO}^-]{\text{i) DMF-SOCl}_2} Ph_2NCHO \qquad (2.206)$$

DMF-POCl$_3$ reacts with p-phenylenediamine to give the bisamidine **337**.[416] Sulfonamides, RSO$_2$NH$_2$, are converted into the N-sulfonyl-formamidines **338**.[477] The amino groups of iodoanilines,[478] amino-pyridines,[479,480] amino-1,2,3-triazolecarboxaldehydes,[481] and naphthyl-amines[482] are also formylated; the amide-POCl$_3$ adducts can be used *in situ*, a ratio of 1:1:2.3 amine:POCl$_3$:acid amide being best.[465] Both N- and C-formylation sometimes occur.[483,484] Bisamidines such as **339** are formed by the reaction of aniline with the adducts of α,ω–bisamides and POCl$_3$.[416]

Reaction of aniline with the appropriate iminium salts gives the amidines **340**.[485] The amino groups of a variety of other (hetero)aromatic compounds have been reacted with amide-POCl$_3$ adducts; the amidine derivatives such as **341**[486] are usually formed,[487-489] although if other suitable adjacent groups are present, the formation of new heterocyclic rings,[490] *e.g.* a fused imidazole[452,453] or a pyrimidine ring[491] may occur. Formamidines of benzo[2,1,3]thiadiazoles and of benzo[2,1,3]selenadiazoles have been prepared in 40-86% yields by reaction of the corresponding amines in benzene with DMF-POCl$_3$.[492]

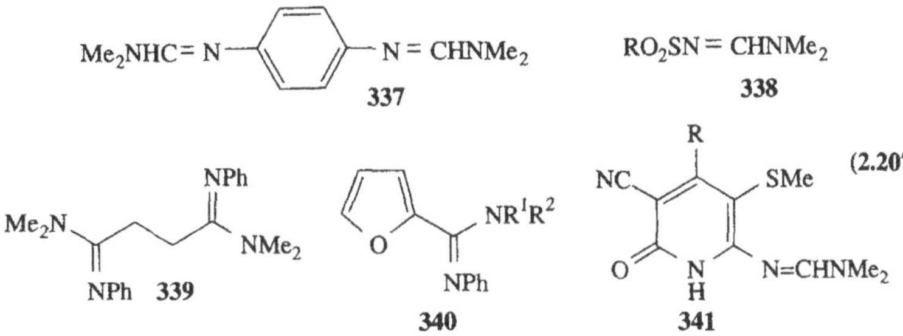

Indole-3-amine was converted by MeCONMe$_2$-POCl$_3$ into **342** (53%) but by *N,N*-diethylformamide-POCl$_3$ into **343** (48%). The indole amidines **344** were prepared in good yields by Vilsmeier reaction of the corresponding aminoindole carboxylate hydrochlorides and R^{2}CON(R^3)$_2$.[493]

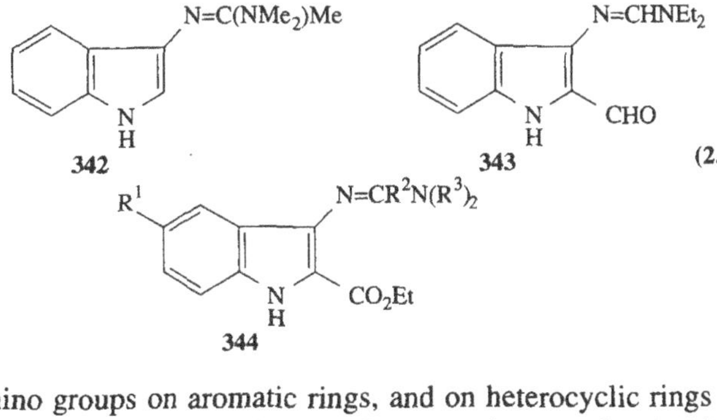

Amino groups on aromatic rings, and on heterocyclic rings can also be converted into amidines using DMF and *p*-toluenesulfonyl chloride.[477b,494] The amidine **346**, rather than the 2-formyl derivative of 3-(acetylamino)-benzo[*b*]thiophene was obtained when amide **345** was reacted with DMF-POCl$_3$.[495]

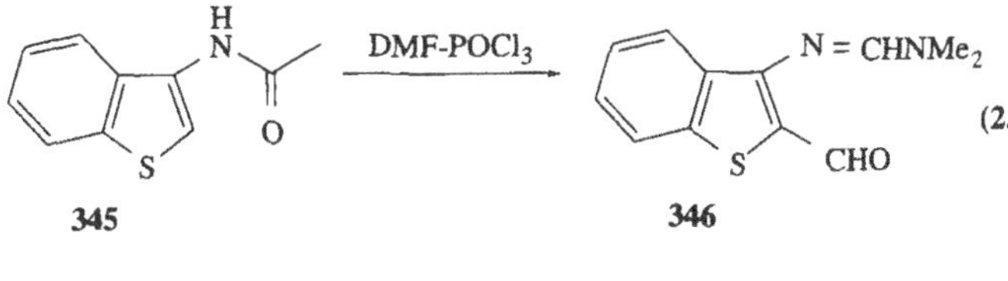

Amidines have been synthesized from halogenating agents by a number of other methods. Thus, formanilide reacts with $POCl_3$ to give *N,N'-*diphenylformamidine.[496] *N* -Arylsulfonyl formamidines undergo self-condensation using $POCl_3$ and pyridine to give *N,N'*-bis(arylsulfonyl) formamidines.[497] Amidines of the holocaine type **347** have also been prepared, as have *N,N,N'*-trisubstituted amidines such as **348**.[498]

$$\underset{\underset{\textbf{347}}{\underset{Me \qquad NHAr}{}}}{\overset{NAr}{\big\|}} \xleftarrow{\quad ArNH_2\text{-}PCl_3 \quad} MeCONHAr \xrightarrow{\quad Et_2NH\text{-}PCl_3 \quad} \underset{\underset{\textbf{348}}{\underset{Me \qquad NEt_2}{}}}{\overset{NAr}{\big\|}} \qquad \textbf{(2.210)}$$

Chloromethyleneiminium chlorides react with hydroxylamine to give aldoxime derivatives (scheme 2.211).[16,499]

$$\underset{Me}{\overset{Me}{>}}\overset{+}{N}\underset{Cl^-}{=}\overset{H}{\underset{Cl}{<}} \quad \xrightarrow{+ H_2NOH} \quad \underset{Me}{\overset{Me}{>}}\overset{+}{N}\underset{Cl^-}{=}\overset{H}{\underset{NHOH}{<}} \quad \longrightarrow \quad \underset{Me}{\overset{Me}{>}}N{-}\overset{H}{\underset{NOH}{<}} \qquad \textbf{(2.211)}$$

Tri-, tetra-, or symmetrically di-substituted ureas react with primary amines in the presence of $POCl_3$ to give guanidines **351**[416]*via* the presumed iminium salts **349** and **350**, believed to exist in equilbrium in the reaction mixture (scheme 2.212).

$$\underset{R^2R^3N}{\overset{R^1RN}{>}}{=}O \xrightarrow{\quad POCl_3 \quad} \underset{\underset{Cl^-}{\underset{\textbf{349}}{}}}{\underset{R^2R^3N}{\overset{R^1RN^+}{>}}}{-}OPOCl_2 \rightleftharpoons \underset{\underset{PO_2Cl_2^-}{\underset{\textbf{350}}{}}}{\underset{R^2R^3N}{\overset{R^1RN^+}{>}}}{-}Cl$$

$$\xrightarrow{\quad R^4NH_2 \quad} \underset{\underset{\textbf{351}}{R^2R^3N}}{\overset{R^1RN}{>}}{=}NR^4 \qquad\qquad \textbf{(2.212)}$$

2.27.2 Amidines from Lactams

The $POCl_3$ adduct of lactams **352** that contain a substituent at nitrogen reacts with primary amines to give lactam imides **353** (or lactamidines).[416]

N-Unsubstituted lactams, after adduct formation, react with primary amines to give one of the tautomers **354** or **355**, or less commonly, a mixture of both. Examples include the lactam imides **356** (from caprolactam-$POCl_3$ and aniline), **357** (from caprolactam-$POCl_3$ and *N*-methylaniline) and **358** (1-methylpyridin-2-one-$POCl_3$ and aniline).   Vinylpyrrolidinone is exceptional in undergoing cleavage of the vinyl group; the $POCl_3$ adduct with aniline forms a mixture of tautomers **354** and **355** (R=Ph).[416]

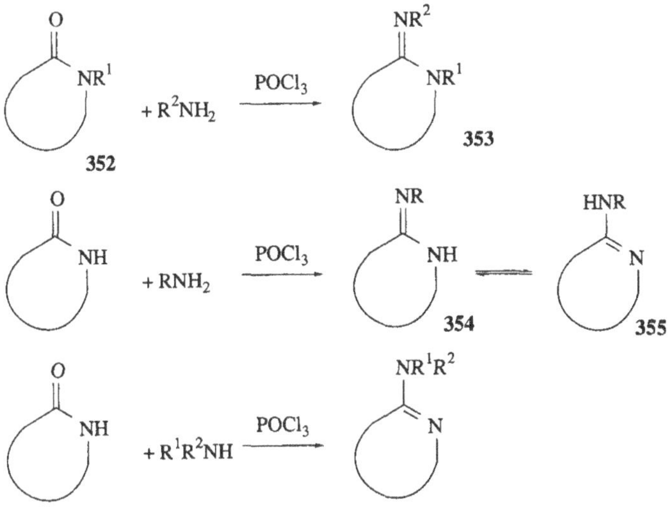

$$(2.213)$$

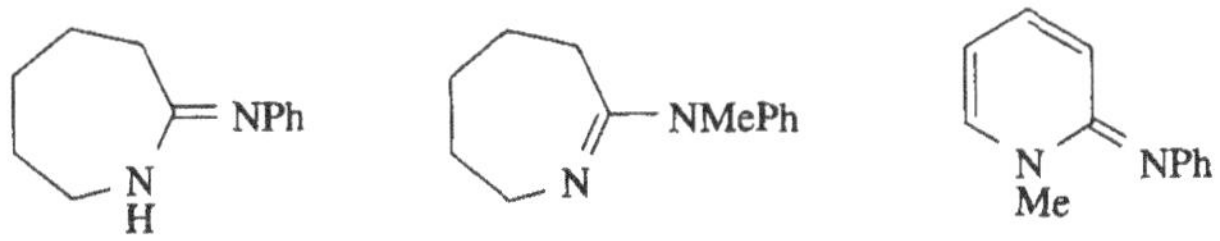

| 356 | 357 | 358 |

Some tertiary amines, including *N,N*-dimethylaniline, react with the pyrrolidinone-POCl$_3$ adduct by means of a cleavage analogous to the von Braun degradation.[416]

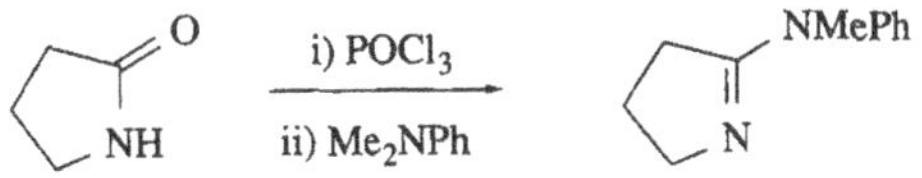

$$(2.214)$$

For certain lactams that contain substituents at, or ring-fusion linked to, the α-position to the nitrogen atom, the above amidines are not formed; instead, a dialkylaminomethylation occurs at the position *alpha* to the carbonyl group, and the amide group becomes transformed into a chloroimine moiety. Thus vinylogous amidines, *e.g.* **359** have been prepared[351,500] from steroidal lactams (scheme 2.215). In general, the vinylogous amidines have tolerable stablity to acids unless aromatization can occur.

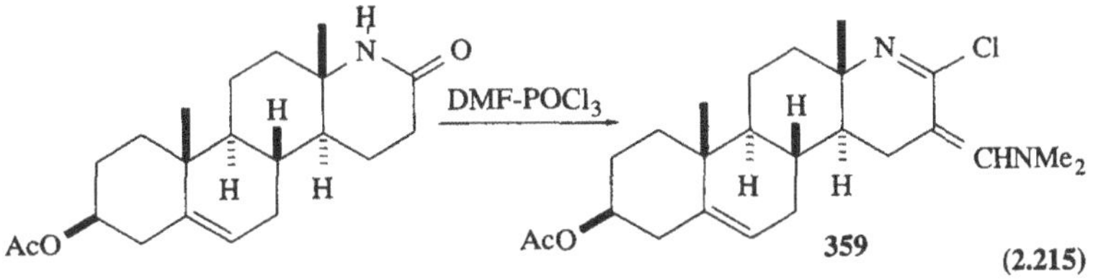

$$(2.215)$$

Stable vinylogous amidines such as **360** can be obtained from lactams in a Vilsmeier reaction (scheme 2.216).[500,501]

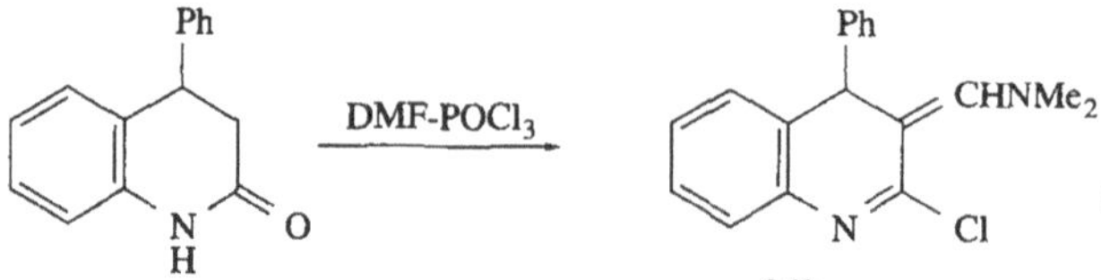

(2.216)

Amidines have been occasionally prepared by ring-cleavage. Thus, isatin β-oxime was converted by DMF-POCl$_3$ into *N,N*-dimethyl-*N*-(*o*-cyanophenyl)formamidine.[502]

2.27.3 Vinylogous Amidines

1-Azafulvene derivatives have been prepared by the action of a Vilsmeier reagent upon 2-bromopyrrole, followed by deprotonation of the resulting iminium species (scheme 2.217).[503]

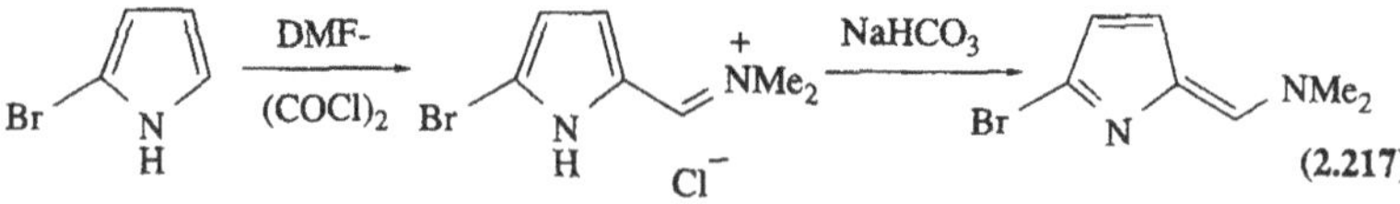

(2.217)

2.28 Amidrazones

The *N*-formylformamidrazones **363** are formed by reacting *N*-mono-substituted hydrazines with DMF-POCl$_3$.[465] By analogy with the reaction of primary amines with amide-POCl$_3$ adducts, the salts of formamidrazones **361** are first formed, and these undergo iminoalkylation to give the salts **362**, which can be hydrolyzed to the amidrazones **363**.

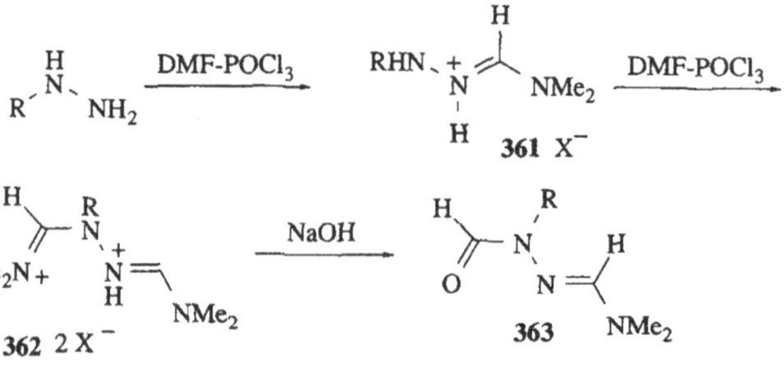

(2.218)

The reactions of *N,N*-disubstituted hydrazines are simpler since only the primary amine group can be formylated. For example, alkaline hydrolysis affords the formamidrazone **364** (scheme 2.219).[465]

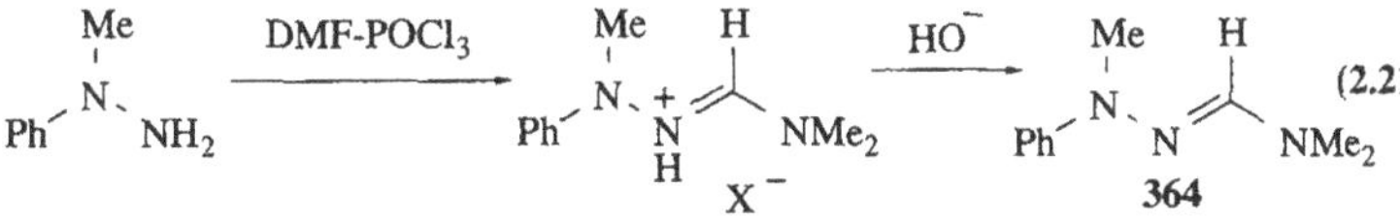

Replacement of a carboxamido group by an iminium group derived from a Vilsmeier reagent can lead to amidrazones.[504]

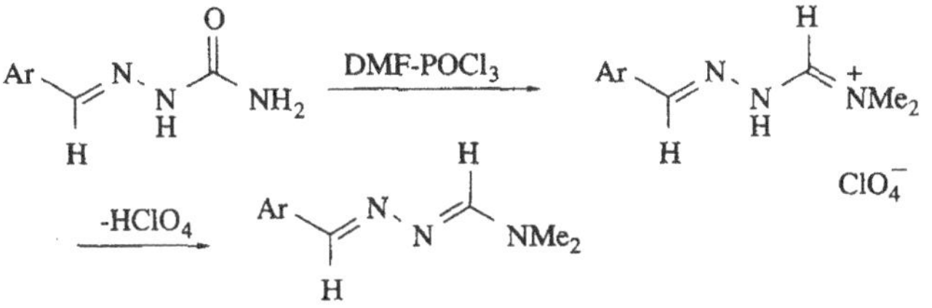

2.29 Iminium Salts

2.29.1 General (including Gold's Reagent)

An exomethylene group that is conjugated with an extended π-system incorporated into a ring can undergo simple monoiminoalkylation with Vilsmeier reagents to give conjugated iminium salts that can be isolated as their perchlorates, such as salts **365**,[75] **366** (from 1,1-dimethyl-4-methylene-1,4-dihydronaphthalene,[75] **367** (from 9,9-dimethyl-10-dimethylene-9,10-dihydroanthracene,[75] and **368** (from methyltropylium perchlorate).[505]

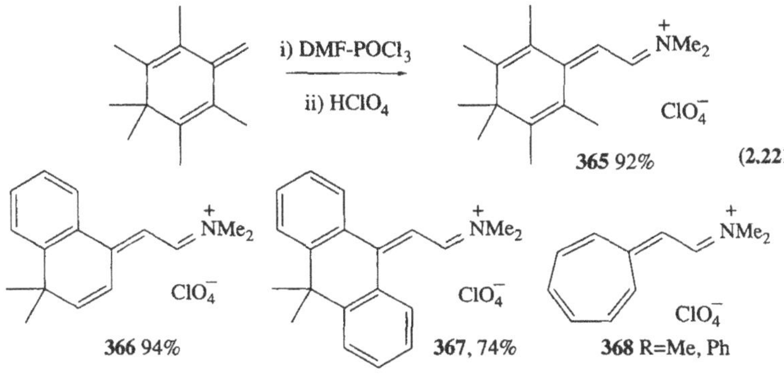

365 92%   (2.221)

366 94%   **367**, 74%   **368** R=Me, Ph

Nitriles react with dimethylchloromethyleneiminium chloride in the presence of HCl to give the salts **369**.[506]

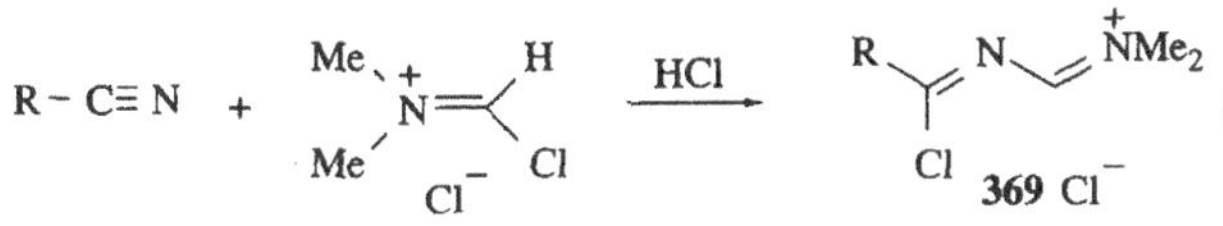

(2.222)

Nitriles of type **370** can undergo repeated iminoalkylation to give the complex iminium salts **371**; acetonitrile forms the salts **372** (scheme 2.223).[507]

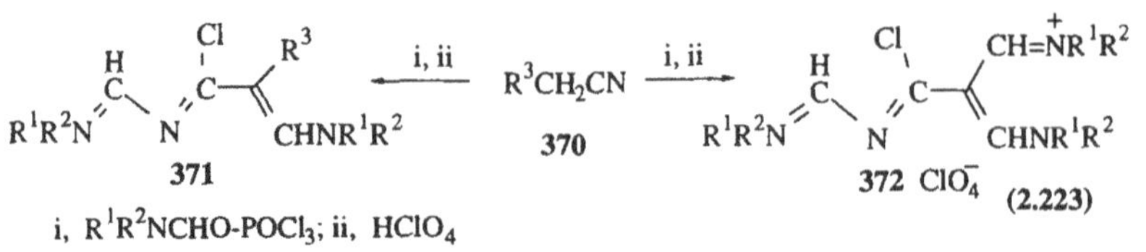

i, R^1R^2NCHO-POCl$_3$; ii, HClO$_4$

Gold's reagent,[508] salt **373**, can be prepared by the reaction of DMF with cyanuric chloride. It reacts with hydrazine to give a 1,2,4-triazole, *e.g.* **374** (R=Ph, 77%),[508] and with amidines to give either 2-monosubstituted triazines or 2,4-disubstituted 1,3,5-triazines **375**, depending on the reaction conditions.[508] Gold's reagent has also been shown to react with anthranilic acid derivatives to

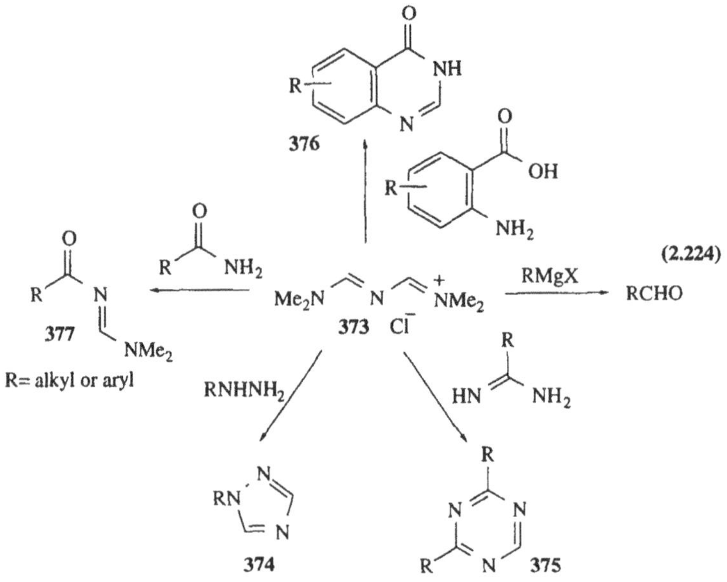

(2.224)

give the quinazolinones **376** and with some acyl- and heteroaryl-amides to give the arylformamidines **377**.[509] A variety of Grignard reagents are converted into aldehydes by Gold's reagent, which acts here as a synthetic equivalent of CHO$^+$.[510]

With Gold's reagent *o*-phenylenediamines afford benzimidazoles, and *o*-aminophenols afford benzoxazoles; *o*-hydroxyacetophenone affords chromone, whereas anthranilamide gives 3*H*-quinazoline-4-one.[511]

3-Chloro-2-propeneiminium salts are 1,3-difunctional electrophiles that have been widely used in synthesis to form new, chiefly heterocyclic, ring systems. The chemistry of 3-chloro-2-propeneiminium salts has been reviewed;[512] they are closely related to (β-chlorovinyl)-aldehydes and -ketones, and are usually prepared by the action of $POCl_3$, PCl_5, $COCl_2$, $SOCl_2$, or $(COCl)_2$ on β-aminovinylcarbonyl compounds.[512] Some can be reduced to 1-substituted-1-chloro-3-*N,N*-dimethylamino-1-propenes.[513,514]

2-Carbethoxy-5-arylpyrroles have been prepared by the reaction of 3-aryl-3-chloropropeniminium salts with esters of α-amino acids.[515]

3-Trifluoromethylsulfonyloxypropeniminium salts have been prepared by the action of Tf_2O on enaminoketones.[516]

A wide variety of iminium salts, including vinylogous amidinium salts, have been prepared by means of the action of Vilsmeier reagents on 1-methoxyalk-1-en-3-ynes (scheme 2.225).[517]

The 3-bromo- and 3-iodo anologues of salt **378** can also be prepared from the enyne and either $DMF-Ph_3P-Br_2$ or $DMF-Ph_3P-I_2$, so that many highly functionalized compounds are accessible.

2.29.2 Amidinium Salts

Much of section 2.27 on amidines is closely related to the present section, since amidinium salts can be generally prepared from amidines and acids; conversely, amidines are generated by treatment of amidinium salts with alkali.[467] At 150-180°C, ammonium chloride, carbonate, or acetate react with chloromethyleneiminium chlorides to give *N,N*-disubstituted form-amidinium chlorides such as **379** (scheme 2.226). This reaction cannot be successfully extended to aliphatic amines, in general, because the amidinium chloride cannot usually be separated from the alkylammonium chloride by recrystallization. However, arylamines do react with acid amide-$POCl_3$ adducts to give the corresponding amidinium salts **3 8 0** (R^3=aryl).[463,464,466,469,480]

$$\underset{\underset{Me}{\overset{Me}{\diagdown}}}{\overset{+}{N}}\!\!=\!\!\underset{Cl^-\ Cl}{\overset{H}{<}} \quad + NH_4Cl \quad \xrightarrow{\ -2\ HCl\ } \quad \underset{\underset{Me}{\overset{Me}{\diagdown}}}{\overset{}{N}}\!\!-\!\!\underset{\overset{+}{N}H_2}{\overset{H}{<}}$$

$$\textbf{379 } Cl^- \qquad\qquad \textbf{(2.226)}$$

$$\underset{\underset{R^2}{\overset{R^1}{\diagdown}}}{\overset{+}{N}}\!\!=\!\!\underset{Cl^-\ OPOCl_2}{\overset{R}{<}} \quad + R^3NH_2 \quad \xrightarrow{\ -\ HOPOCl_2\ } \quad \underset{\underset{R^2}{\overset{R^1}{\diagdown}}}{\overset{+}{N}}\!\!=\!\!\underset{NHR^3}{\overset{R}{<}}$$

$$\textbf{380 } Cl^-$$

When both amidinium nitrogen atoms possess the same groups R^1 and R^2, a carboxylic acid can be condensed with R^1R^2NH in the presence of $POCl_3$, and the resulting amide $RCONR^1R^2$ again condensed with R^1R^2NH and $POCl_3$ to give the amidinium salt.[463, 466] This is a special case of scheme 2.227; the salts were isolated as the perchlorates[466] or as the iodides.[463]

$$\underset{R^1R^2N}{\overset{R}{\diagdown}}\!\!=\!\!O \quad + ArNHR^3 \quad \xrightarrow[\text{ii) HX}]{\text{i) POCl}_3} \quad \underset{R^1R^2N}{\overset{R}{\diagdown}}\!\!=\!\!\overset{+}{N}ArR^3 \quad \underset{X^-}{} \qquad \textbf{(2.227)}$$

In general, arylamines and heteroarylamines react with chloromethyleneiminium chlorides to give good yields of N-aryl substituted amidinium salts.[518] On occasion, adducts of $POCl_3$ and aliphatic amides containing α-methylene groups undergo self-condensation and this interferes with the synthesis of the amidinium salts by scheme 2.227. However, a simple method of suppressing self-condensation is to react $POCl_3$ directly with the free carboxylic acid, and excess amine. Alternatively, $POCl_3$ can be added dropwise to a mixture of the amide and the arylamine, prior to heating at 150°C.[466] Bisamidinium perchlorates derived from glutaric and adipic acids have also been prepared.[466] Aminoanthraquinones have also been converted into amidinium salts by DMF-SOCl$_2$.[16,507]

A variety of amidinium salts result from the reaction of primary, secondary or tertiary amides with primary or secondary aliphatic or aromatic amines,[463,469,519,520] but they are usually converted into the corresponding amidines with alkali (scheme 2.228).

$$\underset{R^1R^2}{\overset{R}{\diagdown}}\!\!=\!\!O \quad + R^3R^4NH \quad \xrightarrow{\ PCl_3\ } \quad R^1R^2\overset{+}{N}\!\!=\!\!\underset{\underset{Cl^-\ NR^3R^4}{}}{\overset{R}{<}} \qquad \textbf{(2.228)}$$

Titanium tetrachloride affords amidinium salts from secondary alkylamines and tertiary acid amides,[521] but the amidinium salt can be difficult to separate from the dialkylamine hydrochlorides. The amide group must not be overly bulky; N,N-dimethylisobutyramide undergoes condensation with dimethylamine, but N,N-dimethylpivalic acid amide does not.

Some amidinium salts can be prepared by the self-condensation of two moles of an amide with thionyl chloride (scheme 2.229).[522,523] Kuehle showed that chloromethyleneiminium chlorides were intermediates.[393] This is confirmed by the reaction of DMF with *N*-phenylchloromethyleneiminium chloride, giving *N,N*-dimethyl-*N´*-phenylformamidinium chloride.[393]

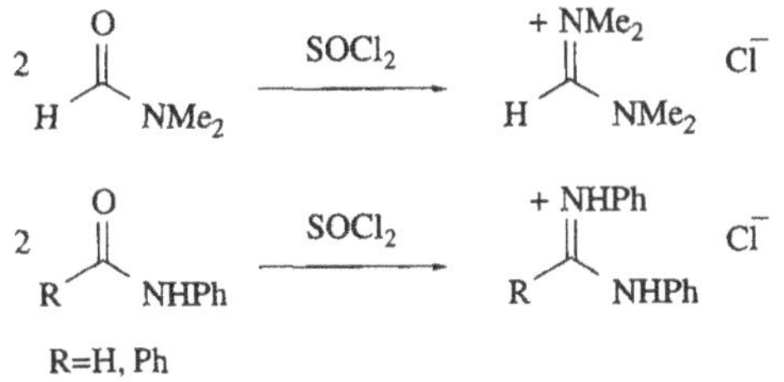

(2.229)

R=H, Ph

N-Monosubstituted formamides react with phosgene to give amidinium salts.[524,525] However, such amides react with chloromethyleneiminium chlorides to give isocyanides.[526]

Urea and its derivatives react with chloromethyleneiminium salts **381** to give amidinium salts **382**.[16,527-529] The salts **381** are generated *in situ* by reacting the amide-urea mixture with thionyl chloride or phosgene. Thioureas,[16] urethanes,[16] biuret,[530] and guanidine carbonate[530] react analogously.

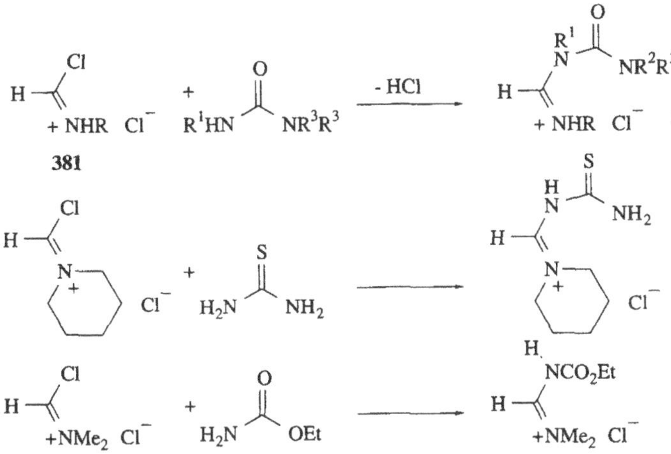

(2.230)

Amidinium bromides can also be prepared from bromomethyleneiminium bromides.[531]

Schiff bases can undergo iminoalkylation at nitrogen to give amidinium salts (scheme 2.231).

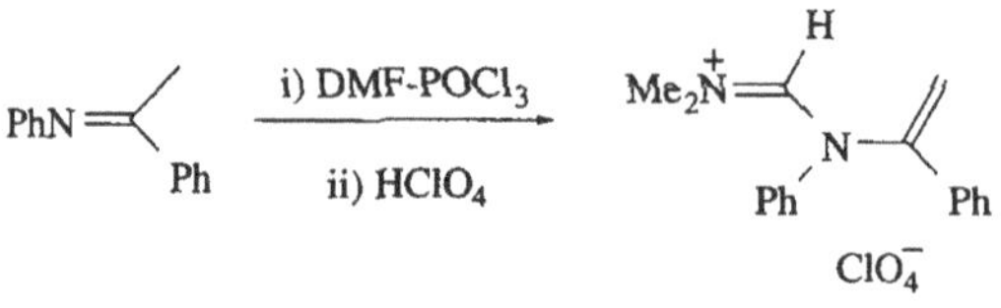

$$(2.231)$$

Acylamidinium salts can be formed by the reaction of nitriles with Vilsmeier reagents in the cold, and with HCl gas bubbling through the mixture;[532] that suggests the formation of imino chlorides which are then alkylated to give salts **383**. This pathway is also consistent with the salt **384** that results from the reaction of malonodinitrile with Vilsmeier reagents.[533]

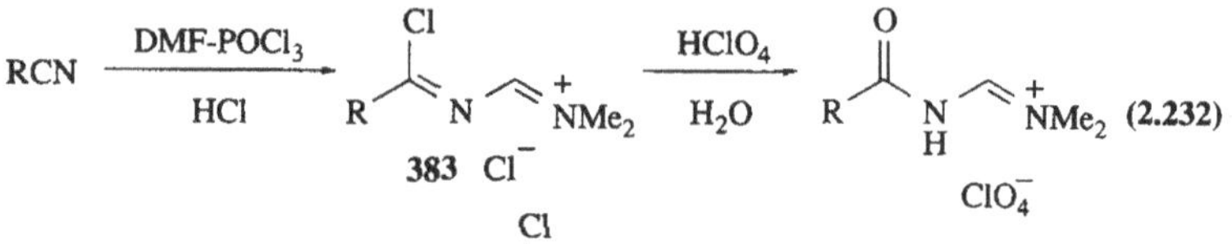

$$(2.232)$$

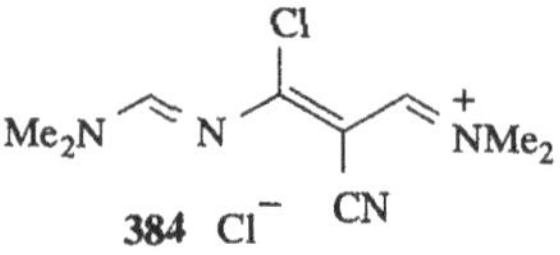

2.29.3 1,5-Diazapentadienium Salts (Vinamidinium or Trimethinium Salts)

The chemistry of vinamidinium salts has been reviewed.[534,535] Vinamidinium salts of the type **388** where R is halo, carboxyl, or an aromatic or heteroaromatic ring, but *not* alkyl, are generally prepared by the reaction of the monosubstituted acetic acid with a Vilsmeier reagent.[159,534,536-538] A plausible mechanism proposes the intermediate ketene **385** being converted into the 3-dimethylaminoacryloyl chloride **386** by the Vilsmeier reagent. The acid chloride is then attacked by DMF to give the presumed cation **387** which by intramolecular substitution affords the vinamidinium salt **388** which can be isolated from the mixture as the perchlorate.[539] That the reaction fails for simple unsubstituted aliphatic carboxylic acids is consistent with the known difficulty in forming simple ketenes **385** from aliphatic acid chlorides. The reaction is effective for many R groups including aryl,[537] chloro and fluoro,[159] and ethoxycarbonyl,[159] but not bromo.[538]

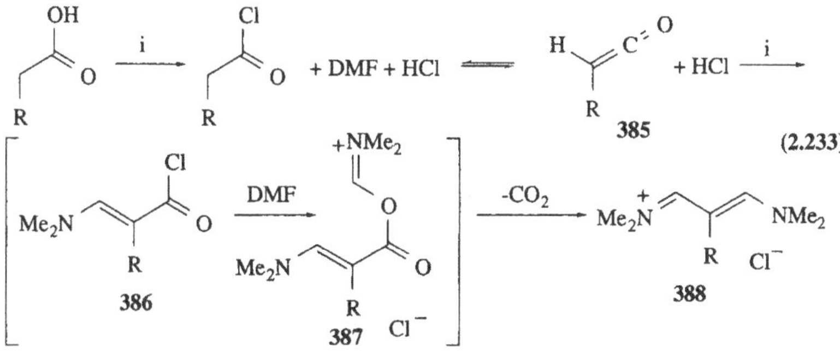

i, Me$_2$N$^+$=CHCl Cl$^-$

Glycine hydrochloride and some of its *N*-substituted derivatives react with DMF-POCl$_3$ to give, after perchloric acid work-up, the diperchlorate **389**. Deprotonation (R=H) affords the synthetically valuable vinamidinium salts **390** with a protected, but readily cleavable amino group.[540]

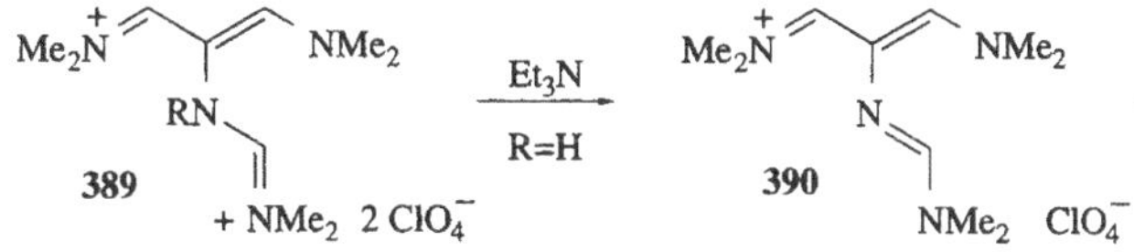

The vinamidinium salt **388**, where R=alkyl, is inaccessible from alkanoic acids, but can be conveniently prepared from the corresponding alkyl malonic acids.[541] Decarboxylation to give the ketene intermediate **385** appears likely. Malonic acid itself undergoes further iminoalkylation to give **388** (R=[CH=NMe$_2$]$^+$).[159] Cyanoacetic acid also affords vinamidinium salts with Vilsmeier reagents.[540]

A convenient route to vinamidinium salts **391** involves *O*-methylation of β-dimethylaminoacrylaldehydes followed by displacement of the ethoxy group with a secondary amine.[542]

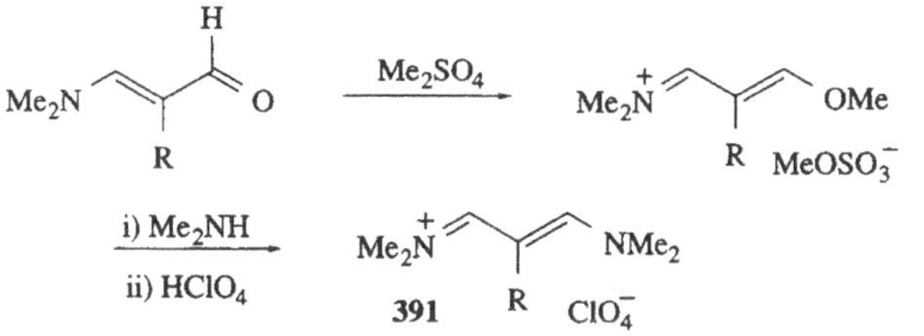

This is an application of Bredereck's procedure for the preparation of amidinium salts.[543-545] The β-dimethylaminoacrylaldehydes themselves are readily prepared by reacting Vilsmeier reagents with either acetals or vinyl ethers (section 2.8.2.2). The reaction of vinamidinium salts with enolates affords functionalized dienones.[546]

The vinamidinium salt **392** (section 1.3.4) is the result of the action of dimethylchloromethyleneiminium chloride upon methylenecyclohexane, although the yield is only 17%.[75]

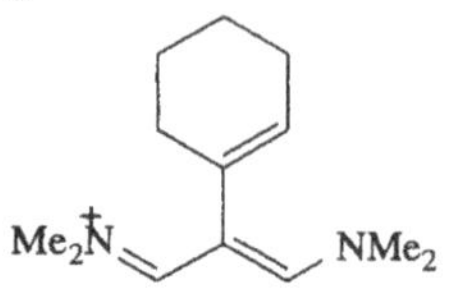

$$(2.236)$$

392 ClO_4^-

Cyclic vinylogous amidinium salts have been generated from the heterocyclic ketone **393**; the extent of iminoalkylation can be controlled by the temperature (scheme 2.237).[547]

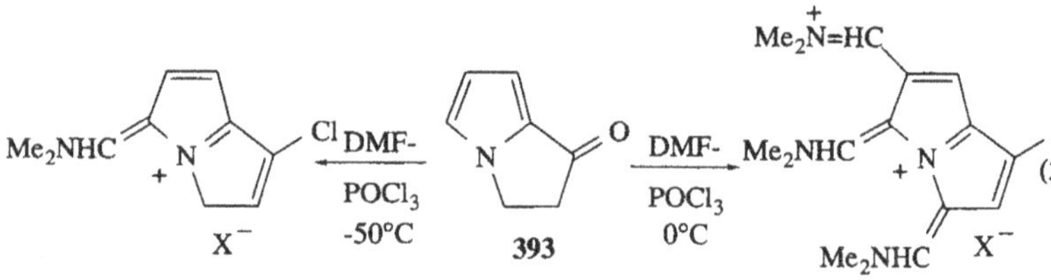

$$(2.237)$$

393

Vinamidinium salts of the type **394** (R^2=Ph) react[548] with aryl Grignard reagents to give after hydrolysis, the 2-alkenals **397**, and with BF_3.THF to give allylamines **396**.[514,,548] The presumed common intermediate in these two sequences is the enamine **395**. Reduction of **394** with $NaBH_4$ also gives an allylamine, $Me_2NCH_2C(R)=CH_2$.[513] α,β-Unsaturated ketones have also been prepared from 1-arylvinamidinium salts and organometallic reagents.[514]

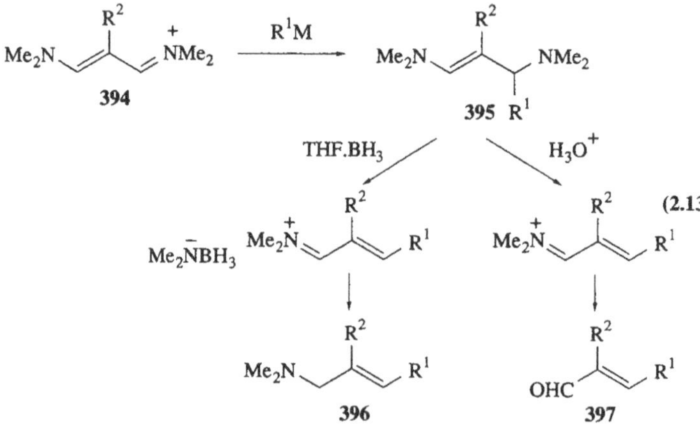

$$(2.138)$$

394

395 R^1

Me_2NBH_3

396   **397**

Some vinamidinium salts react with a variety of amidines to give 2,5-disubstituted pyrimidines; arylhydrazones afford substituted pyrazoles.[549,550]

Reactions of acetonitriles, amides and thioamides afford trimethinium salts of the type **398**; intermediates on the pathway to **398** were indentified.[551] Substituted acetic acids gave trimethinium salts when reacted with DMF-$POCl_3$.[552]

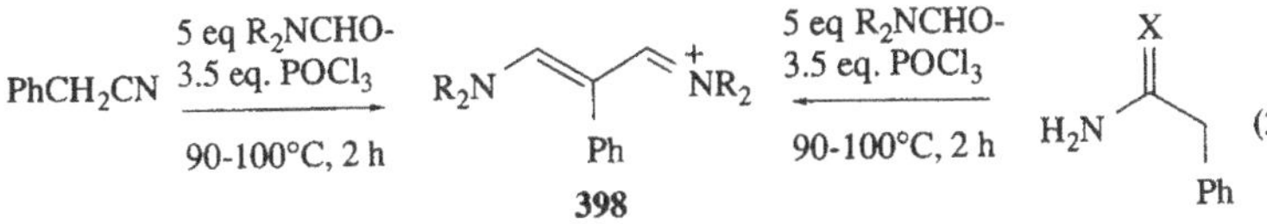

5 eq R₂NCHO-
3.5 eq. POCl₃
PhCH₂CN
90-100°C, 2 h
R₂N
NR₂
Ph
5 eq R₂NCHO-
3.5 eq. POCl₃
90-100°C, 2 h
X
H₂N
Ph

398 (2.239)

2R=-$(CH_2)_n$-, where n=2, 4 and 5; 2R=$O(CH_2CH_2)$ X=O, S

2.29.4 Pentamethinium and Polymethinium Salts

Vinylogs of ethanal react with DMF-$POCl_3$ (that has been deactivated with one equivalent of methanol) to give the chloroiminium salts **399** (n=1 to 4). The salts **399** undergo displacement with dimethylamine to give the methinium salts **400**.[553] The principle can be extended to the synthesis of aryl-substituted pentamethinium salts; **401** is formed *via* the action of a Vilsmeier reagent upon crotonophenone (scheme 2.240).[554]

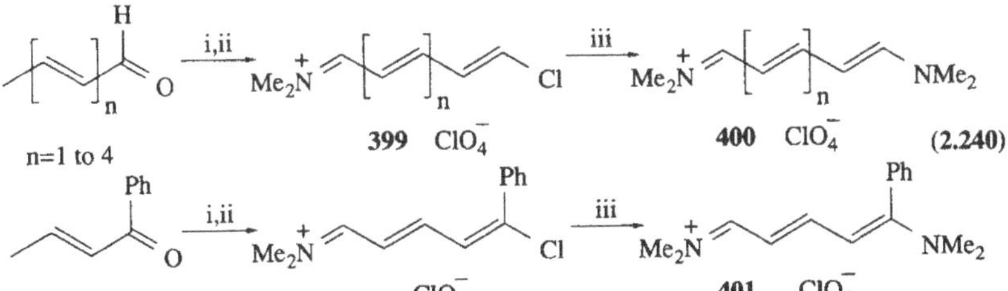

H
O
n
i,ii
Me₂N
n
Cl
iii
Me₂N
n
NMe₂
n=1 to 4
Ph
O
i,ii
Me₂N
Ph
Cl
ClO₄⁻
iii
Me₂N
Ph
NMe₂

399 ClO_4^- **400** ClO_4^- (2.240)

401 ClO_4^-

i, DMF-$POCl_3$ in MeOH; ii, $HClO_4$; iii, Me_2NH

The reaction of acetone **402** (R=H) or methyl ethyl ketone **402** (R=Me) with DMF-$POCl_3$ leads by repeated alkylation a pentamethinium salt **403** that can be isolated as the diperchlorate (90%) in the former case. In the latter case the salt **403** (R=Me) undergoes partial hydrolysis to give the salt **404** (74%).[41]

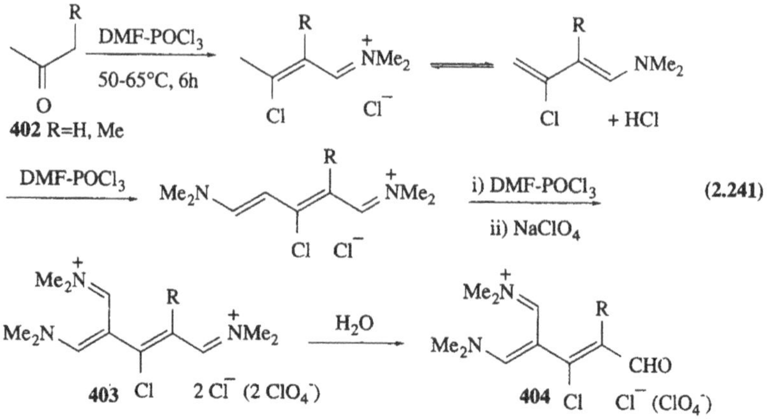

(2.241)

Benzylideneacetone and cinnamylideneacetone are smoothly converted into the respective coloured, crystalline iminium salts **405** and **406**.[555] The reaction of DMF-POCl$_3$ with crotonophenone affords, after treatment with aqueous perchlorate, the salt **407** (65%) (figure 2.242).[554]

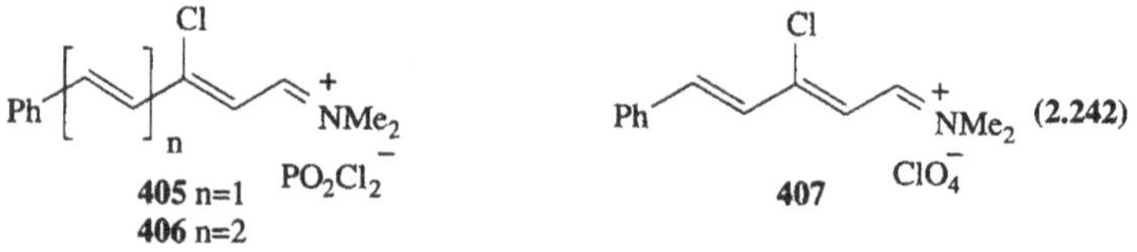

(2.242)

Bistrimethinium salts are obtained by reacting benzene-1,3-diacetic acid and benzene-1,4-diacetic acid with DMF-POCl$_3$; the respective yields are 65% and 82%.[79] Benzene-1,2-diacetic acid behaves differently, forming a benzofulvene (section 4.2.2).

A large excess of the Vilsmeier reagent, followed by addition of sodium perchlorate solution, afforded the 3-chloropentamethinium salts **408** (n=1), which in the case of the cyclohexanone derivative could be hydrolyzed to **409** (n=2; scheme 2.243). Work-up of the reaction mixture from cyclopentanone with aqueous potassium carbonate afforded the salt **410**, which with hot aqueous potassium carbonate is evidently hydrolyzed to **411** and DMF, since a mixture of the amino aldehyde **414** and the amino ketone **415** is obtained.[135] The conversion of **410** into **415** illustrates the important observations of effectively thermal deformylations of intermediates obtained from certain Vilsmeier reactions.

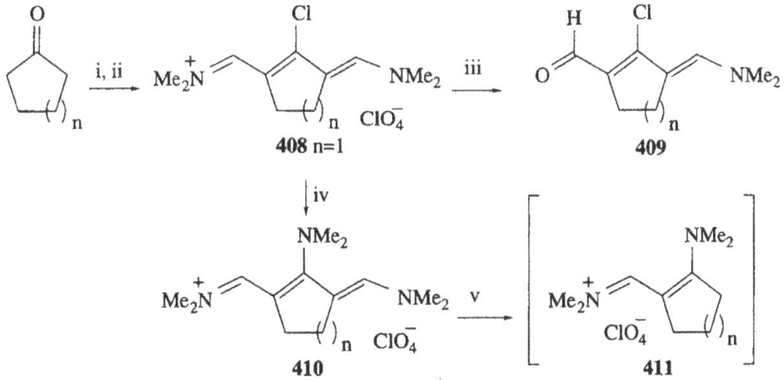

i, DMF-POCl$_3$; ii, aq. NaClO$_4$; iii, aq. NaHCO$_3$; iv, HNMe$_2$; v, aq. K$_2$CO$_3$, heat   (2.243)

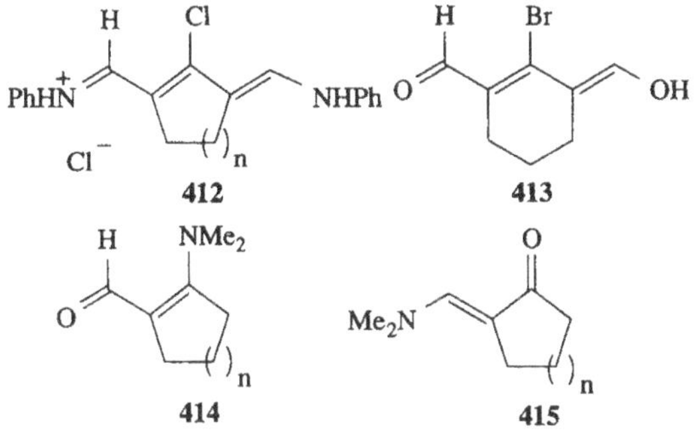

The reaction of cyclohexanone using a bromomethyleneiminium salt afforded a small quantity of the doubly substituted aldehyde **413**.[288] The formation of pentamethinium salts is controlled by the number of alkyl groups present. No other sites of the pentamethinium unit are available for formylation, and its stability evidently prevents isomerization into compounds which might undergo deprotonation and hence further reaction with the Vilsmeier reagent. Scheme 2.244 illustrates a variety of isolable pentamethinium salts.

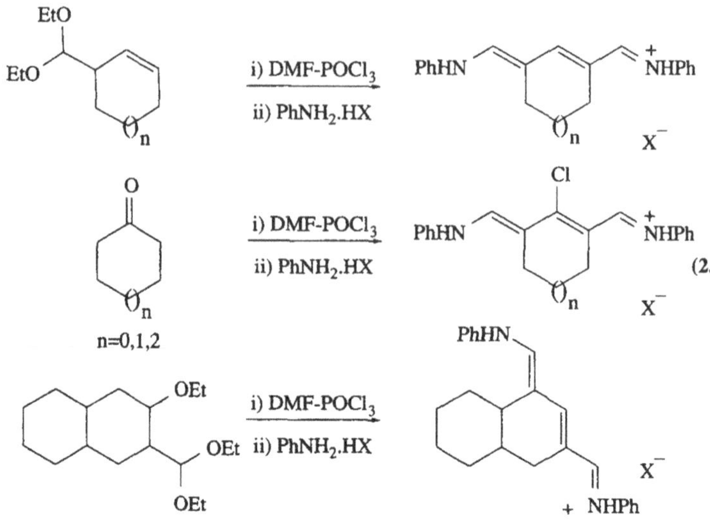

Some interesting cross-conjugated pentamethinium salts have been invoked during the reaction of substituted 2-cyclohexen-1-ones with Vilsmeier reagents (scheme 2.134, section 2.11.3). Cross-conjugated pentamethinium salts are also formed by reacting cyclopentadiene with Vilsmeier reagents (section 2.8.2.1).[81,86,87,556]

2.29.5 Polymethinium Salts

Polymethinium salts as in scheme 2.240 are readily prepared from the appropriate vinylog of ethanal by treatment with a deactivated Vilsmeier reagent, followed by displacement of the chloro group with diethylamine.[553]

Some interesting heptamethinium salts that straddle a cyclohexane ring have been invoked during the reaction of substituted 2-cyclohexen-1-ones with Vilsmeier reagents (scheme 2.134, section 2.11.3).

References

1. D. R. Hepburn and H. R. Hudson, *Chem. Ind.,* 1974, 664.
2. D. R. Hepburn and H. R. Hudson, *J. Chem. Soc., Perkin Trans. 1,* 1976, 754.
3. C. Jutz, in *Advances in Organic Chemistry*, vol. 9, *Iminium Salts in Organic Chemistry*, part 1, E. C. Taylor, (ed.), John Wiley, New York, 1976, pp. 225-342.
4. W. Kantlehner, in *Advances in Organic Chemistry*, vol. 9, *Iminium Salts in Organic Chemistry*, part 2, E. C. Taylor, (ed.), John Wiley, New York, 1976, pp. 65-141.
5. R. R. Koganty, M. B. Shambhue, and G. A. Digenis, *Tetrahedron Lett.,* 1973, 4511.
6 (a) R. F. Dods and J. S. Roth, *Tetrahedron Lett.,* 1969, 165; (b) R. F. Dods and J. S. Roth, *J. Org. Chem.,* 1969, **34**, 1627.
7. (a) M. Petitou, P. Duchaussoy, I. Lederman, J. Choay, and P. Sinay, *Carbohydrate Research,* 1988, **179**, 163; (b) H. Paulsen, M. Heume, and H. Nurnberger, *Carbohydrate Research,* 1990, **200**, 127.
8. (a) G. B. Bachmann and R. W. Dowens, U. S. Pat, 3054829 (1962); *Chem. Abstr.,* 1962, **58**, 3730; (b) G. B. Bachmann, U. S. Pat, 3169150 (1965); *Chem. Abstr.,* 1965, **62**, 14500d.
9. M. Yoshihira, T. Eda, K. Sakakai, and T. Maeshima, *Synthesis,* 1980, 746.
10. M. Pulst, M. Weissenfels, and F. Dietz, *Z. Naturforsch., B: Anorg. Chem., Org. Chem.* 1985, **40B**, 585; *Chem. Abstr.* 1985, **103**, 72600p.
11. I. R. Robertson and J. T. Sharp, *Tetrahedron,* 1984, **25**, 3095.
12. T. L. Gilchrist and R. J. Summersell, *Tetrahedron Lett.,* 1987, **28**, 1469.
13. F. Reicheneder, K. Dury, and P. Dimroth, French Pat., 1413606 (1965).
14. A. Pinner, *Die Iminoaether and ihre Derivative*, Robert Oppenheimer, Berlin, 1892.
15. A. G. Anderson, Jr., N. E. T. Owen, F. J. Freenor, and D. Erickson, *Synthesis,* 1976, 398.
16. H. Eilingsfeld, M. Seefelder, and H. Weidinger, *Angew. Chem.*, 1960, **72**, 836.
17. H. Eilingsfeld, M. Seefelder, and H. Weidinger, *Chem. Ber.,* 1963, **96**, 2671.
18. L. Suranyi and H. Wilk, Ger. Pat., 1150987 (1963).
19. L. Kh. Felidanan, G. S. Senucheva, and N. G. Zhokhovels, *Mechenye Bid. Aktion Veschestv. Sb. Statei,* 1962, 26.
20. W. Ziegenbein and W. Franke, *Chem. Ber.,* 1960, **93**, 1681.
21. W. Ziegenbein and K. H. Hornung, *Chem. Ber.,* 1963, **95**, 2976.

22. W. Ziegenbein and K. H. Hornung, Ger. Pat., 1188570 (1963).

23. H. Priewe and K. Gutsche, Ger. Pat., 1145622 (1963)

24. F. M. Stojanovic and Z. Arnold, *Collect. Czech. Chem. Commun.*, 1967, **32**, 2155.

25. T. Fujisawa, S. Ida, and T. Sato, *Chem. Lett.*, 1984, 1173.

26. M. S. Newman and P. K. Sujeeth, *J. Org. Chem.*, 1978, **43**, 4367.

27. M. Benazza, R. Uzan, D. Beaupère, and G. Demailly, *Tetrahedron Lett.*, 1992, **33**, 4901.

28. R. V. Lemieux, S. Z. Abbas, and B. Y. Chung, *Can. J. Chem.*, 1982, **60**, 58.

29. P. R. Giles and C. M. Marson, unpublished observations.

30. D. T. Kozhich, L. V. Akimenko, A. F. Mironov, and R. P. Evstigneeva, *J. Org. Chem. USSR (Engl. Transl.)*, 1977, **13**, 2418.

31. A. F. Mironov, L. V. Akimenko, V. D. Rumyantseva, and R. P. Evstigneeva, *Khim. Geterotsikl. Soedin.*, 1975, 423; *Chem. Abstr.*, 1975, **83**, 28041b.

32. A. R. Katritzky, Z. Wang, C. M. Marson, R. J. Offerman, A. E. Koziol, and G. J. Palenik, *Chem. Ber.* 1988, **121**, 999.

33. R. S. Al-awar, S. P. Joseph, and D. L. Comins, *Tetrahedron Lett.*, 1992, **33**, 7635.

34. J. Czyzewski, and D. H. Reid, *J. Chem. Soc., Perkin Trans. 1*, 1983, 777.

35. F. B. Dains, *Ber. Dtsch. Chem. Ges.*, 1902, **35**, 2496.

36. I. Matsumoto, *Yakugaku Zasshi*, 1965, **85**, 544.

37. E. V. P. Tao and C. F. Christie, *Org. Prep. Proc. Int.*, 1972, **4**, 73.

38. (a) J. Kuthan, Czech Pat., 109895 (1964); (b) F. Riecheneder, K. Dury, and P. Dimroth, French Pat., 1413603 (1965).

39. Y. Kurasawa, S. Shimabukuro, Y. Okamoto, K. Ogura, and A. Takada, *J. Heterocycl. Chem.* 1985, **22**, 1135.

40. M.-J. Shiao, Li.-M. Shyu, K.-Y. Tarng, and Y.-T. Ma, *Synth. Commun.*, 1990, 2971.

41. Z. Arnold, *Collect. Czech. Chem. Commun.*, 1961, **26**, 1113.

42. J. Zemlicka and Z. Arnold, *Collect. Czech. Chem. Commun.*, 1961, **26**, 2838.

43. K. Bodendorf and R. Meyer, *Chem. Ber.*, 1965, **98**, 3554.

44. K. Bodendorf and P. Kloss, *Angew. Chem., Int. Ed. Eng.*, 1963, **2**, 98.

45. J. Lotzbeyer and K. Bodendorf, *Chem. Ber.*, 1967, **100**, 2620.

46. C. Alexander and W. J. Feast, *Synthesis* , 1992, 735.

47. H. M. Relles, US Pat. 3700743 (1973); *Chem. Abstr.*, 1973, **78**, 85053.

48. V. A. Pattison, J. G. Colson, and R. L. K. Carr, *J. Org. Chem.*, 1968, **33**, 1084.

49. A. N. Grinev and I. N. Nikolaeva, *Arm. Khim. Zh,* 1975, **28**, 1007; *Chem. Abstr.,* 1976, **85**, 21002x.

50. A. P. Shawcross and S. P. Stanforth, *Tetrahedron,* 1989, **45**, 7063.

51. M. Mazaruddin and G. Thyagarajan, *Tetrahedron Lett.,* 1971, 307.

52. M. R. Chandramohan, M. S. Sardessai, S. R. Shah, and S. Seshadri, *Indian J. Chem.,* 1969, **7**, 1006.

53. F. Eiden, *Arch. Pharm. (Weinheim, Ger.),* 1962, **295**, 533.

54. H. Khedija, H. Strzelecka, and M. Simalty, *Bull. Chim. Soc. Fr.,* 1973, 218.

55. K. E. Schulte, J. Reisch, and U. Stoess, *Angew. Chem., Int. Ed. Engl.,* 1965, **4**, 1081.

56. H. von Dobeneck and F. Schnierle, *Tetrahedron Lett.,* 1966, 5327.

57. (a) G. P. Tomakov, T. G. Zemlyanova, and I. I. Grandberg, *Khim. Geterotsikl. Soedin.* 1984, 56; *Chem. Abstr.* 1984, **100**, 209658y; (b) G. M. Tomakov and I. I. Grandberg, *Izv. Timiryazevsk. S-kh. Akad.* 1979, **6**, 151; *Chem. Abstr.* 1980, **92**, 94219c.

58. S. Seshadri, M. S. Sardessai, and A. M. Betrabet, *Indian. J. Chem.,* 1969, **7**, 662.

59. Z. F. Solomko, V. N. Proshkina, N. Ya. Bozhanova, S. V. Loban, and L. N. Babichenko, *Khim. Geterotsikl. Soedin.* 1984, 223; *Chem. Abstr.* 1984, **100**, 209762c.

60. Z. F. Solomko, V. N. Proshkina, V. I. Avramenko, I. A. Plastun, and N. Ya. Bozhanova, *Khim. Geterotsikl. Soedin.* 1984, 1262; *Chem. Abstr.* 1985, **102**, 6436t.

61. U. Sunay, D. Mootoo, B. Molino, and B. Fraser-Reid, *Tetrahedron Lett.,* 1986, **27**, 4697.

62. J. P. Dulcère and J. Rodriguez, *Tetrahedron Lett.,* 1982, **23**, 1887.

63. J. Rodriguez and J. P. Dulcère *J. Org Chem.,* 1991, **56**, 469.

64. T. Fujisawa, T. Mori, T. Tsuge, and T. Sato, *Tetrahedron Lett.,* 1983, **24**, 1543.

65. T. Fujisawa and T. Sato, *Org. Synth.,* 1988, **66**, 121.

66. C. P. Reddy and S. Tanimoto, *Synthesis,* 1987, 575.

67. J. ApSimon, K. E. Fyfe, and A. M. Greaves, in *The Total Synthesis of Natural Products*, J. ApSimon, John Wiley, New York, 1984, vol. 6, pp. 106-107.

68. G. W. Moersch and W. A. Neuklis, *J. Chem. Soc.,* 1965, 788.

69. F. Huet, *Synthesis,* 1985, 496.

70. P. C. Traas, H. J. Takken, and H. Boelens, *Tetrahedron Lett.,* 1977, 2027.

71. A. Hara and M. Sekiya, *Chem. Pharm Bull. Jap.,* 1972, **20**, 309.

72. V. I. Minkin and G.N. Dorofeenko, *Russ. Chem. Rev.,* 1960, **29**, 599.

73. G. A. Olah and S. J. Kuhn, in *Friedel-Crafts and Related Reactions*, G. A. Olah, ed., Wiley Interscience, New York, 1964, vol. 3, part 2, p. 1211.
74. A. R. Katritzky, I. V. Shcherbakova, R. D. Tack, and P. J. Steel, *Can J. Chem.*, 1992, **70**, 2040.
75. C. Jutz and W. Müller, *Chem. Ber.*, 1967, **100**, 1536.
76. C. Jutz, W. Müller, and E. Müller, *Chem. Ber.*, 1966, **99**, 2479.
77. (a) P. C. Traas and H. Boelens, *Rec. Trav. Chim. Pays-Bas*, 1973, **92**, 985; (b) P. C. Traas, H. Boelens and H. J. Trakken, *Rec. Trav. Chim. Pays-Bas*, 1976, **95**, 57.
78. P. C. Traas, H. J. Takken, and H. Boelens, *Tetrahedron Lett.*, 1977, 2129.
79. Z. Arnold, *Collect. Czech. Chem. Commun.*, 1965, **30**, 2783.
80. G. Dauphin, *Synthesis*, 1979, 799.
81. Z. Arnold, *Collect. Czech. Chem. Commun.*, 1960, **25**, 1313.
82. K. Hafner and K. H. Vöpel, *Angew. Chem.*, 1959, **71**, 672.
83. K. Hafner, *Angew. Chem.*, 1960, **72**, 574.
84. K. Hafner and M. Kreuder, *Angew. Chem.*, 1961, **73**, 657.
85. K. Hafner, K. H. Vöpel, G. Ploss, and C. König, *Liebigs Ann. Chem.*, 1963, **661**, 52.
86. K. Hafner, K. H. Häfner, C. König, M. Kreuder, G. Ploss, G. Schulz, E. Sturm, and K. H. Vöpel, *Angew. Chem.*, 1963, **75**, 35.
87. H. Meerwein, W. Florian, N. Schoen and G. Stopp, *Liebigs Ann. Chem.*, 1960, **641**, 1.
88. Z. Arnold and J. Zemlicka, *Collect. Czech. Chem. Commun.*, 1960, **25**, 1302.
89. T. Asao, S. Kuroda, and K. Kato, *Chem. Lett.*, 1978, 41.
90. M. J. Grimwade and M. G. Lester, *Tetrahedron*, 1969, **25**, 4535.
91. R. P. Graber and D. M. Aedo, Spanish Pat. 547158 (1986); *Chem. Abstr.*, 1989, **110**, 115187y.
92. V. V. Gertsev, *Zh. Vses, Khim. O-Va*, 1982, **27**, 341.
93. Z. Zicmanis and M. Klavis, *Latv. PSR Zinat. Akad. Vestis, Kim. Ser.*, 1980, **3**, 354; *Chem. Abstr.*, 1980, **93**, 133030b.
94. L. Cazaux, M. Faher, and P. Tisnes, *J. Chem. Res. (S)*, 1990, 264.
95. M. P. Reddy and G. S. K. Rao, *Synthesis*, 1980, 815.
96. A. Sudalai and G. S. K. Rao, *Indian J. Chem., Sect. B.*, 1989, **28**, 219.
97. M. P. Reddy and G. S. K. Rao, *Tetrahedron Lett.*, 1981, 3549.
98. R. V. Rao and M. V. Bhatt, *Indian J. Chem., Sect. B.*, 1981, **20**, 487.
99. P. A. Reddy and G. S. K. Rao, *Indian J. Chem., Sect. B.*, 1981, **20**, 100.
100. T. P. Velumasy and G. S. K. Rao, *Indian J. Chem., Sect. B.*, 1981, **20**, 351.
101. P. A. Reddy and G. S. K. Rao, *Indian J. Chem., Sect. B.*, 1982, **21**, 885.
102. P. A. Reddy and G. S. K. Rao, *J. Org. Chem.*, 1981, **46**, 5371.

103. T. Shono, Y. Matsumura, K. Tsubata, and Y. Shugihara, *Tetrahedron Lett.,* 1982, 1201.

104. T. Shono, Y. Matsumura, K. Tsubata, Y. Shugihara, S. Yamene, T. Kanazawa, and T. Aoki, *J. Am. Chem. Soc.,* 1982, **104**, 6697.

105. B. Gopalan, K. Rajagopalan, S. Swaminathan, and K. K. Balasubramanian, *Synthesis,* 1976, 752.

106. S. R. Jensen, O. Kirk, and B. J. Nielsen, *Tetrahedron,* 1987, **43**, 1949.

107. N. G. Ramesh and K. K. Balasubramanian, *Tetrahedron Lett.,* 1991, **32**, 3875.

108. N. Langlois and F. Favre, *Tetrahedron Lett.,* 1991, **32**, 2233.

109. F.-W. Ullrich and E. Breitmaier, *Synthesis,* 1983, 641.

110. C. Jutz, *Chem. Ber.,* 1958, **91**, 850.

111. C. Jutz, *Angew. Chem.,* 1958, **70**, 270.

112. G. P. Stepanova and B. I. Stepanov, *J Org. Chem. USSR,* 1971, **7**, 1033.

113. K. Dickore and F. Kroehnke, *Chem. Ber.,* 1960, **93**, 1068.

114. K. Dickore and F. Kroehnke, *Chem. Ber.,* 1960, **93**, 2479.

115. H. Nordmann and F. Kroehnke, *Angew. Chem.,* 1969, **81**, 747.

116. H. Nordmann and F. Kroehnke, *Liebigs Ann. Chem.,* 1970, **731**, 80.

117. Z. Arnold and A. Holy, *Collect. Czech. Chem. Commun.,* 1965, **30**, 40.

118. E. E. Nikolajewski, S. Dähne, D. Leupold, and B. Hirsch, *Tetrahedron,* 1968, **24**, 6685.

119. T. Sugasawa, K. Saskakura and T. Toyoda, *Chem. Pharm. Bull.,* 1974, **22**, 763.

120. D. Burn, G. Cooley, M. T. Davies, J. W. Ducker, B. Ellis, P. Feather, A. K. Hiscock, D. N. Kirk, A. P. Leftwick, V. Petrow, and D. K. Williamson, *Tetrahedron,* 1964, **20**, 597.

121. G. M. Coppola, G. E. Hardtmann, and B. S. Huegi, *J. Heterocycl. Chem.,* 1974, **11**, 51.

122. J. Schmitt, J. J. Panouse, A. Hallot, P.-J. Cornu, H. Pluchet, and P. Comoy, *Bull. Soc. Chim. Fr.,* 1964, 2753.

123. M. G. Lester, V. Petrow, and O. Stephenson, *Tetrahedron,* 1964, **20**, 1407.

124. Z. Arnold, *Collect. Czech. Chem. Commun.,* 1963, **28**, 863.

125. H. Bredereck, G. Simchen, and P. Speh, *Liebigs Ann. Chem.,* 1970, **737**, 46.

126. T. Kato, H. Yamanaka, and H. Hiramimia, *J. Pharm. Soc. Jap.,* 1970, **90**, 870.

127. M. R. Chandramohan and S. Seshadri, *Indian J. Chem.,* 1972, **10**, 573.

128. J. Ciernik, *Collect. Czech. Chem. Commun.,* 1972, **37**, 2273.

129. M. R. Jayanth, H. A. Naik, D. R. Tatke, and S. Seshadri, *Indian J. Chem.,* 1973, **11**, 1112.

130. D. M. Brown and A. Giner-Sorolla, *J. Chem. Soc. (C),* 1971, 128.
131. Z. Arnold, *Collect. Czech. Chem. Commun.,* 1962, **27**, 2993.
132. Z. Arnold and A. Holy, *Collect. Czech. Chem. Commun.,* 1963, **28**, 869.
133. J. Zemlicka and Z. Arnold, *Collect. Czech. Chem. Commun.,* 1961, **26**, 2852.
134. Z. Arnold and J. Zemlicka, *Collect. Czech. Chem. Commun.,* 1960, **25**, 1318.
135. V. Dressler and K. Bodendorf, *Arch. Pharm. (Weinheim, Ger.),* 1970, **303**, 481.
136. B. Eistert and F. Haupter, *Chem. Ber.,* 1959, **92**, 1921.
137. Z. Arnold and J. Zemlicka, *Collect. Czech. Chem. Commun.,* 1959, **24**, 786.
138. Z. Arnold, *Collect. Czech. Chem. Commun.,* 1973, **38**, 1168.
139. C. Reichardt and K. Schagerer, *Angew. Chem., Int Ed. Engl.,* 1973, **12**, 323.
140. P. Knorr, P. Low, P. Hassel, and H. Bronberger, *J. Org. Chem.,* 1984, **49**, 1288.
141. E. A. Jauer, E. Foerster, and J. B. Hirsch, *Signlaufzeichnungsmaterialien,* 1975, **3**, 155; *Chem. Abstr.,* 1975, **83**, 81180x.
142. J. Becher, K. Pluta, N. Krake, K. Brøndum, N. J. Christensen, and M. V. Vinader, *Synthesis,* 1989, 530.
143. R. Brehme, *Chem. Ber.,* 1990, **123**, 2039.
144. M. Weissenfels and M. Pulst, *Tetrahedron Lett.,* 1968, 3045.
145. M. Weissenfels and M. Pulst, *Tetrahedron,* 1972, 5197.
146. M. Weissenfels and M. Pulst, *J. Prakt. Chem.,* 1973, **315**, 873.
147. M. S. Korobov, L. E. Nivorozhkin, and V. I. Minkin, *J. Org. Chem., USSR,* 1973, **9**, 1739.
148. M. Pulst, M. Weissenfels, E. Kleinpeter, and L. Beyer, *Tetrahedron,* 1975, **31**, 1307.
149. B. Schulze, G. Kirsten, S. Kirrbach, A. Rahm, and H. Heingarter, *Helv. Chem. Acta.,* 1991, **74**, 1059.
150. M. Mühlstüdt, R. Braimer, and B. Schulze, *J. Prakt. Chem.,* 1976, **318**, 507.
151. B. Schulze, K. Mütze, D. Selle, and R. Kempe, *Tetrahedron Lett.,* 1993, **34**, 1909.
152. Z. Arnold and J. Zemlicka, *Collect. Czech. Chem. Commun.,* 1959, **24**, 2385.
153. Z. Arnold and F. Sorm, *Collect. Czech. Chem. Commun.,* 1958, **23**, 452.
154. D. H. R. Barton, G. Dressaire, B. J. Willis, A. G. M. Barrett, and M. Pfeffer, *J. Chem. Soc., Perkin Trans. 1,* 1982, 665.

155. V. T. Klimko, T. V. Protopopova, and A. P. Skodinov, *Zh. Obshch. Khim.,* 1964, **34**, 109; *Chem. Abstr.,* 1964, **60**, 10535e.

156. C. Reichardt and K. Halbritter, *Liebigs Ann. Chem.,* 1970, **737**, 99.

157. G. E. Hardtmann, U. S. Pat. 3767650 (1973); *Chem. Abstr.,* 1974, **80**, 3261y.

158. S. M. Makin, O. A. Shavrygina, M. I. Berezhnaya, and T. P. Kolobova, *Zh. Org. Khim.,* 1972, **8**, 1394.

159. Z. Arnold, *Collect. Czech. Chem. Commun.,* 1961, **26**, 3051.

160. N. Roh and G. Kochendoerfer, Ger. Pat 667207 (1937); *Z. Chem.,* 1939, **11**, 3195.

161. O. Bayer, in *Methoden der Organischen Chemie* (Houben-Weyl), Vol VII/1, Sauerstoff-Verbindungen II, Georg Thieme Verlag, Stuttgart, 1954, pp. 29-36.

162. M. R. de Maheas, *Bull. Chim. Soc. Fr.,* 1964, 1989.

163. French Pat., 839359 (1939).

164. H. R. Mueller and M. Seefelder, *Liebigs Ann. Chem.,* 1969, **728**, 88.

165. W. Flitsch, J. Lauterwein, R. Temme, and B. Wibbeling, *Tetrahedron Lett.,* 1988, **29**, 3391.

166. I. M. A. Awad and K. M. Hassan, *Collect. Czech. Chem. Commun.,* 1990, **55**, 2715.

167. H. Normant and G. Martin, *Bull. Chem. Soc. Fr.,* 1963, 1646.

168. Z. Arnold, *Collect. Czech. Chem. Commun.,* 1960, **25**, 1308.

169. S. M. Makin, O. A. Shavrygina, M. I. Berezhnaya, and G. V. Kirillova, *J. Org. Chem., USSR,* 1972, **8**, 682.

170. G. Seitz, *Pharm. Zentrath,* 1968, **107**, 363; *Chem. Abstr.,* 1968, **69**, 76702a.

171. M. Pulst, L. Beyer, and M. Weissenfels, *J. Prakt. Chem.,* 1982, **324**, 292.

172. M. Muraoka and T. Yamamoto, *J. Chem. Soc., Chem. Commun.,* 1985, 1299.

173. M. Muraoka, T. Yamamoto, K. Enomoto, and T. Takeshima, *J. Chem. Soc., Perkin Trans. 1,* 1989, 1241.

174. K. Reimer and F. Tiemann, *Ber. Deut. Chem. Ges.,* 1876, **9**, 824.

175. H. Fischer, H. Berg and A. Schermuller, *Liebigs Ann. Chim.,* 1930, **480**, 153.

176. P. Karrer, *Helv. Chim. Acta,* 1919, **2**, 89.

177. N. Crounse, *Org. React.,* 1949, **5**, 290.

178. J. C. Duff, *J. Chem. Soc.,* 1941, 547.

179. J. C. Duff, *J. Chem. Soc.,* 1945, 276.

180. O. Fischer, A. Muller and A. Vilsmeier, *J. Prakt. Chem.,* 1924, **109**, 69.

181. A. Vilsmeier and A. Haack, *Ber. Deut. Chem. Ges.,* 1927, **60**, 119.

182. J. Jugie, J. A. S. Smith, and G. J. Martini, *J. Chem. Soc., Perkin Trans. 2*, 1975, 2925.
183. S. Alumi, P. Linda, G. Marino, S. Santini, and G. Savelli, *J. Chem. Soc., Perkin Trans. 2*, 1972, 2070.
184. H. Fritz and R. Oehl, *Liebigs Ann. Chem.*, 1971, **749**, 159.
185. Z. Arnold and A. Holy, *Collect. Czech. Chem. Commun.*, 1962, **27**, 2886.
186. G. Martin and M. Martin, *Bull. Chim. Soc. Fr.*, 1963, 1637.
187. A. G. Martinez, R. M. Alvarez, J. O. Barcina, S. de la Moya Cerero, E. T. Vilar, A. G. Fraile, M. Hanack, and L. R. Subramanian, *J. Chem. Soc., Chem. Commun.*, 1990, 1571.
188. K. Hafner and C. Bernard, *Angew. Chem.*, 1957, **69**, 533.
189. K. Hafner and C. Bernard, *Liebigs Ann. Chem.*, 1959, **625**, 108.
190. J. H. Wood and R. W. Bost, *J. Am. Chem. Soc.*, 1937, **59**, 1721.
191. L. F. Fieser and J. L. Hartwell, *J. Am. Chem. Soc.*, 1938, **60**, 2555.
192. C. D. Hurd and C. N. Webb, *Org. Synth., Coll. Vol. I*, 1932, 217.
193. W. Treibs, H. J. Neupert, and J. Hiebsch, *Chem Ber.*, 1959, **52**, 141.
194. M. Rosenblum, *Chem. Ind. (London)*, 1957, 72.
195. H. Vollmann, H. Becker, M. Corell, H. Streeck, *Liebigs Ann. Chem.*, 1937, **531**, 1.
196. N. P. Buu-Hoi and C. T. Long, *Recl. Trav. Chim. Pay-Bas*, 1956, **75**, 1221.
197. Y. Tachibara, K. Obara, Y. Masuyama, Jpn. Kokai Tokkyo Koho JP 62198636 (1987).
198. B. Buylikleev and I. Pozharev, *Dokl. Bolg. Akad. Nauk.*, 1987, **40**, 74; *Chem Abstr.*, 1988, **109**, 149021.
199. N. P. Boi-Hoi, P. Jacquignon, and C. T. Long, *J. Chem. Soc.*, 1957, 505.
200. C. D. Wilson, U. S. patent 2558285; *Chem. Abstr.*, 1952, **46**, 1041.
201. N. P. Buu-Hoi, N. D. Xyong, M. Sy, G. Lejeune, and N. B. Tien, *Bull. Soc. Chim. Fr.*, 1955, 1594.
202. M. Bisagni, N. P. Buu-Hoi, and R. Roger, *J. Chem. Soc.*, 1955, 3693.
203. A. H. Sommers, R. J. Michaels, and A. W. Weston, *J. Am. Chem. Soc.*, 1952, **74**, 5546.
204. N. P. Buu-Hoi and D. Lavit, *J. Chem. Soc.*, 1955, 2776.
205. M. V. Kazankov, L. G. Ginodman, and M. Ya. Mustafina, *J. Org. Chem. USSR, Engl. Transl.*, 1983, **19**, 139.
206. J. Alexander and G. S. K. Rao, *Tetrahedron*, 1971, **27**, 645.
207. O. Pepin-Roussel, P. Jacquignon, and F. Perin, *C. R. Hebd. Seances Acad. Sci., Ser. C*, 1975, **280**, 1315.
208. Yu. N. Porshnev and E. M. Tereshchenko, *J. Org. Chem. USSR, Engl. Transl.*, 1975, **11**, 655.

209. Yu. N. Porshnev and M. I. Cherkashin, *Izv. Akad. Nauk. S.S.S.R., Ser. Khim.,* 1965, 2322.

210. A. Schlözer and J.-H. Fuhrhop, *Angew. Chem., Int. Ed. Engl.,* 1975, **14**, 363.

211. G. V. Ponomarev, B. V. Rozynov, and C. B. Maravin, *Khim. Geterotskl. Soedin.,* 1975, 139.

212. C. Zok and M. S. Wrighton, *J. Am. Chem. Soc.,* 1980, **112**, 7578.

213. J. Frederic and S. Toma, *Collect. Czech. Chem. Commun.,* 1987, **52**, 174.

214. S. Imaki, Y. Takuma, and M. Oushi, Jap. Pat. 6277344 (1987); *Chem. Abstr.,* 1987; **107**, 197790q.

215. D. A. Clark, S. W. Goldstein, R. A. Volkmann, J. F. Eggler, G. F. Holland, B. Hulin, R. W. Stevenson, D. K. Kreutter, E. M. Gibbs, M. N. Krupp, P. Merrigan, P. L. Kelbaugh, E. G. Andrews, D. L. Tickner, R. T. Suleske, C. H. Lamphere, F. J. Rajeckas, W. H. Kappeler, R. E. McDermott, N. J. Hutson, and M. R. Johnson, *J. Med. Chem.,* 1991, **34**, 319.

216. A. C. Shabica, E. E. Howe, J. B. Ziegler, and M. Tischler, *J. Am. Chem. Soc.,* 1946, **68**, 1156.

217. F. T. Tyson and J. T. Shaw, *J. Am. Chem. Soc.,* 1952, **74**, 2273.

218. G. F. Smith, *J. Chem. Soc.,* 1954, 3842.

219. R. E. Walkup and J. Linder, *Tetrahedron Lett.,* 1985, **26**, 2155.

220. D. St. C. Black, N. Kumar, and L. C. H. Wong, *Synthesis,* 1986, 474.

221. D. St. C. Black, A. J. Ivory, P. A. Keller, and N. Kumar, *Synthesis,* 1989, 322.

222. M. V. Trapaidze, Sh. A. Samsoniya, W. A. Kuprashvili, L. M. Mamaladze, and N. N. Suorov, *Khim. Geterotsikl. Soedin.* 1988, 603; *Chem Abstr.* 1988, **110**, 114709.

223. D. A. Partsvaniya, R. N. Akhvlediani, V. E. Zhiguchev, E. N. Gordeev, C. N. Kuleshove, and V. N. Suvorov, *Khim. Geterotsikl. Soedin.* 1987, 919; *Chem Abstr.* 1987, **108**, 150385s.

224. J.-C. Lancelot, D. Ladureé, and M. Robba, *Chem. Pharm. Bull. Jap.,* 1985, **33**, 4242.

225. C. F. Candy, R. A. Jones, and P. H. Wright, *J. Chem. Soc.(C),* 1970, 2563.

226. E. Ju-Hwa Chu and T. C. Chu, *J. Org. Chem.,* 1954, **19**, 266.

227. R. Kreher, G. Vogt, and M.-L. Schultz, *Angew. Chem., Int. Ed. Engl.,* 1975, **14**, 821.

228. J. M. Muchowski and R. Naef, *Helv. Chem. Acta,* 1984, **67**, 1168.

229. J. White and G. McGilluray, *J. Chem. Soc., Perkin Trans. 2,* 1982, 259.

230. A. Guzmán, M. Romero, and J. M. Muchowski, *Can. J. Chem.,* 1990, **68**, 791.

231. N. Ono, E. Muratami, and T. Ogewa, *J. Heterocycl. Chem.,* 1991, **28**, 205.

232. B. L. Bray and J. M. Muchowski, *Can. J. Chem.,* 1990, **68**, 1305.

233. R. Chong, P. S. Clezy, A. J. Liepa, and A. W. Nichol, *Aust. J. Chem.,* 1969, **22**, 229.

234. H. H. Inhoffen, J.-H. Fuhrop, H. Voigt, and H. Broekmann, Jr., *Liebigs Ann. Chem.,* 1966, **695**, 133.

235. M. Gosmann and B. Franck, *Angew. Chem., Int. Ed. Eng.,* 1986, **25**, 1100.

236. M. G. H. Vicente and K. M. Smith, *J. Org. Chem.,* 1991, **56**, 4407.

237. V. N. Eraksina, L. B. Shagalov, and N. N. Surorov, *Khim. Geterotskl. Soedin.,* 1975, 1257.

238. J. W. Buchler, C. Dreher and G. Herget, *Liebigs Ann. Chem.,* 1988, 43.

239. G. V. Ponomarev, G. V. Kurilliva, L. B. Lazukova, and T. A. Babushkina, *Khim. Geterosikl. Soedin.,* 1982, 1507.

240. H. Ogoshi, K. Saita, K. Sakurai, T. Watanabe, H. Toi, Y. Aoyama, and Y. Okamoto, *Tetrahedron Lett.,* 1986, **27**, 6365.

241. G. V. Ponomarev and A. M. Shul'ga, *Khim. Geterotskl. Soedin.,* 1987, 922.

242. W. H. Traynelis, J. J. Miskel, Jr., and J. R. Sowa, *J. Org. Chem.,* 1957, **22**, 1269.

243. J. P. Marquet, E. Basigni and J. A. Louisfert, *Bull. Chim. Soc. Fr.,* 1973, 2323.

244. B. L. Feringa, R. Hulst, R. Rikers, and L. Brandsma, *Synthesis,* 1988, 316.

245. W. J. King and F. F. Nord, *J. Org. Chem.,* 1948, **13**, 635.

246. W. J. King and F. F. Nord, *J. Org. Chem.,* 1949, **14**, 405.

247. W. J. King and F. F. Nord, *J. Org. Chem.,* 1949, **14**, 638.

248. S. Kato and M. Ishizaki, Jap. Pat. 85/135874 (1985); *Chem. Abstr.,* 1986, **106**, 231755u.

249. A. Tsubouchi, N. Matsumura, H. Inoue and K. Yanagi, *J. Chem. Soc., Perkin Trans. 1,* 1991, 909.

250. D. L. Comins and J. J. Herrick, *Heterocycles,* 1987, **26**, 2159.

251. D. L. Comins and N. B. Mantlo, *J. Org. Chem.,* 1986, **51**, 5456.

252. D. L. Comins and Y. C. Myoung, *J. Org. Chem.,* 1990, **55**, 292.

253. M. Natsume, S. Kumadaki, Y. Kanda, and K. Kiuchi, *Tetrahedron Lett.,* 1973, **26**, 2335.

254. A. Guzmán, M. Romero, M. L. Maddox and J. M. Muchowski, *J. Org. Chem.,* 1990, **55**, 5793.

255. S. Senda, K. Hirota, G.-N. Yang, and M. Shirahashi, *Yakugaku Zasshi,* 1971, **91**, 1372; *Chem. Abstr.,* 1972, **76**, 126915q.

256. N. M. Cherdantseva, V. M. Nesterov, and T. S. Safonova, *Khim. Geterotsikl. Soedin.* 1983, 834; *Chem Abstr.* 1983, **99**, 139895h.

257. D. Prajapati, P. Bhuyan, and J. S. Sandhu, *J. Chem. Soc., Perkin Trans. 1*, 1988, 607.

258. H. Bredereck, G. Simchen, H. Wagner, and A. A. Santos, *Liebigs Ann. Chem.*, 1972, **766**, 73.

259. J. Negrillo, M. Nogueras, A. Sánchez and M. Melgarejo, *Chem. Pharm. Bull. Jap.*, 1988, **36**, 386.

260. M. Y. Yeh, H. J. Tien, L. Y. Huang, and M. H. Chen, *J. Chin. Chem. Soc. (Taipei)*, 1983, **30**, 29.

261. Y. Murakami, Y. Yokoyama, and N. Okuyama, *Tetrahedron Lett.*, 1983, **24**, 2189.

262. Y. Murakami and H. Ishii, *Chem. Pharm. Bull. Jap.*, 1981, **29**, 699.

263. J. Haüfel and W. Breitmaier, *Angew. Chem., Int. Ed. Engl.*, 1974, **13**, 604.

264. I. L. Finar and G. H. Lord, *J. Chem. Soc.*, 1957, 331.

265. I. L. Finar and G. H. Lord, *J. Chem. Soc.*, 1959, 1819.

266. L. A. Kutulya, A. E. Shevchenko, Y. Surov, and N. Surov, *Khim. Geterotskl. Soedin.*, 1975, 250.

267. B. S. Holla and S. Y. Ambekar, *J. Ind. Chem. Soc.*, 1974, **51**, 965.

268. A. N. Grinev, V. I. Shedov, N. K. Chizov, and T. F. Vlasova, *Khim. Geterotskl. Soedin.*, 1975, 1250.

269. O. Fuentes and W. N. Paudler, *J. Heterocycl. Chem.*, 1975, **12**, 379.

270. D. M. Brown and G. A. R. Koy, *J. Chem. Soc.*, 1948, 2147.

271. D. Farquhar, T. T. Gough, and D. Leaver, *J. Chem. Soc., Perkin Trans. 1*, 1976, 341.

272. W. Flitsch, A. Gurke, and B. Mueter, *Chem. Ber.*, 1975, **108**, 2969.

273. J.-C. Lancelot, D. Laureé, and M. Robba, *Chem. Pharm. Bull. Jap.*, 1985, **33**, 3122.

274. T. Mukai, T. Kumagai, and S. Tanaka, *Jpn. Kokai Tokkyo Koho*, JP 62207275; *Chem Abstr*. 1988, **108**, 186728v.

275. A. Horváth and I. Hermecz, *J. Heterocycl. Chem.*, 1986, **23**, 1295.

276. G. Roma, M. Di Braccio, A. Balbi, M. Mazzei, and A. Ermili, *J. Heterocycl. Chem.*, 1987, **24**, 329.

277. (a) S. Mackenzie and D. H. Reid, *J. Chem. Soc., Chem. Commun.*, 1966, 401; (b) S. Mackenzie and D. H. Reid, *J. Chem. Soc. C*, 1970, 145.

278. R. K. Mackie, S. Mackenzie, D. H. Reid and R. G. Webster, *J. Chem. Soc., Perkin Trans. 1*, 1973, 657.

279. P. D. Croce, C. La Rosa, and R. Ritieni, *Synthesis*, 1989, 783.

280. J. White and G. McGillivray, *J. Org. Chem.*, 1977, **42**, 4248.

281. H. Akimoto, A. Kawai, and H. Nomura, *Bull. Chem. Soc. Jap.*, 1985, **58**, 123.

282. S. Pennanen, *Acta. Chem. Scand.*, 1973, **27**, 3133.

283. (a) R. D. Youssefyeh, *Tetrahedron Lett.,* 1964, 2161;
     (b) R. D. Youssefyeh, *J. Am. Chem. Soc.,* 1963, **85,** 3901.
284. W. Ziegenbein, *Angew. Chem.,* 1965, **77,** 380.
285. P. Caigniant and G. Kirsch, *C. R. Hebd. Seances Acad. Sci, Ser., C,* 1976, **282,** 465.
286. J. S. Pizey, *Synthetic Reagents,* Vol. 1, 1974, pp. 1-99.
287. M. Pulst and M. Weissenfels, *Z. Chem.,* 1976, **16,** 337.
288. Z. Arnold and A. Holy, *Collect. Czech. Chem. Comun.,* 1961, **26,** 3059.
289. J. Schmitt, J. J. Panouse, P.-J.Cornu, H. Pluchet, A. Hallot, and P. Commoy, *Bull. Chim. Soc. Fr.,* 1964, 2760.
290. R. A. S. Chandraratna, A. L. Bayerque, and W. H. Okamura, *J. Am. Chem. Soc.,* 1983, **105,** 3588.
291. A. Carpita, A. Lezzi, R. Rossi, F. Marchetti, and S. Merlino, *Tetrahedron,* 1985, **41,** 621.
292. W. Ziegenbein and W. Franke, *Angew. Chem.,* 1959, **71,** 573.
293. L. A. Paquette, B. A. Johnson, and F. M. Hinga, *Org. Synth.,* 1966, **46,** 18.
294. H. Schellhorn, S. Hauptmann, and H. Frischleder, *Z. Chem.,* 1973, **13,** 97.
295. W. Ziegenbein and W. Franke, *Angew. Chem.,* 1959, **71,** 628.
296. W. Ziegenbein and W. Lang, *Chem. Ber.,* 1960, **93,** 2743.
297. M. Weissenfels, H. Schurig, and G. Huesham, *Z. Chem.,* 1966, **6,** 471.
298. M. Weissenfels, M. Pulst, and P. Schneider, *Z. Chem.,* 1973, **13,** 175.
299. J. M. F. Gagan, A. G. Lane, and D. Lloyd, *J. Chem. Soc. C,* 1970, 2484.
300. M. A. Volodina, A. P. Terent'ev, L. G. Roshcupkina, and V. G. Mishina, *Zh. Obshch. Khim.,* 1964, **39,** 469.
301. G. Alvernhe, B. Langlois, A. Laurent, I. Le Drean, A. Selmi, and M. Weissenfels, *Tetrahedron Lett.,* 1991, **32,** 643.
302. C. M. Beaton, N. B. Chapman, K. Clarke, and J. M. Willis, *J. Chem. Soc., Perkin Trans. 1,* 1976, 2355.
303. (a) M. D. Rausch and A. Siegal, *J. Organomet. Chem.,* 1969, **17,** 117; (b) M. Rosenblum, N. Brown, J. Papenmeier, and M. Applebaum, *J. Organomet. Chem.,* 1966, **6,** 173.
304. Yeu, *Ann. Chem. (Paris),* 1962, **7,** 785.
305. V. I. Shvedo and I. N. Fedorova, *Zh. Org. Khim.,* 1991, **27,** 247; *Chem. Abstr.,* 1991, **115,** 158150b.
306. Z. Arnold and J. Zemlicka, *Proc. Chem. Soc.,* 1958, 227.
307. H. A. Brandman, E. Heilweil, J. A. Virgilio, and T. F. Wood, Ger. Offen. 2,851,024 (1979); *Chem. Abstr.,* 1979, **91,** 107808a.
308. H. M. Relles, U.S. Pat. 3700743 (1972); *Chem. Abstr.,* 1973, **78,** 85033w.
309. M. Sreenivasulu and G. S. K. Rao, *Indian J. Chem., Sect. B.,* 1989, **28B,** 494.

310. M. Wessienfels, M. Pulst, M. Haase, U. Pawlowski, and H.-F. Uhlig, *Z. Chem.*, 1977, **17**, 56.
311. K. Bodendorf and R. Mayer, *Chem. Ber.*, 1965, **98**, 3565.
312. W. R. Benson and A. E. Pohland, *J. Org. Chem.*, 1965, **30**, 1126.
313. M. Weissenfels, *J. Prakt. Chem.*, 1979, **321**, 671.
314. J.-M. Magar, J.-F. Muller, and D. Cagniant, *C. R. Hebd. Seances Acad. Sci., Ser. C*, 1978, **286**, 241.
315. J. O. Karlsson and T. Frejd, *J. Org. Chem.*, 1983, **48**, 1921.
316. R. Sciaky and U. Pallini, *Tetrahedron Lett.*, 1964, 1839.
317. G. W. Moersch and W. A. Neuklis, *J. Chem. Soc.*, 1965, 788.
318. A. R. Katritzky and C. M. Marson, *J. Org. Chem.*, 1987, **52**, 2726.
319. P. C. Traas, H. Boelens, and H. J. Takken, *Recl. Trav. Chim. Pays-Bas*, 1976, **95**, 308.
320. P. C. Traas, *U.S. Pat.* 4152355 (1979); *Chem. Abstr.*, 1979, **91**, 140428z.
321. J. Schmitt, J. J. Panouse, H. Pluchet, A. Hallot, P.-J. Cornu, and P. Comoy, *Bull. Soc. Chim. Fr.*, 1964, 2768.
322. R. Sciaky and F. Mancini, *Tetrahedron Lett.*, 1965, 137.
323. H. Laurent and R. Wiechert, *Chem. Ber.*, 1968, **101**, 2393.
324. H. Laurent, G. Schulz, and R. Wiechert, *Chem. Ber.*, 1966, **99**, 3057.
325. A. R. Katritzky and C. M. Marson, *J. Am. Chem. Soc.*, 1983, **105**, 3279.
326. A. R. Katritzky, C. M. Marson, S. S. Thind, and J. Ellison, *J. Chem. Soc., Perkin Trans. 1*, 1983, 487.
327. P. Cagniant and G. Kirsch, *C. R. Hebd. Seances Acad. Sci., Ser. C*, 1975, **281**, 393.
328. P. Cagniant, P. Perin, G. Kirsch, and D. Cagniant, *C. R. Hebd. Seances Acad. Sci., Ser. C*, 1973, **277**, 37.
329. L. A. Paquette, U. S. Pat. 3129257 (1964); *Chem. Abstr.*, 1964, **61**, 1814a.
330. P. A. Reddy and G. S. K. Rao, *Proc. Indian Acad. Sci., [Ser.]: Chem. Sci.*, 1980, **89**, 435; *Chem. Abstr.* 1981, **94**, 208585t.
331. T. Koyama, T. Hirota, F. Yagi, S. Ohmori, and M. Yamamoto, *Chem. Pharm. Bull. Jap.*, 1975, **23**, 3151.
332. G. Kalischer, A. Scheyer, and K. Keller, Ger. Pat. 514415 (1927); *Chem. Abstr.*, 1931, **25**, 1536.
333. M. Opmane, A. Ya. Strakov, and E. Gudriniece, *Latv. PSR Zinat. Akad. Vestis, Kim. Ser.* 1976, **6**, 701; *Chem. Abstr.*, 1977, **86**, 171322m.
334. D. T. Drewry and R. M. Scrowston, *J. Chem. Soc. (C)*, 1969, 2750.
335. (a) P. R. Giles and C. M. Marson, *Tetrahedron Lett.*, 1990, **31**, 5227;
     (b) P. R. Giles and C. M. Marson, *Tetrahedron*, 1991, **47**, 1303.
336. C. H. Chen and G. A. Reynolds, *J. Org. Chem.*, 1979, **44**, 3144.

337. J. Andrieux, J.-P. Battioni, M. Girard, and D. Mohlo, *Bull. Soc. Chim. Fr.,* 1973, 2093.
338. J. Smith and R. H. Thompson, *J. Chem. Soc.,* 1960, 346.
339. P. E. Brown, W. Y. Marcus, and P. Anastasis, *J. Chem. Soc., Perkin Trans. 1,* 1985, 1127.
340. M. V. Naidu and G. S. K. Rao, *Synthesis,* 1979, 708.
341. T. Eszenyi and T. Timar, *Synth. Commun.,* 1990, **20**, 3219.
342. C. G. Rimbault, Brit. U. K. Pat. Appl. GB 2145719 (1985); *Chem. Abstr.,* 1986, **104**, 88354t.
343. G. Litkei, T. Patonay, L. Szilagyi, and Z. Dinya, *Org. Prep. Proc. Int.,* 1991, **23**, 741.
344. P. Cagniant and A. Deluzarche, *C. R. Hebd. Seances Acad. Sci., Ser. C,* 1946, **223**, 677.
345. V. A. Kortunenko, D. Greif, V. V. Ischenko, and M. Weissenfels, *Ukr. Khim. Zh.,* 1988, **54**, 775; *Chem. Abstr.,* 1988, **110**, 192631r.
346. E. Lippmann, P. Strauch, E. Tenor, and T. Eckhard, Ger. (East) DD 264,439 (1988); *Chem. Abstr.* 1989, **111**, 115203w.
347. T. Ya. Mozhaeva, O. L. Samsonova, and V. L. Savel'ev, *Khim. Geterotsikl Soedin.,* 1988, 1287; *Chem. Abstr.,* 1989, **110**, 212552w.
348. G. A. Mironova, E. N. Kirillova, V. N. Kuklin, N. A. Smorygo, and B. A. Ivin, *Khim. Geterotsikl Soedin.,* 1984, **10**, 1328; *Chem. Abstr.,* 1985, **102**, 149205c.
349. I. P. Yakovlev, G. A. Mironova, M. M. Timoshenko, Yu. V. Chizhov, and B. A. Ivin, *Zh. Org. Khim.,* 1987, **23**, 1556; *Chem. Abstr.,* 1988, **108**, 111713c.
350. O. Aki and Y. Nakagawa, *Chem. Pharm. Bull. Jap.,* 1972, **20**, 1325.
351. H. Singh and D. Paul, *Indian J. Chem.,* 1974, **12**, 1210.
352. M. Weissenfels and S. Kaubisch, *Z. Chem.,* 1981, **21**, 259.
353. M. Weissenfels and S. Kaubisch, *Z. Chem.,* 1982, **22**, 23.
354. T. Aubert, M. Farnier, I. Meunier, and R. Guilard, *J. Chem. Soc., Perkin Trans. 1,* 1989, 2095.
355. W. Flitsch, J. Koszinowski, and P. Witthake, *Chem. Ber.,* 1979, **112**, 2465.
356. D. Ramesh, B. G. Chatterjee, and J. K. Ray, *Indian J. Chem., Sect. B,* 1986, **25B**, 964.
357. C. M. Marson, PhD Thesis, University of East Anglia, 1982.
358. G. M. Coppola, *J. Heterocycl. Chem.,* 1981, **18**, 845.
359. J. A. Virgilio and E. Heilwell, *Org. Prep. Proc. Int.,* 1982, **14**, 9.
360. N. Ben Hassaine-Coniac, G. Hazerbroucq, and J. Gardent, *Bull. Chem. Soc. Fr.,* 1971, 3985.

361. M. E. Kuehne, W. G. Bornmann, W. G. Earley, and I. E. Markó, *J. Org. Chem.*, 1986, **51**, 2913.
362. (a) P. G. Willard, and S. E. de Laszlo, *J. Org. Chem.*, 1983, **48**, 1123; (b) P. G. Willard and S. E. de Laszlo, *J. Org. Chem.*, 1985, **50**, 3738.
363. I. A. Strakova, A. Ya. Strakov, and E. Gudriniece, *Latv. PSR Zinat. Akad. Vestis, Kim. Ser.*, 1972, **5**, 627; *Chem. Abstr.*, 1973, **78**, 29667u.
364. R. E. Mewshaw, *Tetrahedron Lett.*, 1989, **30**, 3753.
365. W. A. Gregory, U. S. Pat. 288456 (1959).
366. Hoechst, Belgian Pat. 553871 (1957).
367. H. H. Bosshard and H. Zollinger, *Angew. Chem.*, 1959, **71**, 375.
368. (a) J. P. Maffraud, M. Amoros, D. Frehel, and F. Eloy, *Chem. Ther.*, 1974, **9**, 150; (b) J. P. Maffraud, M. Amoros, D. Frehel, and F. Eloy, *Cheminform*, 1974, **50-206**, p. 70.
369. E. F. Godefroi, C. A. M. van der Eycken, and C. van de Westeringh, *J. Org. Chem.*, 1964, **29**, 3707.
370. M. Zaoral and Z. Arnold, *Tetrahedron Lett.*, 1960, 9.
371. H. Ulrich, B. Tucker and A. A. R. Sayigh, *J. Org. Chem.*, 1967, **32**, 4052.
372. I. Hoppe and U. Schollkopf, *Chem. Ber.*, 1976, **109**, 482.
373. (a) L. Brehm, P. Kroogsgaard, and H. Hjeds, *Acta Chem. Scand., Ser. B*, 1974, **28**, 308; (b) L. Brehm, P. Kroogsgaard, and H. Hjeds, *Cheminform*, 1974, **32-297**, p. 121.
374. B. J. R. Nicolaus, E. Bellasio, and E. Testa, *Helv. Chem. Acta*, 1962, **45**, 717.
375. E. Testa, B. J. R. Nicolaus, L. Mariani, and G. Pagani, *Helv. Chem. Acta*, 1963, **46**, 766.
376. W. Reid and G. Neidhardt, *Liebigs Ann. Chem.*, 1963, **666**, 148.
377. T. Fujisawa and T. Sato, *Org. Synth.*, 1988, **66**, 116.
378. P. E. Bratchanski and G. Kommitsarova, *Zh. Prikl. Khim.*, 1974, **47**, 239; *Chem. Abstr.*, 1974, **80**, 107958x.
379. H. H. Bosshard, R. Mory, M. Schmid, and H. Zollinger, *Helv. Chem. Acta*, 1959, **42**, 1653.
380. Z. Arnold and A. Holy, *Collect. Chem. Soc. Commun.*, 1962, **27**, 2886.
381. F. Hallman, *Ber. Deut. Chem. Ges.*, 1876, **9**, 846.
382. R. N. McDonald and R. A. Kreuger, *J. Org. Chem.*, 1963, **28**, 2542.
383. Y.-W. Chang, Fr. Pat., 1385595 (1965); *Chem. Abstr.*, 1965, **62**, 14702h.
384. F. Eiden and G. Bachmann, *Arch. Pharm. (Weinheim, Ger.)*, 1973, **306**, 401.
385. Belgian Pat., 618751 (1962).
386. Z. Arnold, *Collect. Czech. Chem. Commun.*, 1963, **28**, 2047.
387. T. Kotscheff, F. Wolf, and G. Wolter, *Z. Chem.*, 1966, **6**, 148.

388. A. Roedig and W. Wenzel, *Angew. Chem., Int. Ed. Engl.,* 1969, **8**, 71.
389. H. Boehme and F. Soldam, *Chem. Ber.,* 1961, **94**, 3112.
390. E. Kuehle, *Angew. Chem.,* 1962, **74**, 861.
391. H. Ulrich, B. Tucker, and A. A. Sayigh, *Angew. Chem., Int. Ed. Engl.,* 1967, **6**, 636.
392. J. C. Sheenan, P. A. Cruickshank, and G. L. Boshart, *J. Org. Chem.,* 1961, **26**, 2525.
393. E. Kuehle, *Angew. Chem., Int. Ed. Engl.,* 1962, **1**, 647.
394. S. Christopherson and P. Carlsen, *Tetrahedron Lett.,* 1973, 211.
395. J. Barluenga, P. J. Campos, E. Gonzalez-Nunez, and G. Asensio, *Synthesis,* 1985, 426.
396. K. Morita, S. Noguchi, and M. Nishukawa, *Chem. Pharm. Bull. Jap.,* 1959, 7, 896.
397. V. K. Brel, L. A. Chepakova, N. F. Kartavtseva, and I. V. Martynov, *Izv. Akad. Nauk. S.S.S.R., Ser. Khim.,* 1989, 1423.
398. S. Morimura, H. Horiuchi, and K. Murayama, *Bull. Chem. Soc. Jap.,* 1977, **50**, 2189.
399. Z. Arnold, *Collect. Czech. Chem. Commun.,* 1961, **26**, 1723.
400. R. C. K. Boeckman, Jr. and B. Ganem, *Tetrahedron Lett.,* 1974, 913.
401. P. A. Stadler, *Helv. Chem. Acta,* 1978, **61**, 1675.
402. F. M. Stoyanovich, B. P. Federov, and G. M. Adrianova, *Dokl. Akad. Nauk. SSSR,* 1962, **145**, 584; *Chem. Abstr.,* 1963, **58**, 4448.
403. T. Kato, T. Chiba, and T. Okada, *Chem. Pharm. Bull. Jap.,* 1979, **27**, 1186.
404. F. Cramer, S. Rittner, W. Reinhard, and P. Desai, *Chem. Ber.,* 1966, **99**, 2252.
405. H. Bredereck, R. Gompper, H. Rempfer, K. Klemm, and H. Keck, *Chem. Ber.,* 1959, **92**, 329.
406. D. Konwar, R. C. Boruah, and J. S. Sandhu, *Tetrahedron Lett.,* 1987, **28**, 955.
407. M. Lacora, T. Y. Nguyen, J. J. Halgas, *Stud. Org. Chem. (Amsterdam),* 1988, **35**, 384; *Chem. Abstr.,* 1990, **113**, 78219z.
408. A. Speziale and L. R. Smith, *J. Org. Chem.,* 1963, **28**, 3492.
409. O. Wallach, *Ber. Dtsch. Chem. Ges.,* 1876, **9**, 1212.
410. O. Wallach and P. Pirath, *Ber. Dtsch. Chem. Ges.,* 1879, **12**, 1063.
411. A. Bernthesen, *Ber. Dtsch. Chem. Ges.,* 1877, **10**, 1238.
412. H. Leo, *Ber. Dtsch. Chem. Ges.,* 1877, **10**, 2133.
413. G. Seitz H. Morck, K. Mann, and R. Schmiedel, *Chem.-Ztg.,* 1974, **98**, 459.
414. W. Walter and G. Maerten, *Liebigs Ann. Chem.,* 1963, **669**, 66.

415. (a) M. H. Brown, U. S. Pat. 3092637 (1963); *Chem. Abstr.*, 1963, **59**, 12674g; (b) H. Bredereck and K. B. Bredereck, Swiss Pat. 384564 (1965); *Chem. Abstr.*, 1965, **62**, 16135h.

416. H. Bredereck and K. Bredereck, *Chem. Ber.,* 1961, **94**, 2278.

417. H. H. Bosshard, E. J. Jenny, and H. Zollinger, *Helv. Chim. Acta,* 1961, **44**, 1203.

418. (a) H. Eilingsfeld, M. Seefelder, and H. Weidinger, Ger. Pat. 1119872 (1961); *Chem. Abstr.*, 1961, **56**, 14083; (b) Ciba Ltd., Br. Pat. 911475 (1962); *Chem. Abstr.*, 1962, **58**, 13852x.

419. H. Eilingsfeld, M. Seefelder, and H. Weidinger, *Chem. Ber.*, 1963, **96**, 2899.

420. R. Buijle, A. Halleux, and H. G. Viehe, *Angew. Chem., Int. Ed. Engl.,* 1966, **5**, 584.

421. L. Ghosez, B. Haveaux, and H. G. Viehe, *Angew. Chem.,* 1969, **81**, 468.

422. H. Bredereck, R. Gompper, and K. Klemm, *Chem. Ber.,* 1959, **92**, 1456.

423. M. Zaoral and Z. Arnold, Czech Pat., 99291 (1960); *Chem. Abstr.,* 1961, **56**, 8836g.

424. T. M. Jacob and H. G. Khorana, *J. Am. Chem. Soc.,* 1964, **86**, 1630.

425. M. Ikehara and H. Umo, *Chem. Pharm. Bull. Jap.,* 1964, **12**, 742.

426. J. Toye and L. Ghosez, *J. Am. Chem. Soc.,* 1975, **97**, 2276.

427. D. Habeck and W. J. Houlihan, *J. Heterocycl. Chem.,* 1976, **13**, 897.

428. A. J. Speziale and L. R. Smith, *J. Org. Chem.,* 1962, **27**, 4361.

429. R. G. Glushkov, V. G. Shminova, K. A. Zaitseva, N. A. Novitskaya, M. D. Mashovskii, and G. N. Pershin, *Khim. Farm. Zh.,* 1974, **8**, 14.

430. A. J. Speziale and R. C. Freeman, *J. Am. Chem. Soc.,* 1960, **82**, 903.

431. A. J. Speziale and R. C. Freeman, *J. Am. Chem. Soc.,* 1960, **82**, 909.

432. M. A. Kira, Z. M. Nofal, and K. Z. Gadalla, *Tetrahedron Lett.,* 1970, 4215.

433. (a) D. St. C. Black, M. C. Bowyer, A. Choy, D. C. Craig, and N. Kumar, *J. Chem. Soc, Perkin Trans. 1,* 1989, 200; (b) A. J. Ivory, Ph. D. Thesis, University of New South Wales, 1992.

434. (a) J. von Braun and A. Heymons, *Ber. Dtsch. Chem. Ges.,* 1929, **62**, 409; (b) J. von Braun, F. Jostes, and W. Muench, *Liebigs Ann. Chem.,* 1926, **453**, 113.

435. J. von Braun, *Angew. Chem.,* 1934, **47**, 611.

436. J. von Braun, *Ber. Dtsch. Chem. Ges.,* 1904, **37**, 2812.

437. H. Gross and J. Gloede, *Chem. Ber.,* 1963, **96**, 1387.

438. J. von Braun, *Ber. Dtsch. Chem. Ges.,* 1904, **37**, 2915.

439. G. Fodor, J. J. Rayan. and F. Letourneau, *J. Am. Chem. Soc.,* 1969, **91**, 7768.

440. B. A. Phillips, G. Fodor, J. Gal, F. Letourneau, and J. J. Ryan, *Tetrahedron*, 1973, **29**, 3309.

441. P. Kurtz, in Houben-Weyl-Mueller, *Methoden der Organischen Chemie*, Georg Thieme Verlag, Stuttgart, 1952, Vol. 8, p. 330.

442. J. C. Thurman, *Chem. Ind.*, 1964, 752.

443. E. A. Lawton and D. D. McRitchie, *J. Org. Chem.*, 1959, **24**, 26.

444. C. S. Marvel and M. M. Martin, *J. Am. Chem. Soc.*, 1958, **80**, 6600.

445. J.-P. Dulcère, *Tetrahedron. Lett.*, 1981, **22**, 1599.

446. K. Wallenfels, D. Hofmann, and R. Kern, *Tetrahedron*, 1965, **21**, 2231.

447. M. El-Kerdawy, M. N. Tolba, and A.-G. El-Agamey, *Acta. Pharm. Jugosl.*, 1976, **26**, 141.

448. Z. Arnold and M. Svoboda, *Collect. Czech. Chem. Commun.*, 1977, **42**, 1175.

449. G. H. Barnett, H. J. Anderson, and C. E. Loader, *Can. J. Chem.*, 1980, **58**, 409.

450. J. Liebscher and U. Bechstein, *Z. Chem.*, 1984, **24**, 56.

451. (a) J. Ugi and R. Meyr, *Angew. Chem.*, 1958, **70**, 702; (b) J. Ugi and R. Meyr, *Chem. Ber.*, 1960, **93**, 239; (c) J. Ugi, U. Fetzer, U. Eholzer, H. Knupfer, and K. Offermann, *Angew. Chem.*, 1965, **77**, 492.

452. F. Yoneda, M. Higuchi, T. Matsumura, and K. Senga, *Bull. Chem. Soc. Jap.*, 1973, **46**, 1837.

453. C. V. Z. Smith, R. K. Robins, and R. L. Tolman, *J. Chem. Soc., Perkin Trans. 1*, 1973, 1855.

454. A. Hantzsch, *Ber. Dtsch. Chem. Ges.*, 1931, **64**, 667.

455. K. Hantke, U. Schoellkopf, and H. H. Hausberg, *Liebigs Ann. Chem.*, 1975, 1531.

456. P. Jacobsen, *Acta. Chem. Scand., Ser. B*, 1976, **30**, 995.

457. K. Bartel and W. P. Fehlhammer, *Angew. Chem., Int. Ed. Engl.*, 1974, **13**, 599.

458. G. Skorna and I. Ugi , *Angew. Chem., Int. Ed. Engl.*, 1977, **16**, 259.

459. W. P. Fehlhammer and A. Mayr, *Angew. Chem., Int. Ed. Engl.*, 1975, **14**, 757.

460. T. Ignasiak, J. Suzuko, and B. Ignasiak, *J. Chem. Soc., Perkin Trans. 1*, 1975, 2122.

461. A. R. Katrizky, L. Xie, and W.-Q. Fan, *Synthesis*, 1993, 45.

462. T. Fujisawa and K. Sakai, *Tetrahedron Lett.* 1976, 3331.

463. C. G. Raison, *J. Chem. Soc.*, 1949, 3319.

464. H. Bredereck, F. Effenberger, H. Botsch, and H. Rehn, *Chem. Ber.*, 1965, **98**, 1081.

465. H. Bredereck, R. Gompper, K. Klemm, and H. Rempfer, *Chem. Ber.*, 1959, **92**, 837.
466. C. Jutz and H. Amschler, *Chem. Ber.*, 1963, **96**, 2100.
467. R. L. Shriner and F. W. Neumann, *Chem. Rev.*, 1944, **35**, 351.
468. E. E. Bures and M. Kundera, *Cas. Ceks. Lek.*, 1934, **14**, 272.
469. T. L. Davis and W. E. Yelland, *J. Am. Chem. Soc.*, 1937, **59**, 1998.
470. E. Beckmann and E. Fellrath, *Liebigs Ann. Chem.*, 1893, **273**, 1.
471. J. von Braun and W. Pinkernelle, *Ber. Dtsch. Chem. Ges.*, 1934, **67**, 1218.
472. J. von Braun and K. Weissbach, *Ber. Dtsch. Chem. Ges.*, 1932, **65B**, 1574.
473. J. von Braun and G. Lemke, *Ber. Dtsch. Chem. Ges.*, 1922, **55**, 3526.
474. J. von Braun, F. Jostes, and A. Heymons, *Ber. Dtsch. Chem. Ges.*, 1927, **60**, 92.
475. J. Arient and J. Podstata, *Collect. Czech. Chem. Commun.*, 1974, **39**, 3177.
476. M. D. Scott and H. Spedding, *J. Chem. Soc. C*, 1968, 1603.
477. (a) E. Enders, Ger. Pat. 949285 (1956); *Chem. Abstr.*, 1959, **53**, 95250e; (b) G. R. Pettit and R. E. Kadunce, *J. Org. Chem.*, 1962, **27**, 4566.
478. K. Dierbach, E. Ger. Pat 26918 (1964); *Chem. Abstr.*, 1964, **61**, 13236h.
479. W. Klocher and M. Herberz, *Montash. Chem.*, 1965, **96**, 1567.
480. E. C. Taylor and R. W. Morrison, Jr., *Angew. Chem., Int Ed. Engl.*, 1965, **4**, 868.
481. A. Albert and H. Taguchi, *J. Chem. Soc., Perkin Trans. 1*, 1973, 2037.
482. E. M. Roberts, J. M. Grisar, R. D. McKenzie, G. P. Claxton, and T. R. Blohm, *J. Med. Chem.*, 1972, **15**, 1270.
483. I. Ya. Kvitko and T. M. Loginova, *J. Org. Chem. USSR, Engl. Trans.*, 1974, **10**, 1101.
484. S. B. Barnela, R. S. Pandit, and S. Seshadri, *Indian J. Chem., Sect. B*, 1976, **14**, 668.
485. M. Grdinic and V. Hahn, *J. Org. Chem.*, 1965, **30**, 2381.
486. F. Pochat, *Tetrahedron*, 1986, **42**, 4461.
487. (a) T. Tsuji and Y. Kamo, *Chem. Letters*, 1972, 641; (b) T. Tsuji, *Bull. Chem. Soc. Jap.*, 1974, **22**, 471.
488. C. H. Foster and E. U. Elam, *J. Org. Chem.*, 1976, **41**, 2646.
489. W. Schulze, P. Held, and A. Junov, *Z. Chem.*, 1975, **15**, 184.
490. M. Ichiba, K. Senga, S. Nishigaki, and F. Yoneda, *J. Heterocycl. Chem.*, 1977, **14**, 175.
491. S. Kwon, F. Ikeda, and K. Izegawa, *Nippon Kagaku Kaishi*, 1973, 1944.
492. E. K. D'yachenko, V. G. Pesin, A. A. Smirnova, A. I. Kapitan, N. D. Rozhkova, and M. P. Papirnik, *Zh. Org. Khim.*, 1987, **23**, 2450.

493. S. V. Simakov, V. S. Velezheva, T. A. Kozik, and N. N. Suvorov, *Khim. Geterosikl. Soedin.,* 1985, 76; *Chem. Abtsr.,* 1985, **103**, 53907c.
494. C. L. Dickinson, W. J. Middleton, and V. A. Engelhardt, *J. Org. Chem.,* 1962, **27**, 2470.
495. A. Ya. Bushkov, O. I. Lantsova, V. A. Bren, and V. I. Minkin, *Khim. Geterotsikl. Soedin.,* 1985, 565; *Chem. Abstr.,* 1985, **103**, 37451u.
496. S. Seshadri, M. S. Sardessai and M. A. Betrabet, *J. Ind. Chem. Soc.,* 1969, **7**, 667.
497. I. Hagedorn, H. Etling, and K. E. Lichtel, *Chem. Ber.,* 1966, **99**, 520.
498. S. Kobayashi, *Bull. Chem. Soc. Jap.,* 1973, **46**, 2835.
499. H. Eilingsfield, M. Seefelder, and H. Weidinger, *Chem. Ber.,* 1963, **96**, 2672.
500. M. R. Chandramohan and S. Seshadri, *Indian J. Chem.,* 1973, **11**, 1108.
501. M. R. Chandramohan and S. Seshadri, *Indian J. Chem.,* 1974, **12**, 940.
502. M. N. Deshpandi and S. Seshadri, *Indian J. Chem.,* 1973, **11**, 538.
503. B. L. Bray, P. Hess, J. M. Muchowski, and M. E. Scheller, *Helv. Chem. Acta,* 1988, **71**, 2053.
504. (a) M. A. Kira, A. Bruckner-Wilhelm, F. Ruff, and J. Borsi, *Acta Chim. Acad. Sci. Hung.,* 1968, **56**, 189; (b) M. A. Kira, M. O. Abdel-Rahman, and K. Z. Gadalla, *Tetrahedron Lett.,* 1969, 109.
505. C. Jutz, *Chem. Ber.,* 1964, **97**, 2050.
506. (a) S. Yanagida, T. Fujita, M. Ohoka, J. Katagari, and S. Komori, *Bull. Chem. Soc. Jap.,* 1973, **46**, 292; (b) S. Yanagida, T. Fujita, M. Ohoka, J. Katagari, M. Miyake, and S. Komori, *Bull. Chem. Soc. Jap.,* 1973, **46**, 303.
507. J. Liebscher and H. Hartmann, *Collect. Czech. Chem. Commun.,* 1976, **41**, 1565.
508. H. Gold, *Angew. Chem.,* 1960, **72**, 956.
509. J. T. Gupton, C. Colon, C. R. Harrison, M. J. Lizzi, and D. E. Polk, *J. Org. Chem.,* 1980, **45**, 4522.
510. J. T. Gupton and D. E. Polk, *Synth. Commun.,* 1981, **11**, 571.
511. (a) J. T. Gupton, K. F. Correia, and B. S. Foster, *Synth. Commun.,* 1986, **16**, 365; (b) J. T. Gupton, K. F. Correia, and G. Hertel, *Synth. Commun.,* 1984, **14**, 1013.
512. J. Liebscher and H. Hartmann, *Synthesis,* 1979, **9**, 241.
513. J. T. Gupton, M. A. M. Moebus, and T. Buck, *Synth. Commun.,* 1986, **16**, 1561.
514. J. T. Gupton, J. Coury, M. A. M. Moebus, and S. Fitzwater, *Synth. Commun.,* 1986, **16**, 1575.
515. J. T. Gupton, D. A. Krolikowski, R. H. Yu, P. Vu, J. A. Sikorski, M. L. Dahl, and C. R. Jones, *J. Org. Chem.,* 1992, **57**, 5480.

516. B. Singer and G. Maas, *Chem. Ber.,* 1987, **120**, 485.
517. C. Nolte, G. Schaefer, and C. Reichardt, *Liebigs Ann. Chem.,* 1991, 111.
518. (a) Z. Csuros, R. Soós, J. Palinkas, and I. Bitter, *Acta Chim. Acad. Sci. Hung.,* 1970, **63**, 215; (b) Z. Csuros, R. Soós, J. Palinkas, and I. Bitter, *Acta Chim. Acad. Sci. Hung.,* 1971, **68**, 397; (c) Z. Csuros, R. Soós, J. Bitter, and I. Palinkas, *Acta Chim. Acad. Sci. Hung.,* 1972, **72**, 59.
519. E. Bamberger and J. Lorenzen, *Ann. Chem.,* 1893, **273**, 269.
520. (a) A. J. Hill and I. Rabinowitz, *J. Am. Chem. Soc.,* 1926, **48**, 732; (b) A. J. Hill and M. V. Cox, *J. Am. Chem. Soc.,* 1926, **48**, 3214.
521. J. D. Wilson, C. F. Hobbs, and H. Weingarten, *J. Org. Chem.,* 1970, **35**, 1542.
522. W. Kantlehner and P. Speh, *Chem. Ber.,* 1971, **104**, 3714.
523. W. H. Warren and F. E. Wilson, *Ber. Dtsch. Chem. Ges.,* 1935, **68**, 957.
524. W. Jentsch, *Chem. Ber.,* 1964, **37**, 1361.
525. A. A. R. Sayigh and H. Ulrich, *J. Chem. Soc.,* 1963, 3146.
526. H. M. Walborsky and G. E. Niznik, *J. Org. Chem.,* 1972, **37**, 187.
527. Z. Arnold and J. Zemlicka, *Chem. Listy,* 1958, **52**, 459.
528. W. Jentsch and M. Seefelder, *Chem. Ber.,* 1965, **98**, 274.
529. Z. Csuros, R. Soós, I. Bitter, and E. A. Karpati, *Acta Chim. Acad. Sci. Hung.,* 1972, **73**, 239.
530. K. H. Beyer, H. Eilingsfeld, and H. Weidinger, Ger. Pat., 1110625 (1961); *Chem. Abstr.,* 1961, **56**, 3363b.
531. R. Walther and R. Grossman, *J. Prakt. Chem.,* 1908, **78**, 478.
532. J. Liebscher and H. Hartmann, *Z. Chem.,* 1974, **14**, 358.
533. C. Jutz and W. Müller, *Angew. Chem., Int. Ed. Engl.,* 1966, **5**, 1042.
534. D. Lloyd and H. McNab, *Angew. Chem., Int. Ed. Engl.,* 1976, **15**, 459.
535. L. Berlin, and O. Reister, in Houben-Weyl, *Methodenden Organischen Chemie,* ed. E. Mueller, Georg Thieme Verglag, Stuttgart, Vol 5/1d, 1972, p. 234.
536. D. Lloyd, K. S. Tucker, and D. R. Marshall, *J. Chem. Soc., Perkin Trans. 1,* 1981, 726.
537. C. Jutz, R. Kirchlechner, and H.-J. Seidel, *Chem. Ber.,* 1969, **102**, 2301.
538. Z. Arnold, *Collect. Czech. Chem. Commun.* 1965, **30**, 2125.
539. C. Reichardt and K. Halbritter, *Liebigs Ann. Chem.,* 1970, **737**, 99.
540. Z. Arnold, J. Sauliova, and V. Krchnak, *Collect. Czech. Chem. Commun.,* 1973, **38**, 2633.
541. J. Kucera and Z. Arnold, *Collect. Czech. Chem. Commun.,* 1967, **32**, 3792.
542. Z. Arnold and A. Holy, *Collect. Czech. Chem. Commun.,* 1963, **28**, 2040.
543. H. Bredereck, F. Effenberger, and G. Simchen, Ger. Pat., B65348 IVb/120 (1961).

544. H. Bredereck, F. Effenberger, and D. Zeyfang, *Angew. Chem., Int. Ed. Engl.*, 1965, **4**, 242.

545. H. Bredereck, F. Effenberger, and G. Simchen, *Chem. Ber.*, 1963, **96**, 1350.

546. V. Nair and C. S. Cooper, *J. Org. Chem.*, 1981, **46**, 4759.

547. R. A. Jones in *Comprehensive Heterocyclic Chemistry*, A. R. Katritzky and C. W. Rees (Eds.), CRC Press, New York, 1984, Vol. 4, p. 223.

548. J. T. Gupton, B. Norman, and E. Wysong, *Synth. Commun.*, 1985, **15**, 1305.

549. J. T. Gupton, J. E. Gall, S. W. Resinger, S. Q. Smith, K. M. Bevirt, J. A. Sikorski, M. L. Dahl, and Z. Arnold, *J. Heterocyl. Chem.*, 1991, **28**, 1281.

550. J. T. Gupton, S. W. Riesinger, A. S. Shah, J. E. Gall, and K. M. Bevirt, *J. Org. Chem.*, 1991, **56**, 976.

551. J. Liebscher, A. Knoll, H. Hartmann, and S. Anders, *Collect. Czech Chem. Commun.*, 1987, **52**, 761.

552. I. Muller, W. Himmerling, R. Wingen, Ger. Pat. DE. 3842062 (1991); *Chem. Abstr.*, 1991, **114**, 61678.

553. H. E. Nikolajewski, S. Dühne, and B. Hirsch, *Chem. Ber.*, 1967, **100**, 2616.

554. Z. Arnold and A. Holy, *Collect. Czech. Chem. Commun.*, 1965, **30**, 47.

555. J. Loetzberger and K. Bodendorf, *Chem. Ber.*, 1967, **100**, 2620.

556. S. M. Makin, L. I. Boiko, and O. A. Shavrygina, *J. Org. Chem. USSR (Engl. Transl.)*, 1977, **13**, 1093.

chapter three

Ring Formation by Aromatization and Heteroaromatization

3.1 Ring Formation by Aromatization

3.1.1 Derivatives of Benzene

Although cyclizations to form a benzene ring are discussed in section 4.2.1, it is useful to compare and contrast the functional groups and their orientation when Vilsmeier reagents induce either (a) cyclization to form a benzene ring or (b) aromatization of an existing ring. Some allylic alcohols give a mixture of di- and tri-formyl benzenes, although the yields are poor (scheme 3.1).[1-3] Each of the allylic alcohols **1-4** were converted into the same dialdehyde **5** in 22-28% yields.[4] All of these reactions are consistent with an initial dehydration step to give a 1,3-diene, followed by polyimino-alkylation, and subsequent ring closure.

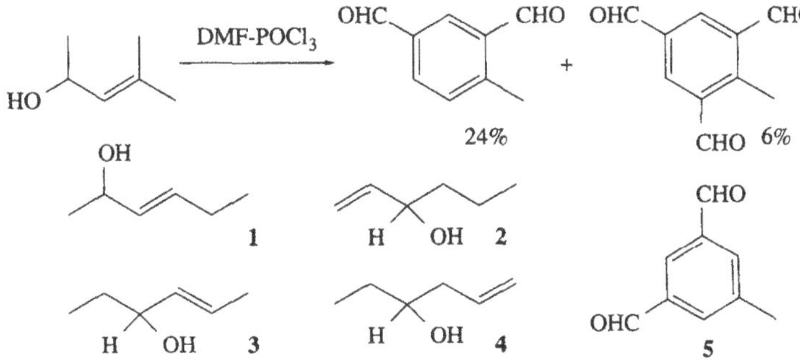

Some acyclic 2,4-dienoic acids can be converted into di- and tri-formyl-benzenes (scheme 3.2).[5]

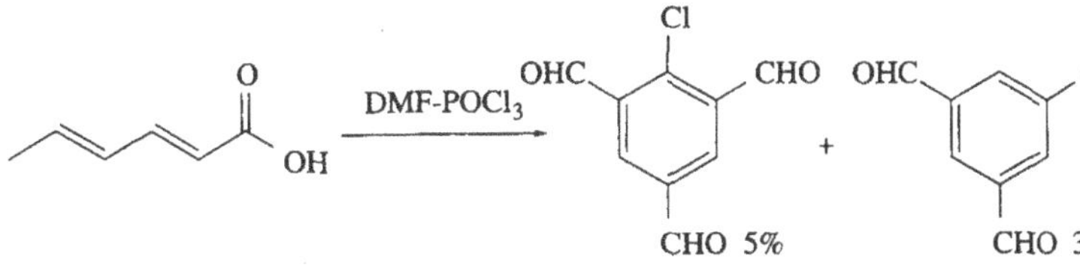

A mixture of α,β-unsaturated ketones afforded the chloroformylbenzene **6** (scheme 3.3).[6] Some acyclic precursors also afford chloroformyl-benzenes.[7]

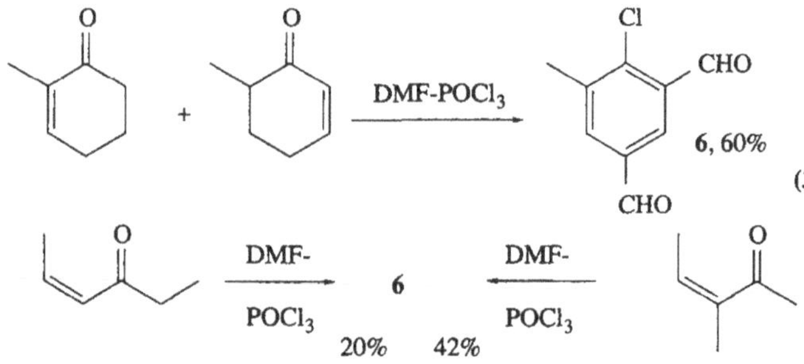

$$(3.3)$$

The initial *C*-iminoalkylation of substituted 2-cyclohexen-1-ones proceeds at C-6, *via* a presumed enolic intermediate (scheme 2.134, section 2.11.3). Depending on the amount of Vilsmeier reagent used and the temperature of the reaction, further iminoalkylation can occur. The observed sequences are complex, but relatively well understood. Oxidation, presumably aerially, can result in monoformylbenzenes and triformyl-benzenes (scheme 2.134, section 2.11.3).

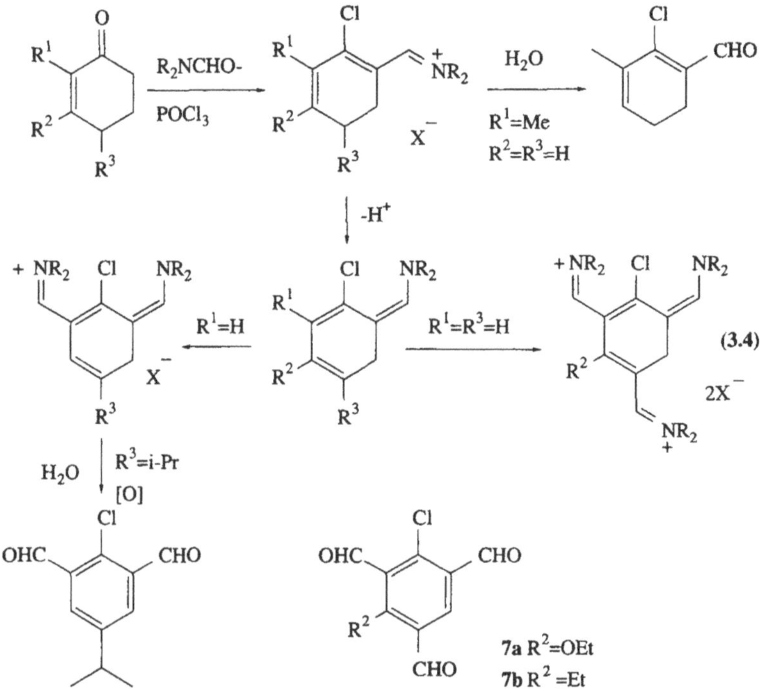

$$(3.4)$$

A unified pathway[8] illustrating typical processes involving 2-cyclohexen-1-ones lacking a methyl group at C-3 is given in scheme 3.4. For 3-methyl-2-cyclohexen-1-one and its 5-substituted derivatives an entirely different course is followed (scheme 2.137).[8-10] Although it is uncertain whether Y=Cl or an oxygenated moiety in the exocyclic alkene **211**, several exocyclic alkenes analogous to **211** (scheme 2.137) are known to be thermo-dynamically favoured over their endocyclic isomers.

Trialdehydes **7** have been obtained by the Vilsmeier reaction of cyclo-hexenones prepared by Birch reduction. The trialdehyde **8** was formed from cyclohexenone in a Vilsmeier reaction, and the dialdehyde **9** from 4-methyl-2-cyclohexen-1-one.[6]

(3.5)

The cross-conjugated dialdehyde **207a** (scheme 2.134, section 2.11.3) was formed (24%) by treating cyclohexane-1,3-dione with DMF-POCl$_3$ at 20°C.[11] The same reaction was also observed for the 3-halo-2-cyclohexen-1-ones **195b** and **195c** (scheme 2.134). The pathway by which dialdehydes **207** are formed is outlined in scheme 2.134. The methinium species is thought to be the final product of the reaction prior to hydrolysis. The stability of alkenes **203** and **207** towards aromatization has been investigated;[12] heating the reaction mixture under reflux afforded the pentasubstituted benzene **10a** in 20% yield. MNDO calculations indicated an energy difference between **207a** and **10b** of 13 kcal mol^{-1} in favour of the arene **10b**. Reaction of cyclohexane-1,3-dione with DMF-POCl$_3$ i n chloroform at 50°C afforded 2,4-dichloro-1-formylcyclohexa-1,3-diene **199** (scheme 2.134). This formation of a less substituted aldehyde at higher temperatures suggests that thermal decomposition of **203** (known to be formed at 20°C) can occur (presumably by nucleophilic attack, probably involving addition of HCl to the dication **203**) to give **200**, and hence the aldehyde, **199** (scheme 2.134). A related thermal deformylation was referred to in section 2.29.4.

(3.6)

10a R^1=CHO, R^2=CHNMe$_2$
10b R^1=CHNMe$_2$, R^2=CHO

Cyclohexane-1,4-dione reacts with *N*-formylmorpholine-POCl$_3$ to give the phenol **11** (scheme 3.7).[13] The probable intermediates are not stabilized by strongly electron-withdrawing groups, as are the methinium species depicted in scheme 2.134, so that pathways leading to aromatization are favored.

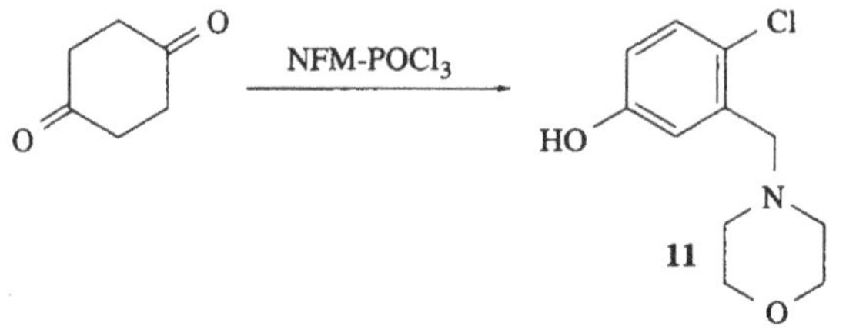

(3.7)

1,4-Dihydrobenzoic acids react with DMF-POCl$_3$ to give polyformylated benzenes, possibly *via* decomposition of intermediate **12**.[14] In general, formylation occurs at the 3- and the 5-positions, unless they already bear substituents, in which case the unsubstituted positions are formylated. Thus, 3,5-dimethyl-1,4-dihydrobenzoic acid affords 3,5-dimethylbenzaldehyde (45%). A substituent at the 2- or 4-position is tolerated, but the yields are poor; 1,4-dihydro-4-methylbenzoic acid affords 2,4,6-triformyl-1-methyl-benzene in 17% yield.

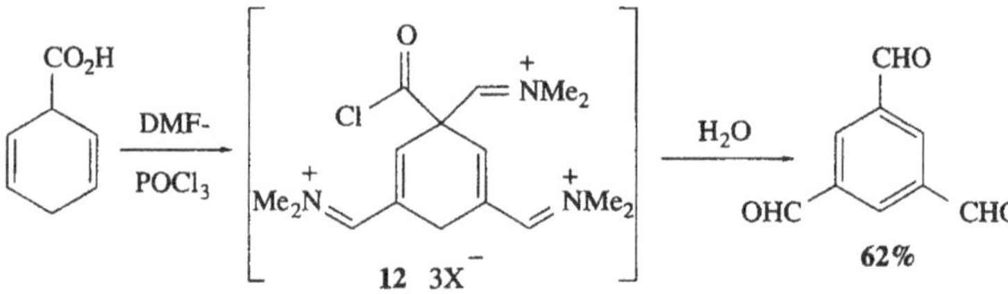

(3.8)

A related study of the reaction of 1-methoxy-4-methyl-1,4-cyclo-hexadienes with DMF-POCl$_3$. A triformylbenzene results, usually, though not always one in which the methoxy group has undergone replacement by chloro. However, 1,4-dimethoxy-1,4-cyclohexadiene afforded 2,5-diformyl-1,4-dimethoxybenzene (14%).[15] Which intermediate undergoes oxidation, and whether aerial oxidation is a requirement, has not been established with certainty. Similar results were obtained from simple alkyl substituted 1,4-cyclohexadienes, but the yields were poor.[16]

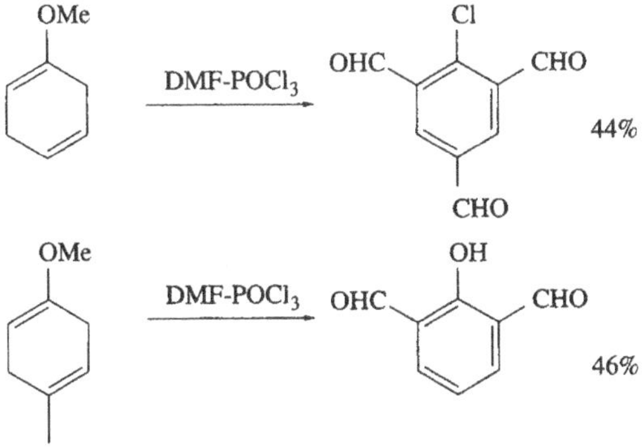

$$(3.9)$$

Reaction of vinyl- and thienyl-cyclohexenes with chloromethylene-iminium salts can provide an aromatic ring, and hence an approach to anthracyclines.[17]

3.1.2 Derivatives of Benzo-fused Cycloalkanes

A mixture of α,β-unsaturated and β,γ-unsaturated ketones **13** afforded the indane **14**.[18]

$$(3.10)$$

Rao has reported the synthesis of a variety of benzenoid systems from acyclic precursors. In this manner, a number of indane, tetralin, and benzosuberone mono- and di-carboxaldehydes was obtained from cyclic homoallylic alcohols **15**, although in poor yields.[19]

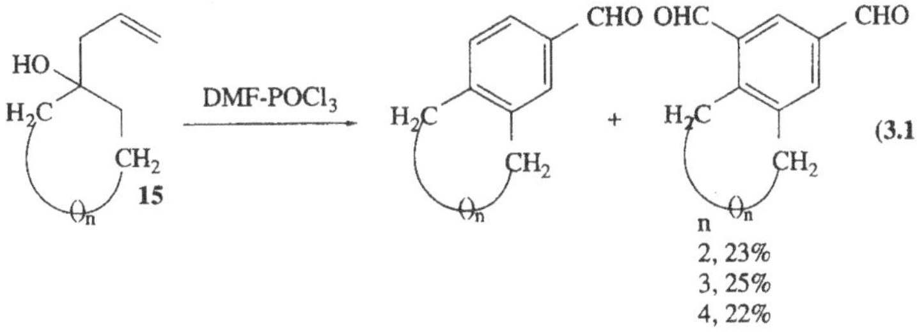

$$(3.11)$$

n
2, 23%
3, 25%
4, 22%

3.1.3 Derivatives of Naphthalene

In the same way that 1,4-dihydrobenzoic acids react with DMF-POCl₃ to give benzene mono- di-, and tri-carboxaldehydes, the analogous acids **16** afford naphthalene-1,3-dicarboxaldehyde (scheme 3.12).[14]

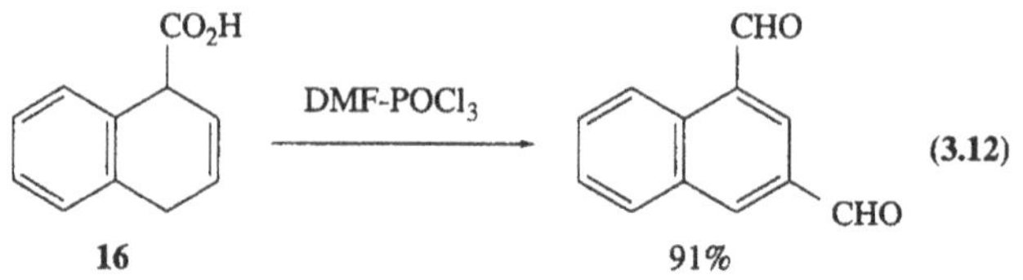

$$\text{(3.12)}$$

16   **91%**

β-Tetralones have been converted into 1,3-diformyl-2-chloro-naphthalenes. Paquette [20] obtained only the monoformyl derivatives, with no aromatization. However, the use of excess DMF-POCl₃ with DMF as solvent and prolonged reaction times enabled the isolation of the 1,3-diformyl compounds **17a** and **17b** in respective yields of 54% and 30%.[21]

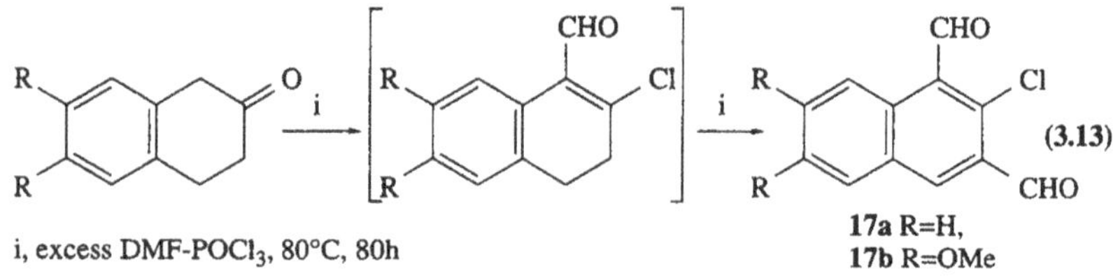

$$\text{(3.13)}$$

i, excess DMF-POCl₃, 80°C, 80h

17a R=H,
17b R=OMe

3.1.4. Derivatives of Anthracene

The aromatization of anthrone by Vilsmeier reagents to give 10-chloro-9-anthracenecarboxaldehyde has been long known,[22] and proceeds *via* the cation **18** which imparts an intense deep red colour to the reaction mixture.

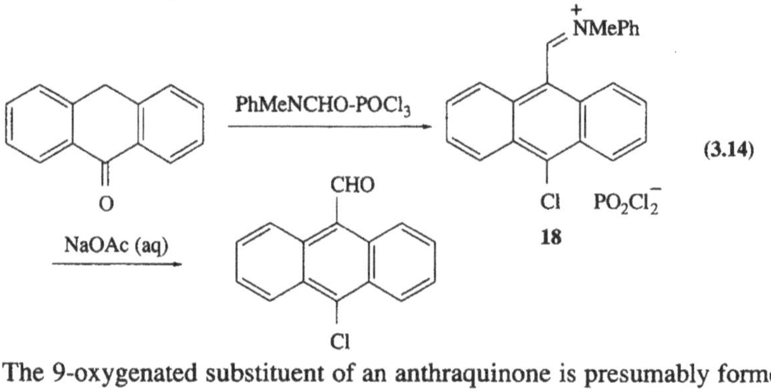

$$\text{(3.14)}$$

18

The 9-oxygenated substituent of an anthraquinone is presumably formed and activates the 10-position towards iminoalkylation; the chloride ion then displaces the 9-oxygenated substituent.

3.1.5 Derivatives of Carbazoles

The effects of temperature and stoichiometry of Vilsmeier reagents can be crucial to the outcome to the reaction of tetrahydrocarbazoles. 9-Benzyl-1,2,3,4-tetrahydrocarbazole is formylated in the 1-position at low temperatures in 90% yield (scheme 3.15). This formylation presumably proceeds *via* deprotonation of an iminium salt to give the enamine endocyclic to the cyclohexane ring; the enamine would then undergo acylation followed by reformation of the pyrrole ring by deprotonation. At higher temperatures, aromatization of the product occurs yielding the 3-formylcarbazole **19**, and minor quantities of the 3-formyl-1-(1,1-dimethyl-amino)methyl derivative **20**. These derivatives were converted into ellipticine and olivicine.[23]

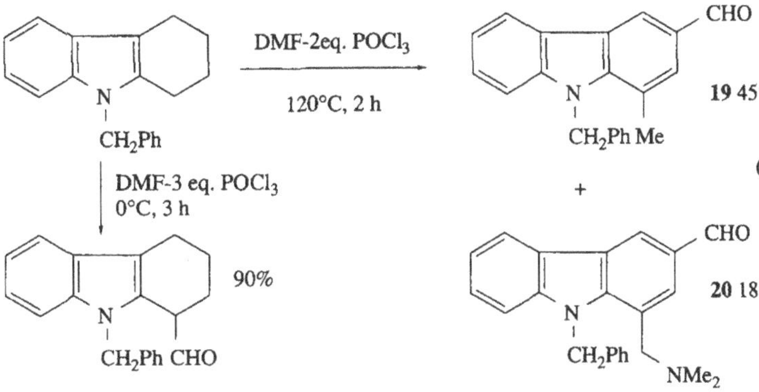

$$(3.15)$$

3.2 Rings by Heteroaromatization

3.2.1 Furans

3-Chloro-2-formylbenzo[*b*]furan **21** is formed by the reaction of 3-coumarinone with DMF-POCl₃.[24]

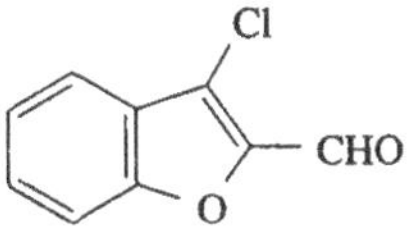

$$(3.16)$$

3.2.2 Thiophenes

Aromatization of the keto-ester **22** to thiophenes is temperature-dependent.[25] Some 4-oxo-4,5-dihydrothiophenes such as **23** are converted by Vilsmeier reagents into 4-chloro-5-formylthiophene derivatives **24** which are useful intermediates for preparing thiophene azo dyes (scheme 3.17).[26]

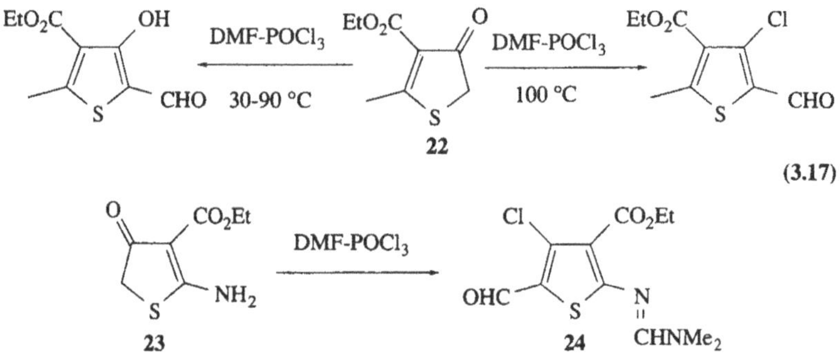

(3.17)

The aldehyde **25** can be prepared in good yield by reacting 3-methoxy-benzo[b]thiophene with DMF-POCl$_3$ [27]

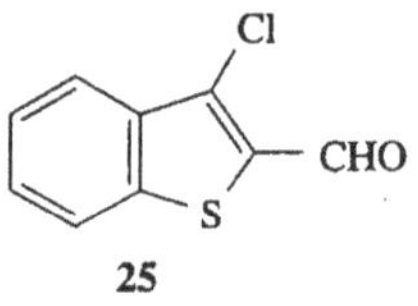

3.2.3 Pyrroles

2-Chloropyrrole-3-carboxaldehydes have been prepared by treating Δ^4-pyrrolidin-2-ones with DMF-POCl$_3$ in chloroform.[28,29] The reaction of 4-methoxycarbonyl-5-methyl-Δ^4-pyrrolin-2-one with chlorinated and brominated Vilsmeier reagents has been studied.[30] A wide variety of Δ^3-pyrrolin-2-ones **26** are smoothly converted into the functionalized pyrroles **27** (scheme 3.19).[31-32]

The aldehyde **28** has been prepared from the corresponding pyrrolone and DMF-POCl$_3$.[33]

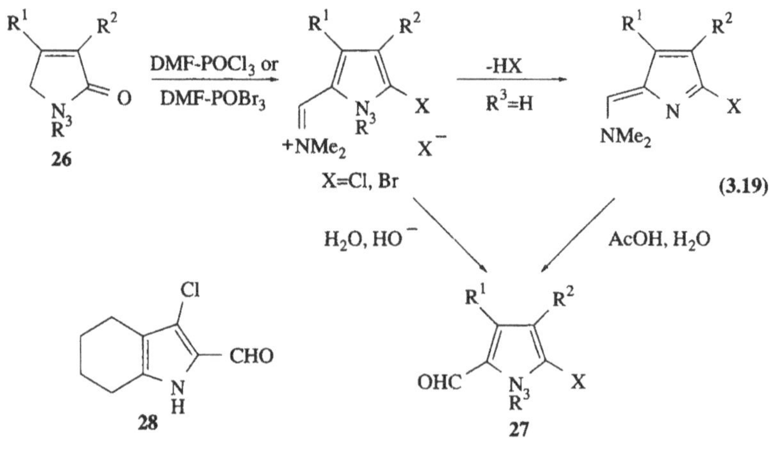

(3.19)

3.2.4 Pyrazoles

5-Pyrazolones usually react with Vilsmeier reagents to give formylated pyrazoles.[34] A variety of other functionalities may be produced, depending upon the substrate and reaction conditions. Formylpyrazole derivatives **29** have been prepared by treating 5-pyrazolones with DMF-POCl$_3$ (scheme 3.20).

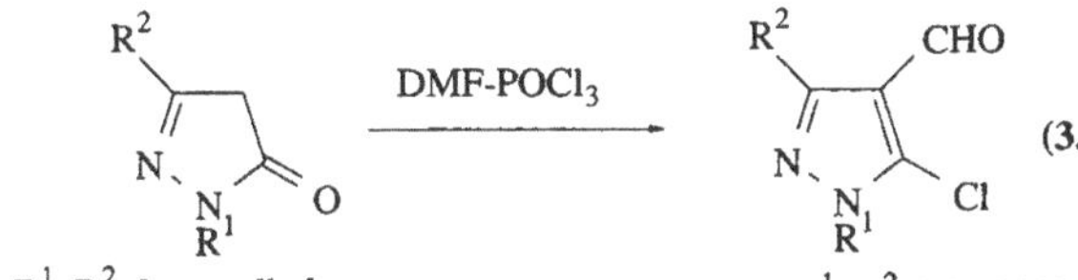

(3.20)

R^1, R^2=lower alkyl   **29** R^1=R^2=Me, 91%

3-Methyl-1-phenyl-5-pyrazolone **30** afforded the pyrazole carbox-aldehyde **31** when the reaction mixture was added to water and slowly neutralized.[35] However, when the mixture was poured into aqueous K$_2$CO$_3$, some of the hydroxymethylene derivative **32** was also obtained. Working at lower temperatures, the formation of the aminomethylenepyrazolone **33** has been observed.

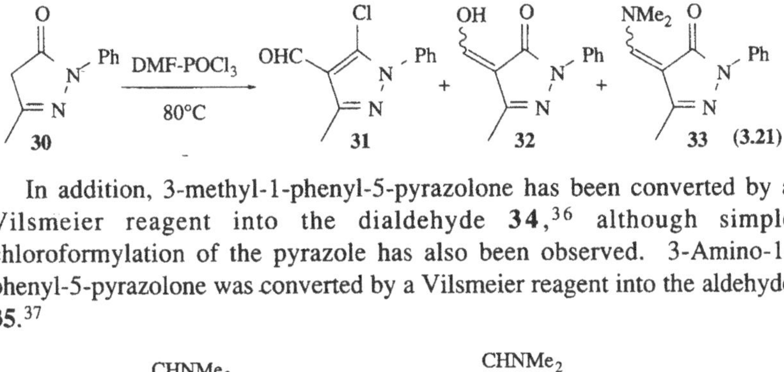

(3.21)

In addition, 3-methyl-1-phenyl-5-pyrazolone has been converted by a Vilsmeier reagent into the dialdehyde **34**,[36] although simple chloroformylation of the pyrazole has also been observed. 3-Amino-1-phenyl-5-pyrazolone was converted by a Vilsmeier reagent into the aldehyde **35**.[37]

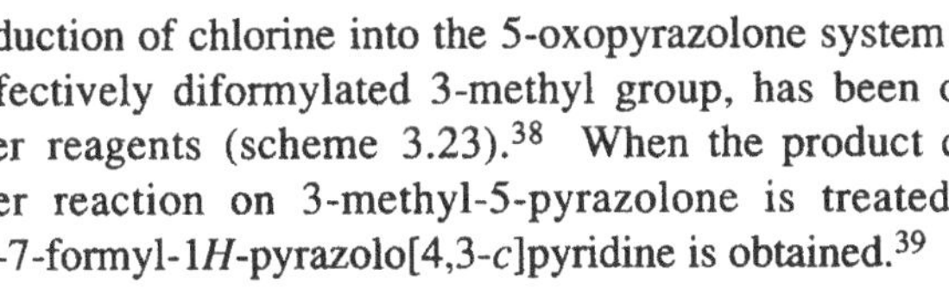

(3.22)

Introduction of chlorine into the 5-oxopyrazolone system **36**, in addition to an effectively diformylated 3-methyl group, has been observed using Vilsmeier reagents (scheme 3.23).[38] When the product derived from a Vilsmeier reaction on 3-methyl-5-pyrazolone is treated with NH$_4$Cl, 3-chloro-7-formyl-1*H*-pyrazolo[4,3-*c*]pyridine is obtained.[39]

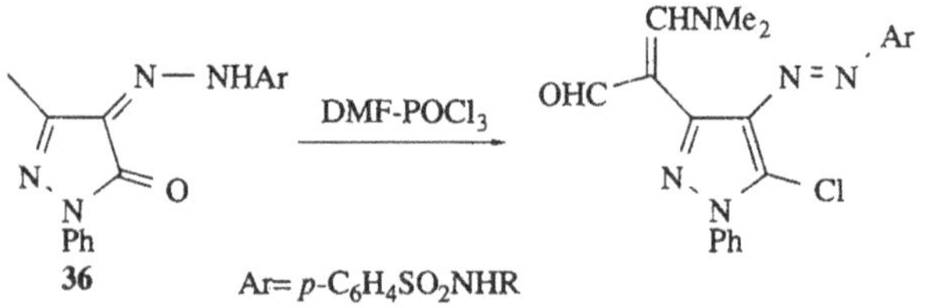

(3.23)

Ar= *p*-C₆H₄SO₂NHR

The 2-pyrazolin-5-one **36** was diformylated to give amino acrolein derivatives **37**.[40]

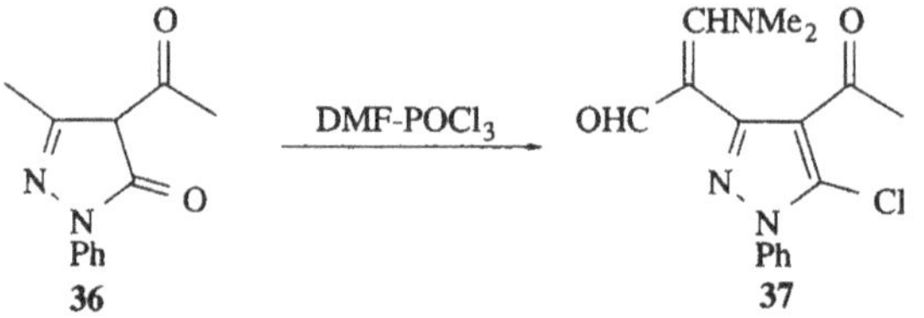

(3.24)

3.2.5 Indoles

Early work showed that indolin-2-one (oxindole) **38** (R=H) was converted by DMF-POCl₃ into the 2-chloro-3-formylindole **39a** (scheme 3.25).[41] The reaction also proceeds with a variety of 1-substituted oxindoles.[42] The remarkable conversion of 1-methyloxindole into 3-chloro-1-methyl-2-quinolinone (79%) has been reported.[43]

The Vilsmeier reagent converted 1-acyloxindoles **40** into the 3-(dimethylamino)methylidene derivatives **41a** and **41b**. The products **42** had to be prepared by acylation of 2-chloro-3-formylindole.[44] Some oxindoles substituted at either the 5- or 7-position were converted into the corresponding 2-chloro-3-formylindoles;[45] however, other substituted oxindoles reacted differently.[41]

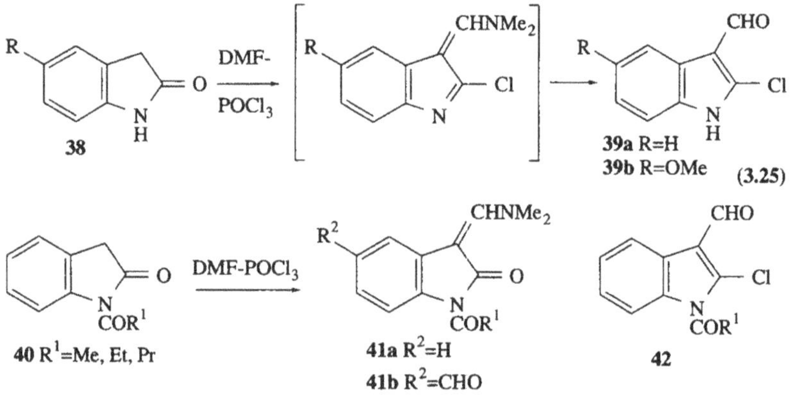

(3.25)

A general approach to 2,2′-, 2,3′- and 2,7′-bi-indolyls using Vilsmeier type intermediates has been described.[46]   For example, addition of

2-phenylindole in chloroform to a mixture of oxindole and $POCl_3$ affords 2,3′-bi-indolyls (60%), together with a terindolyl (19%). The latter yield in increased to 60% by using an excess of the Vilsmeier reagent. By using indoles blocked at both the 2- and 3-positions, and activated in the benzene ring by methoxy groups, the reaction can be made to occur at the 7-position. For example, **43**, is obtained in 56% yield using the Vilsmeier reagent derived from oxindole and $POCl_3$; by using $(CF_3SO_2)_2O$ as the condensing agent, a quantitative yield of **43** was obtained.[47]

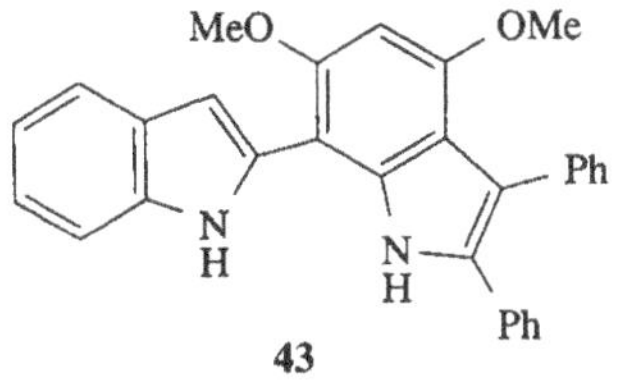

(3.26)

4,6-Dimethoxyindolin-2-one undergoes self-condensation across the 2- and 3-positions, when reacted with $POCl_3$, to give low yields of trimeric products.[47]

3.2.6 Pyrimidines

Barbituric acid reacts with DMF-$POCl_3$ to give the aldehyde **44** (R=Cl); other intermediates **44** (R=Cl, NMe_2, NEt_2, Ph) for dyes have been similarly prepared using Vilsmeier reagents.[48]

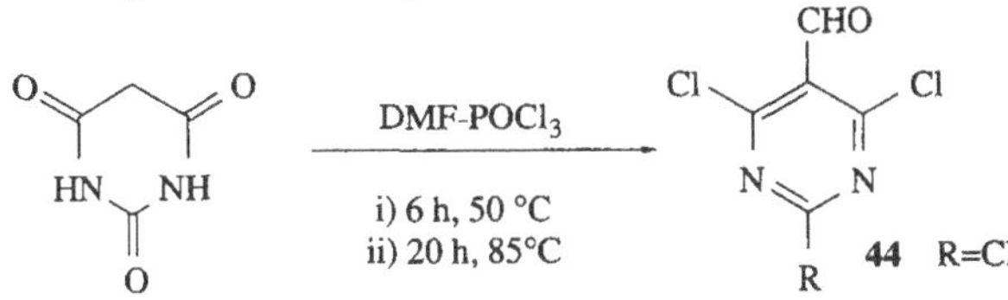

(3.27)

3.2.7 Isoquinolines

Dihydroisoquinolin-3-ones **45** have been converted into 3-chloro-4-formylisoquinolines **46** by the action of Vilsmeier reagents followed by oxidation (scheme 3.28).[49] The sequence of the Vilsmeier-Haack reaction followed by oxidation can also be applied to other dihydroisoquinolinone derivatives including **47** and **48** (scheme 3.28); blocking the nitrogen atom afforded 3-chloro-1,2-dihydro-1-phenylisoquinoline-4-aldehyde (25%). Blocking the 1-position of the isoquinolone enabled the spirocyclic chloroketone **49** to be obtained (scheme 3.29).[49]

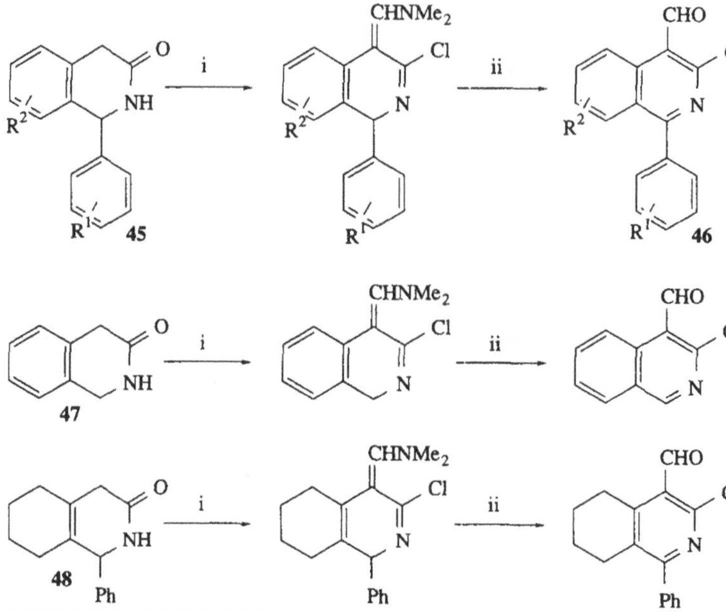

i, DMF-POCl$_3$; ii, H$_2$SO$_4$-KMnO$_4$

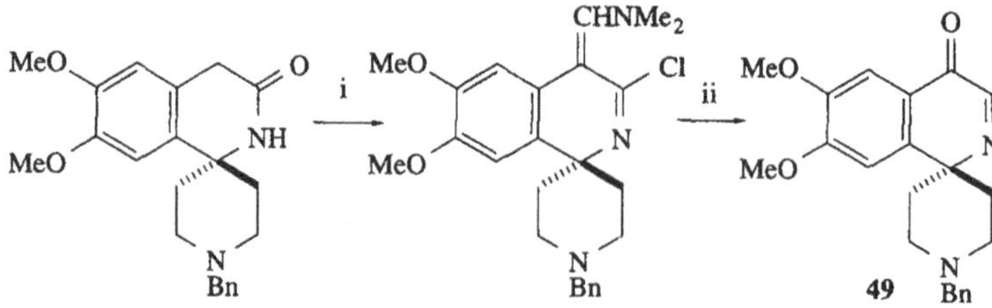

i, DMF-POCl$_3$; ii, H$_2$SO$_4$-KMnO$_4$

3.2.8 Miscellaneous Ring Systems

Vilsmeier-Haack reactions on 5(4H)-isoxazolones **50** led to dichloromethylisoxazoles.[50]

The five-membered heterocycles **51** and **52** were prepared by reacting the appropriate heterocyclic amides with Vilsmeier reagents (scheme 3.31).[51]

(3.31)

51 X=N, CMe   52

Rhodanine **53** was converted into 4-dimethylaminoformylidene rhodanine **54**. 2-Phenyliminothiazolidin-4-one is converted by a Vilsmeier reagent into the versatile derivatives **55** (scheme 3.32) from which several 5,5-fused heterocycles have been made.[52] Heterocycle **56** is an interesting example of a chloroformylation together with aromatization and formation of an amidine moiety (scheme 3.33).[53]

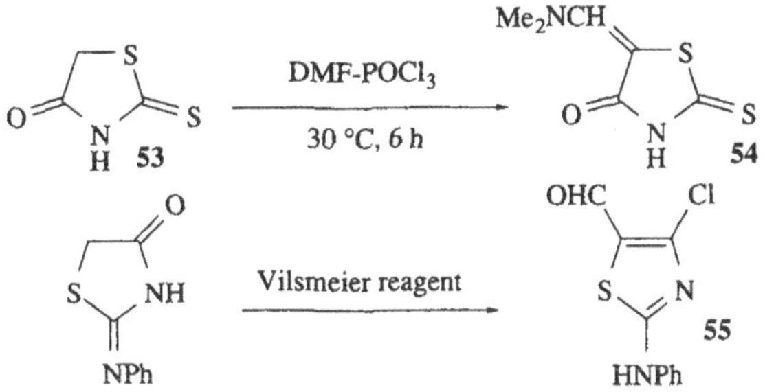

(3.32)

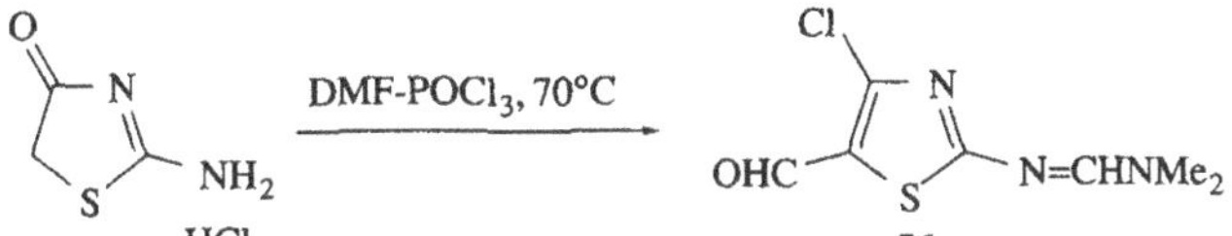

(3.33)

The aldehyde **57** was obtained from a Vilsmeier reaction of 4-azaazulen-5-one. It was subsequently transformed into a [2,3,4]cyclazine.[54]

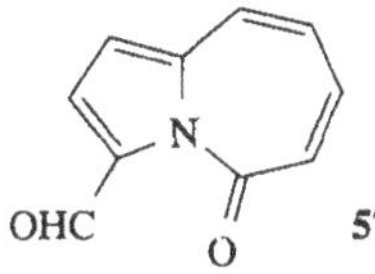

(3.34)

OHC   O   **57**

References

1. M. S. C. Rao and G. S. K. Rao, *Indian J. Chem., Sect. B*, 1988, **27B**, 213.
2. M. Sreenivasulu and G. S. K. Rao, *Indian J. Chem., Sect. B*, 1987, **26B**, 581.

3. M. Sreenivasulu and G. S. K. Rao, *Indian J. Chem., Sect. B*, 1987, **26B**, 1187.
4. M. Sreenivasulu, M. S. C. Rao, and G. S. K. Rao, *Indian J. Chem., Sect. B*, 1987, **26B**, 173.
5. B. Raju and G. S. K. Rao, *Indian J. Chem., Sect. B*, 1987, **26B**, 175.
6. B. Raju and G. S. K. Rao, *Indian J. Chem., Sect. B*, 1987, **26B**, 177.
7. M. S. C. Rao and G. S. K. Rao, *Indian J. Chem., Sect. B*, 1989, **28B**, 494.
8. A. R. Katritzky, Z. Wang, C. M. Marson, R. J. Offerman, A. E. Koziol, G. J. Palenik, *Chem. Ber.* 1988, **121**, 999.
9. P. C. Traas, H. Boelens, and H. J. Takken, *Recl. Trav. Chim. Pays-Bas*, 1976, **95**, 308.
10. W. Hoffmann and E. Mueller, Ger. Pat., 2152193 (1973); *Chem. Abstr.*, 1973, **79**, 18212.
11. A. R. Katritzky and C. M. Marson, *Tetrahedron Lett.*, 1985, **26**, 3753.
12. A. R. Katritzky, C. M. Marson, G. Palenik, A. E. Koziol, H. Luce, M. Karelson, B.-C. Chen, and W. Brey, *Tetrahedron*, 1988, **44**, 3209.
13. A. R. Katritzky and C. M. Marson, *J. Org. Chem.*, 1987, **52**, 2726.
14. B. Raju and G. S. K. Rao, *Synthesis*, 1987, 197.
15. B. Raju and G. S. K. Rao, *Synthesis*, 1985, 779.
16. B. Raju and G. S. K. Rao, *Indian J. Chem., Sect. B*, 1987, **26B**, 892.
17. T. Iqbal, *Diss. Abstr. Int. B.*, 1983, **43**, 3598; *Chem. Abstr.*, 1983, **99**, 87304.
18. B. Raju and G. S. K. Rao, *Indian J. Chem., Sect. B*, 1987, **26B**, 1185.
19. M. S. C. Rao and G. S. K. Rao, *Indian J. Chem., Sect. B*, 1988, **27B**, 660.
20. L. A. Paquette, US Pat. 3129257 (1964); *Chem. Abstr.*, 1964, **61**, 1814a.
21. S. L. Evans, H. A. Lloyd, D. LeBeau, and E. B. Sokoloski, *Org. Prep. Proc. Int.*, 1990, **22**, 764.
22. G. Kalischer, A. Scheyer, and K. Keller, Ger. Pat. 514415 (1927); *Chem. Abstr.*, 1931, **25**, 1536.
23. Y. Yokoyama, N. Okuyama, S. Iwadate, T. Momoi, and Y. Murakami, *J. Chem. Soc., Perkin Trans. 1*, 1990, 1319.
24. Y. Anmo, Y. Tsurata, S. Ito, and K. Noda, *Yakugashi Zasshi*, 1963, **83**, 807; *Chem. Abstr.*, 1963, **59**, 15239c.
25. V. I. Shuedov, X. Vasil'eva, and G. N. Grinev, *Khim. Geterotsikl. Soedin.*, 1972, 427; *Chem Abstr.*, 1972, **77**, 88356u.
26. R. Egli, and B. Henzi, Ger. Offen. 3529831 (1986); *Chem. Abstr.*, 1987, **106**, 6402m.
27. A. Ricci, D. Balucani, and N. P. Buu-Hoï, *J. Chem. Soc. (C)*, 1967, 779.
28. A. Monge, I. Aldana, I. Lezamiz, and E. Fernandez-Alvarez, *Synthesis*, 1984, 160.

29. K. E. Schulte, J. Reisch, and U. Stoess, *Arch. Pharm. (Weinheim, Ger.)*, 1972, **305**, 523.
30. T. Messerschmitt, U. von Specht, and H. von Dobeneck, *Liebigs Ann. Chem.*, 1971, **751**, 50.
31. H. von Dobeneck and T. Messerschmitt, *Liebigs Ann. Chem.*, 1971, **751**, 32.
32. F. Schnierle, H. Reinhard, N. Dieter, E. Lippacher, and H. von Dobeneck, *Liebigs Ann. Chem.*, 1968, **715**, 90.
33. K. E. Schulte, R. Reisch, and U. Stoess, *Angew. Chem., Int. Ed. Engl.*, 1965, **4**, 1081.
34. M. Kishida, H. Hanaguchi, and T. Akita, Jpn. Kokai Tokkyo Koho JP 63,267,762 (1988); *Chem. Abstr.* 1989, **111**, 57728h.
35. M. R. Chandramohan, M. S. Sardessai, S. R. Shah, and S. Seshadri, *Indian J. Chem.*, 1969, **7**, 1006.
36. S. B. Barnela, R. S. Pandit, and S. Seshadri, *Indian J. Chem., Sect. B*, 1976, **14B**, 665.
37. S. B. Barnela, R. S. Pandit, and S. Seshadri, *Indian J. Chem., Sect. B*, 1976, **14B**, 668.
38. I. M. A. Awad and K. M. Hassan, *Phosphorus, Sulfur Silicon Relat. Elem.* 1989, **44**, 135; *Chem. Abstr.* 1990, **112**, 178773y.
39. S. V. Thiruvikraman and S. Seshadri, *Indian J. Chem., Sect. B*, 1984, **23B**, 768.
40. I. M. A. Awad and K. M. Hassan, *J. Chin. Chem. Soc. (Taipei)*, 1990, **37**, 599; *Chem. Abstr.*, 1991, **114**, 122153.
41. S. Seshadri, M. S. Sardessai, and A. M. Betrabet, *Indian J. Chem.*, 1969, **7**, 662.
42. L. Marchetti and A. Andreani, *Ann. Chim. (Rome)*, 1973, **53**, 681; *Chem. Abstr.*, 1975, **82**, 72723d.
43. A. Andreani, D. Bonazzi, and M. Rambaldi, *Boll. Chim. Farm.*, 1976, **115**, 732; *Chem. Abstr.*, 1978, **89**, 24107d.
44. A. Andreani, D. Bonazzi, M. Rambaldi, G. Mungiovino, and L. Greci, *Farmaco, Ed. Sci.*, 1978, **33**, 781; *Chem. Abstr.*, 1979, **90**, 38743r.
45. A. Andreani, M. Rambaldi, D. Bonazzi, A. Guarnieri, and L. Greci, *Boll. Chim. Farm.*, 1977, **116**, 589; *Chem. Abstr.*, 1978, **88**, 169876t.
46. D. St. C. Black and N. Kumar, *J. Chem. Soc., Chem. Commun.*, 1984, 441.
47. A. J. Ivory, Ph. D. Thesis, University of New South Wales, 1992.
48. J. Dehnert, Ger. Offen. 3603797 (1987); *Chem. Abstr.*, 1988, **108**, 7511z.
49. W. Bartmann, E. Konz, and W. Rüger, *Synthesis*, 1988, 680.
50. R. K. M. R. Kallury and P. S. U. Devi, *Tetrahedron Lett.*, 1977, 3655.
51. J. Becher, P. H. Olesen, N. A. Knudsen, and H. Toftlund, *Sulfur Lett.*, 1986, **4**, 175.

52. R. A. Pawar and A. P. Rajput, *Indian J. Chem., Sect. B*, 1989, **28B**, 866.
53. R. Egli, Ger. Offen. 3015121 (1980); *Chem. Abstr.*, 1981, **94**, 158317j.
54. W. Flitsch, *Adv. Heterocycl. Chem.*, 1988, **43**, 35.

chapter four

Ring-forming Reactions using either Vilsmeier Reagents or β-Chlorovinylaldehydes

4.1 Non-Benzenoid Systems

4.1.1 Cyclopentenes

Two examples of ring-closure to give derivatives of cyclopentene have been reported; a suitably placed alkyl substituent that can be deprotonated is a requirement, as is a 1,3-diene that undergoes iminoalkylation, directed by an electron-rich alkoxy or amino group (schemes 4.1, 4.2). The perchlorate salt of **1** was isolated in 55% yield.[1] The diamine salt **2** was obtained in 64% yield (scheme 4.2).

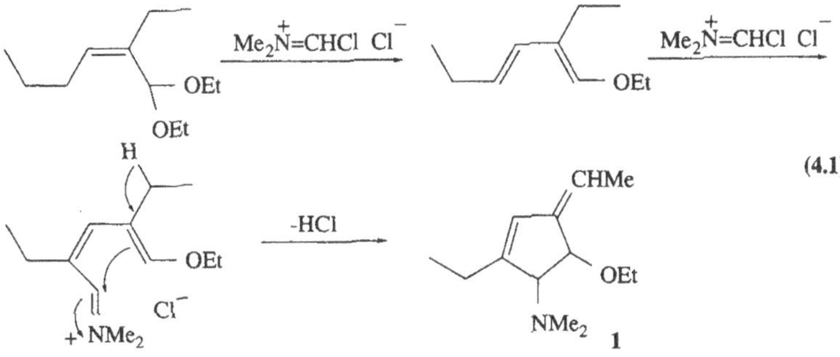

(4.1)

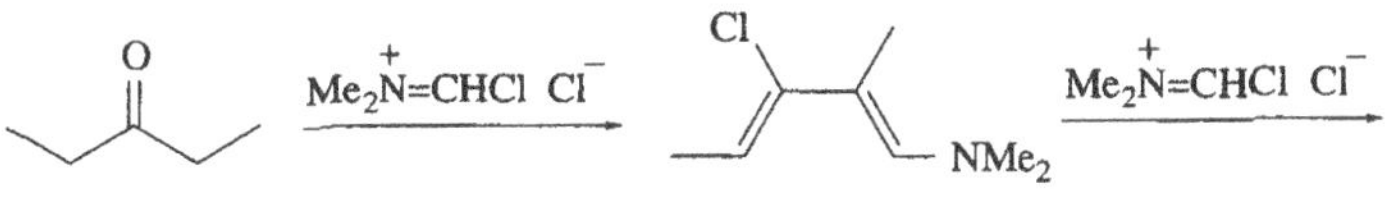

(4.2)

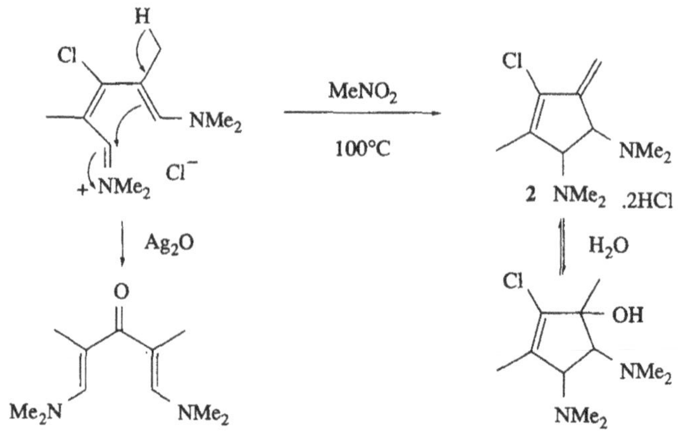

(4.2)

4.1.2 Fulvenes

Although fulvenes can be obtained from cyclopentadiene by a Vilsmeier formylation (section 2.8.2.1), they have also been obtained from 1,4-diaryl-butadienes which undergo iminoalkylation with Vilsmeier reagents (section 1.3.4) to give, after hydrolysis, fulvenecarboxaldehydes.[2] The mechanism of ring-closure is similar to that given in scheme 4.1.

4.1.3 Fused Ring Systems

The electron-rich five-membered ring of azulene undergoes reaction with Vilsmeier reagents to give iminium salts (section 2.8.4). In addition to this, ring-closure can occur by deprotonation of an adjacent methyl group (scheme 4.3). The overall yields of the fused systems **3** and **4** from 4,6,8-trimethyl-azulene are 27 and 55% respectively.[3]

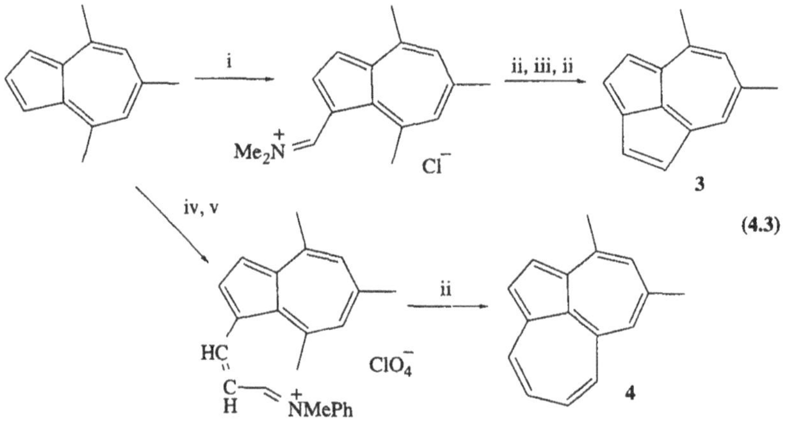

(4.3)

i, DMF-POCl$_3$; ii, NaOMe; iii, MeI; iv, Ph(Me)NCH=CHCHO-POCl$_3$; v, Na ClO$_4$

The [2.2.4]cyclazinium salts **5** were obtained under Vilsmeier conditions from 5-methylene-4-azaazulenes.[4]

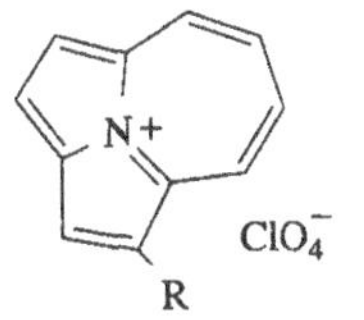

$$(4.4)$$

5 R=H, CN

4.2 Benzenoid Systems

4.2.1 Benzene Rings

A few cyclizations to form a benzene ring have been referred to in section 3.1.1. Homoallylic benzylic tertiary alcohols undergo terminal imino-alkylation as well as elimination. Electrocyclic ring closure of a presumed aminohexatriene intermediate, followed by elimination with aromatization, provides an elegant route to biphenyl and derivatives of biphenyls with substituents in a variety of positions.[5] By using α-tetralols, ring-closure to give several 9,10-dihydrophenanthrenes has been achieved.

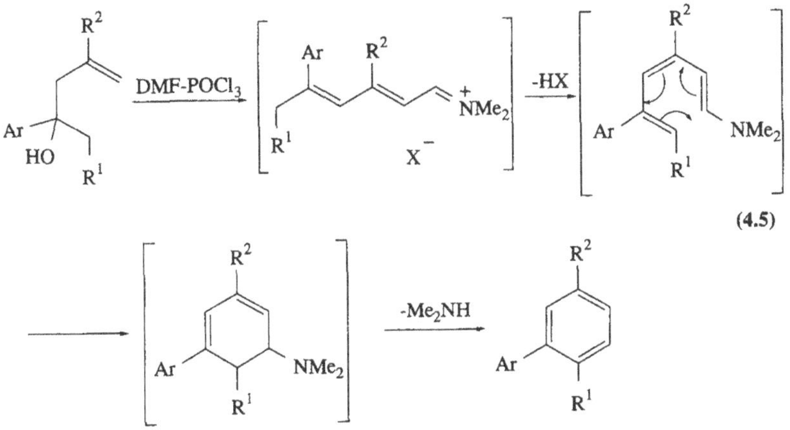

$$(4.5)$$

Reaction of heptamethinium salt **6** with Vilsmeier reagents leads to polymethinium species **7**. Although this could undergo ring closure by an enamine-iminium type attack (as shown for **7a**) there is some evidence that suggests the mechanism involves a 6π electrocyclic ring closure (as shown for **7b**). The resulting cation **8**, after elimination and hydrolysis, affords triformylbenzene **9**.[6-9]

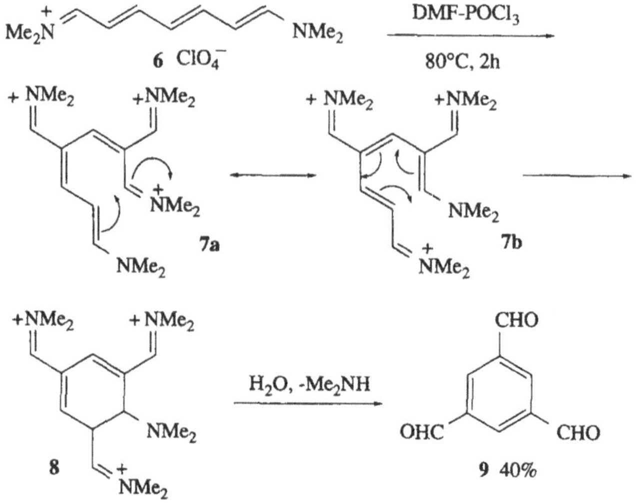

$$(4.6)$$

Pentane-2,4-dione **10** (R=OH) reacts in a similiar manner, although the heptamethinium cation must bear chloro substituents prior to ring-closure. Also, ring-closure takes place on the monocation, so that only one formyl group is found in the benzene ring (scheme 4.7).[10]

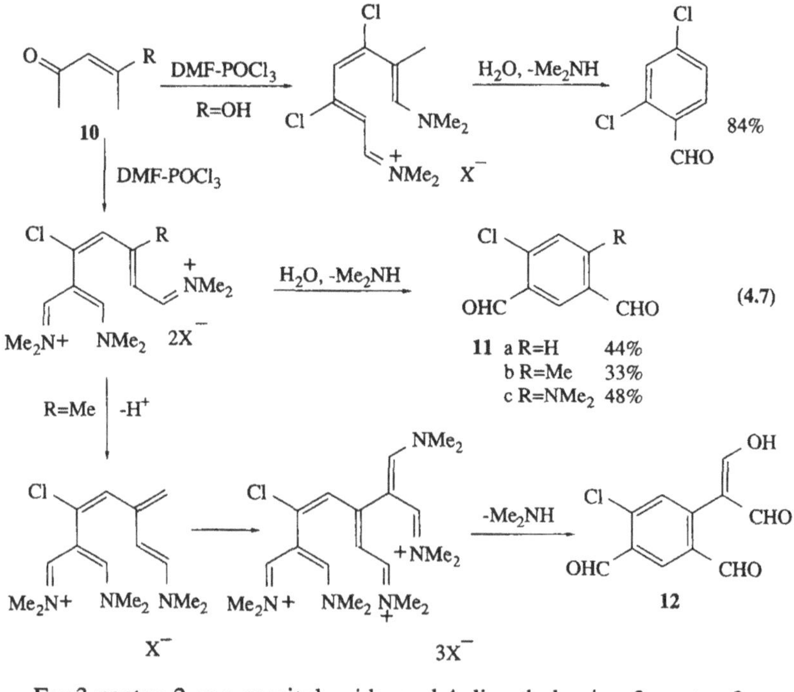

$$(4.7)$$

For 3-penten-2-one, mesityl oxide, and 4-dimethylamino-3-penten-2-one, iminoalkylation proceeds further, so that after ring-closure, elimination and hydrolysis, the corresponding dialdehydes **11a**, **11b**, and **11c** are obtained respectively.[10] However, for the methyl ether of pentane-2,4-dione (**10**, R=OMe), ring closure proceeds after a single iminoalkylation; 3-chloro-anisole is obtained (42%) (scheme 4.8). The dialdehyde is also obtained (23%) in a manner that presumably mirrors the introduction of chlorine into the ketone **10** (R=OH) as well as the further iminoalkylation that also occurs on ketones **10**.

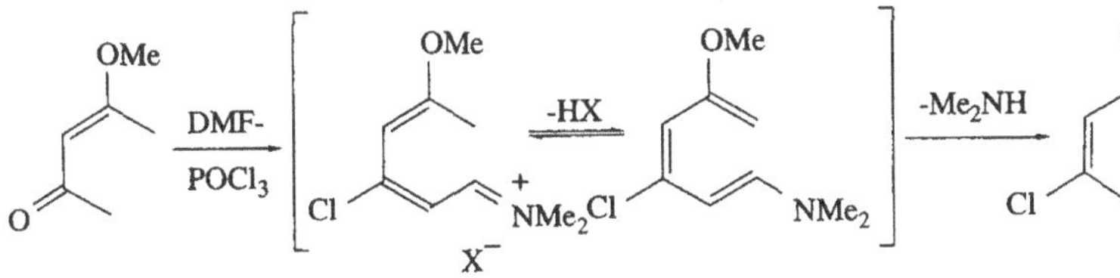

(4.8)

Pentamethinium salts can also react with Vilsmeier reagents to give substituted benzenes (scheme 4.9). The salt evidently undergoes a double iminoalkylation prior to ring-closure. These and other examples imply that the fine balance of both steric and electronic effects dictates the constitution of the products in these ring-closures to arenes. In these ring-closures, from one to five carbon atoms can be derived from the Vilsmeier reagent, of which from zero to three may be formally represented as iminium cations.

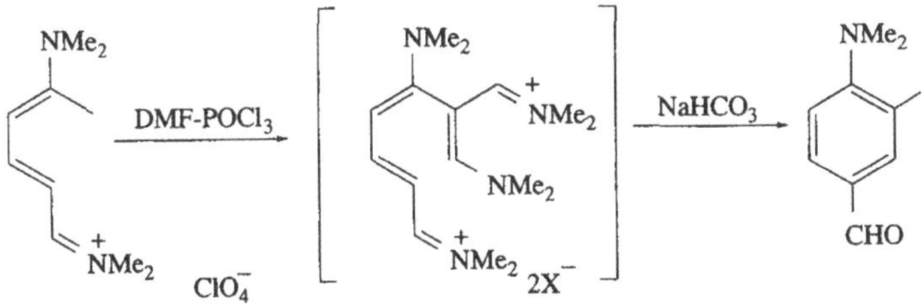

(4.9)

Extensions to the above ring-closures have been reported. Several α,β-unsaturated alkenones have been converted by Vilsmeier reagents into chlorobenzene mono-, di-, and tri- carboxaldehyde. Conversion of 2-hexen-2-one into aldehyde **13** is illustrative (scheme 4.10).[11]

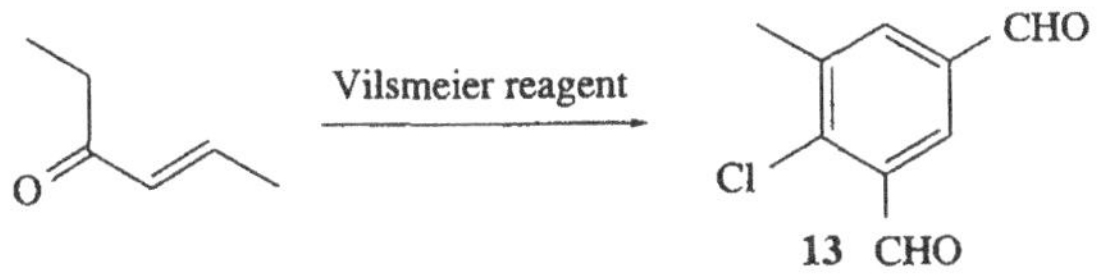

(4.10)

Such reactions are of synthetic value for several reasons; the products may not be readily obtained by other synthetic methods; where more than one product is formed, they are usually readily separable; the procedure is adaptable to large scales, is convenient and also economical.

Holy and Arnold[10] showed that treatment of acetylacetone with DMF-POCl₃ afforded 2,4-dichlorobenzaldehyde (84%; scheme 4.11). The route involves the heptamethinium species **14** which undergoes ring-closure, probably in a pericyclic process, although a closure of the enamine-Exo-6-Exo-Trig type cannot be excluded. Katritzky and Marson[12] showed that the course of the reaction depends on the nature of the dialkylformamide; 4,6-dichloroisophthalaldehyde **16** was the major product when the Vilsmeier complex derived from *N*-formylmorpholine-POCl₃ was used. The relative bulk of the R group when R₂N=morpholino presumably decreases the rate of ring-closure of the cation **14**, so that further iminoalkylation can occur giving the dicationic species **15** which then yields the dialdehyde **16**.

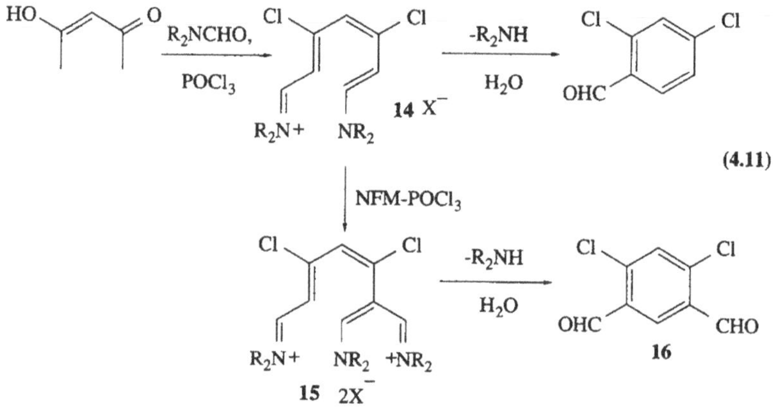

(4.11)

A seminal study[13] of the action of Vilsmeier reagents on acyclic β-diketones showed that dichlorobenzaldehydes (*e.g.* **17**, **18**, and **19**) were formed, although steric hindrance markedly decreased the yields when the pentasubstituted benzene **19** was formed (scheme 4.12). A terminal acetyl group appears to be necessary for cyclization to 2,4-dichlorobenzaldehydes. Replacement of one of the alkyl groups by phenyl necessarily prevented the formation of a substituted benzene; the case of benzoylacetone has been investigated (section 4.3.1).[13]

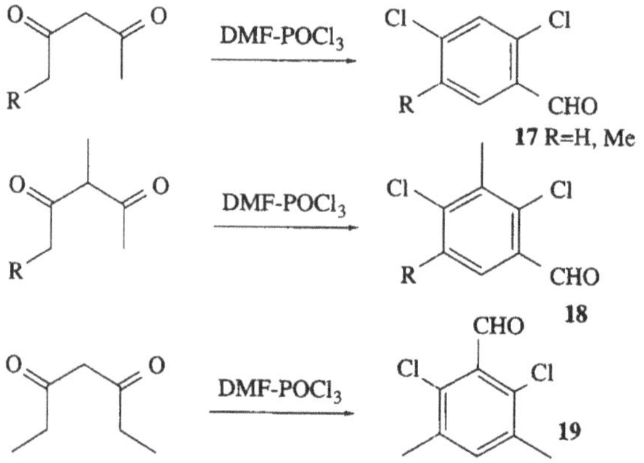

(4.12)

A comparative study of the ligand reactivity at the 3-position of trisacetylacetonates of Cr(III) and Co(III) by means of Vilsmeier-Haack reactions has been made.[15]

Formylation of 2,4-dienoic acids under Vilsmeier conditions gave low yields of isophthalaldehyde derivatives, with poor selectivity. However, 3,5-xylenol was prepared in 45% yield from 3-methylhexadienoic acid.[16]

The vinylindoles 20 have been converted into carbazoles by Vilsmeier reagents at high temperatures. At ambient temperatures, the 3-formyl derivatives are formed in high yields.[17,18]

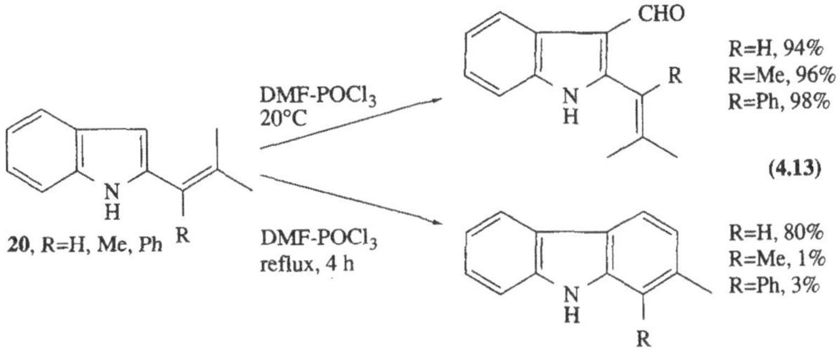

$$\text{(4.13)}$$

4.2.2 Indenes and Benzofulvenes

Whereas 3,4-dimethoxyacetophenone 21a undergoes both mono- and di-formylation,[19] propioveratrone 21b affords the indene 22.[20]

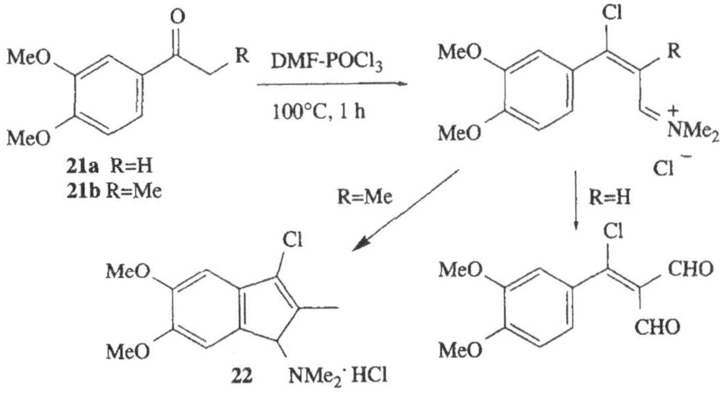

$$\text{(4.14)}$$

Treatment of aryl benzyl ketones 23 with DMF-POCl$_3$ afforded indanes 25 in 20-37% yield. With desoxyveratroin, 23 (X=Y=OMe), the β-chlorovinylaldehyde 24 was formed in 59% yield; at elevated temperatures 23 afforded 25 (X=Y=OMe) in 35% yield. The β-chlorovinylaldehyde 24 was converted into 25 (X=Y=OMe) by reaction with DMF-POCl$_3$ at 90°C in 57% yield.[21]

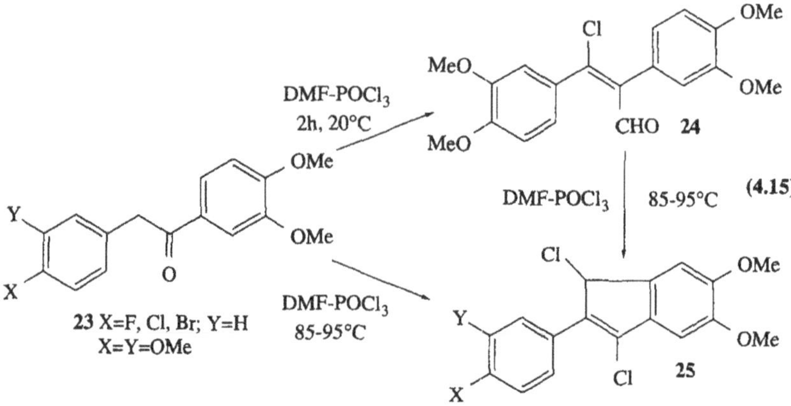

DMF-POCl₃ 2h, 20°C → structure 24

23 X=F, Cl, Br; Y=H
X=Y=OMe

DMF-POCl₃ 85-95°C

DMF-POCl₃   85-95°C   (4.15)

25

A benzofulvene can be prepared in low yield from benzene-1,2-diacetic acid (scheme 4.16).[22]

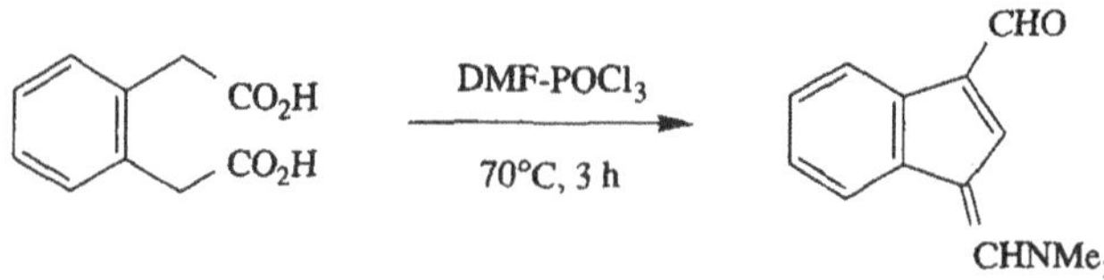

DMF-POCl₃
70°C, 3 h
(4.16)

Indenes are formed in yields of 22-31% by the reaction of DMF-POCl₃ with benzalacetophenones (scheme 4.17).[23]

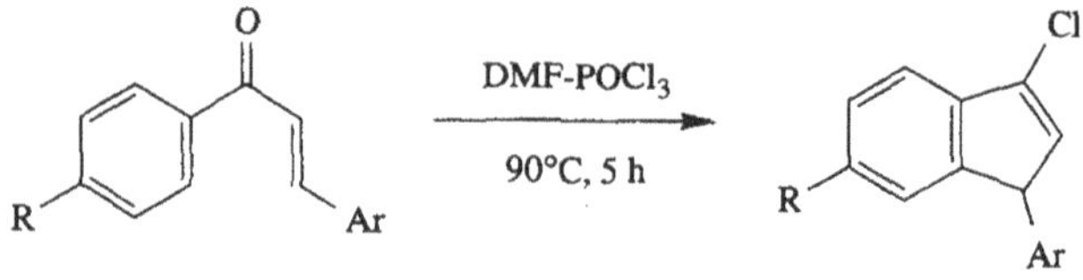

DMF-POCl₃
90°C, 5 h
(4.17)

4.2.3 Dihydronaphthalenes

The ring-closure of certain allylbenzenes to formyldihydronaphthalenes under Vilsmeier conditions has been reported (schemes 4.18 and 4.19). Analysis of products suggests that nuclear formylation is probably the first step in these reactions. This cyclisation fails when the *ortho*-position (R) to the allyl group is not activated; only formylation at the arene ring occurs.[24]

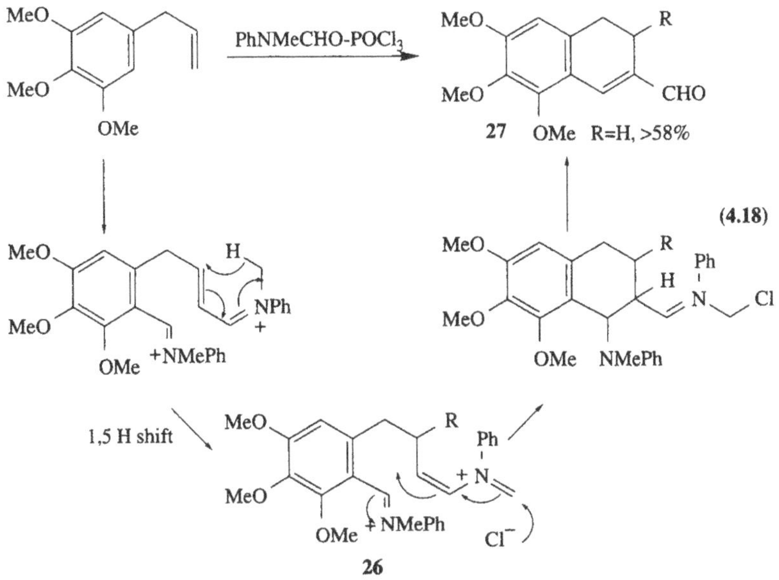

(4.18)

A proposed mechanism involves a [1,5] hydride shift to give the diiminium cation **26.** Such a hydride transfer was confirmed by using PhN(CHO)CD$_3$ which gave only **27** (R=D) as the product. The use of PHN(CDO)CH$_3$ gave only the 1-deuterio-3,4-dihydroaldehyde (CDO). Additionally, no deuterium was incorporated when D$_2$O was used to hydrolyze the reaction.[24a]

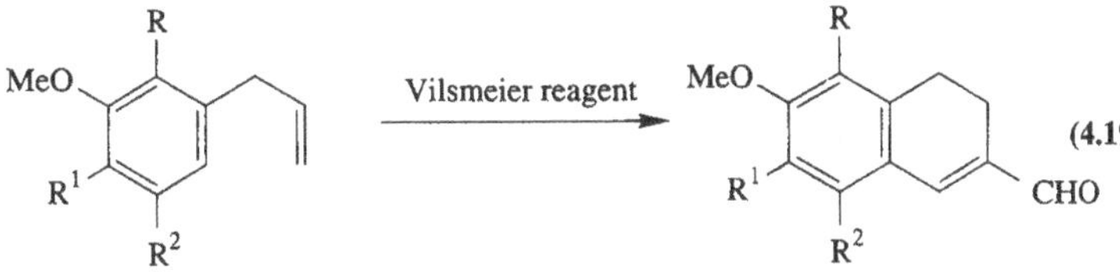

(4.19)

4.2.4. Dithienobenzenes

Dithienobenzene can be isolated by reacting 3,3′-dithienylmethane with a Vilsmeier reagent.[25]

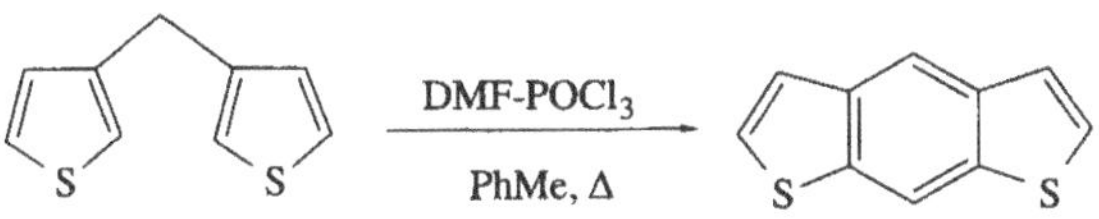

(4.20)

4.3 Synthesis of Non-Aromatic Heterocycles

4.3.1 Pyrans, Pyrones and their Derivatives

A double iminoalkylation of dibenzylketone occurs with Vilsmeier reagents; the isolated 4-pyrone **30** is considered to be formed by a 6π-electro-cyclic ring-closure of the pentadienal **28** and subsequent hydrolysis of the pyrylium salt **29** (scheme 4.21).[26]

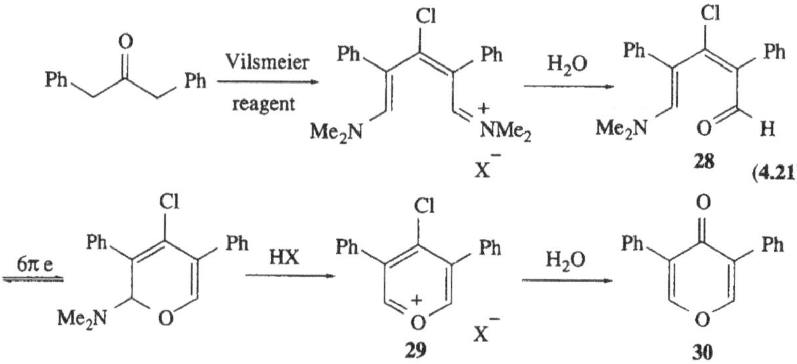

$$(4.21)$$

The pyrone **32** is formed together with the β-chlorovinylketone **33** when benzoylacetone **31** is reacted with Vilsmeier reagents (scheme 4.22).[13] In such reactions, the absence of a two enolizable carbonyl groups that can give rise to iminoalkylation at both the C-1 and C-5 positions means that the alternative ring-formation leading to a substituted benzene (section 4.2.1) cannot occur.

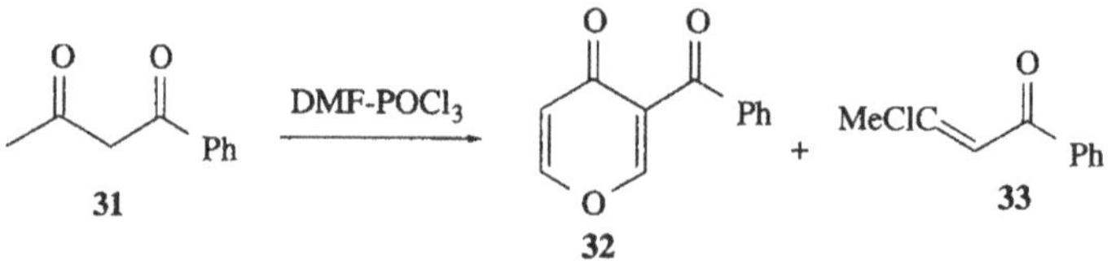

$$(4.22)$$

A cross-conjugated and fused pyran system can be isolated from the reaction of isophorone with a large excess of DMF-POCl$_3$ (scheme 2.137, section 2.11.3).[27,28]

The pyran-2,4-dione **34**, a useful intermediate in the preparation of dyes and pharmacologically active compounds, was prepared by treating triacetic acid lactone with DMF and POCl$_3$ under cooling.[29]

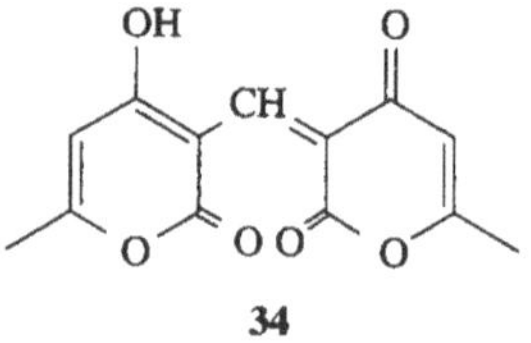

$$(4.23)$$

β-Chloro-α-methylcinnamaldehyde gives the pyran-2-one **35** with malonic acid in the presence of one molar equivalent of pyridine.[30]

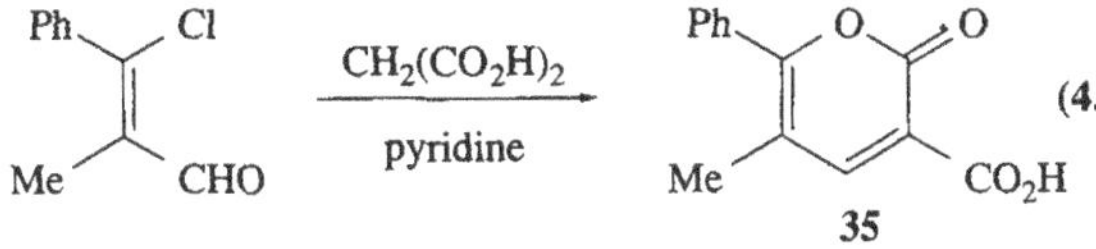

(4.24)

Condensation of β-chlorovinylaldehydes with *o*-hydroxyacetophenone under basic conditions affords benzopyrans. α-Substituted chloroaldehydes give 2-formylmethylene-2*H*-benzopyrans **36**; using 1-chlorocyclohexene-2-carboxaldehyde, the tricyclic pyran **37** is isolated (scheme 4.25).[31]

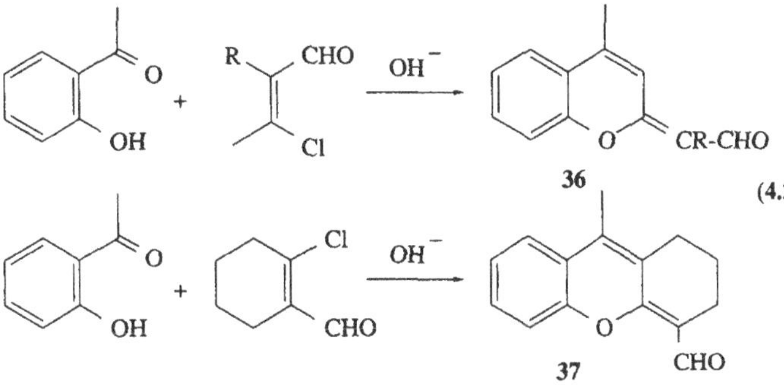

(4.25)

The reaction of DMF-POCl₃ with 1,5-diketones **38** to give the 4*H*-pyrans **39** probably proceeds by conversion of one carbonyl group into the corresponding chloromethyleneiminium salt, attack of the other carbonyl group, resulting in ring-closure, then attack at the 5-position of the *O*-heterocycle by another molecule of the Vilsmeier reagent.[32]

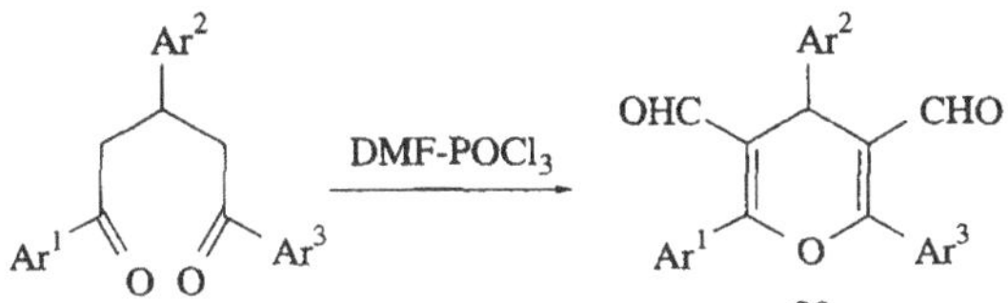

(4.26)

4*H*-Pyrans can be formylated under Vilsmeier conditions (see also section 4.4.2.1). Thus, 2,4,6-triphenyl-4*H*-pyran and DMF-POCl₃ gave the corresponding 3-carboxaldehyde.[33]

4.3.2 Thiopyrans, Thiopyranones and their Derivatives

Compounds containing an activated methylene group, including malono-nitrile, a phenacyl cyanamide, cyanamide, or a cyanocarboxylic acid, react with β-chlorovinylaldehydes to give the condensation products **40**. Subsequent reaction with aqueous sodium sulfide affords thiapyran-2-ones **41** (scheme 4.27).[34]

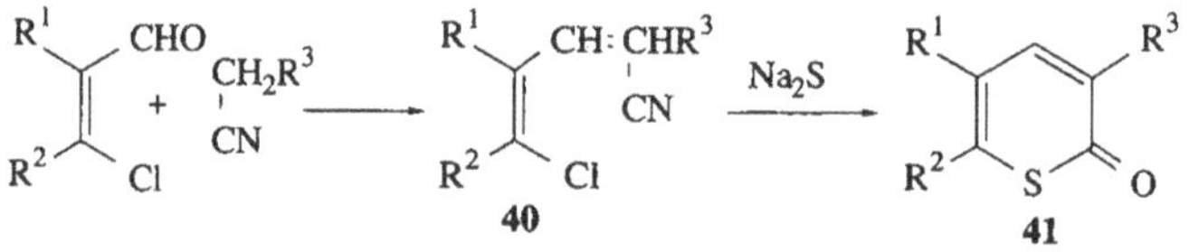

$$(4.27)$$

Weissenfels[35,36] and others,[37] found that the stoichiometry of reagents also influenced the course of the reaction. If a 1:2 molar ratio of sodium sulfide to β-chlorovinylaldehyde is used, then a bis-(β-formylvinyl) sulphide **42** is formed. Owing to the activated nature of the β-methyl group, base-catalyzed condensation is possible to give the 2-formylmethylene-2H-thiopyran **43** (scheme 4.28). Certain steroidal systems have been shown to undergo the same transformation.[37]

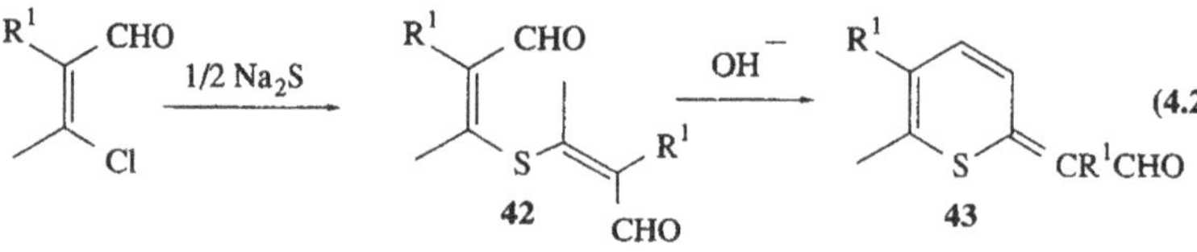

$$(4.28)$$

The reaction of substitued cyclopentadienes with DMF-POCl$_3$ was shown by Hafner and co-workers to lead to cross-conjugated iminium salts of type **44**.[38] This reaction has been used to prepare cyclopenta[c]thiopyrans **45**, albeit in poor yields.[39]

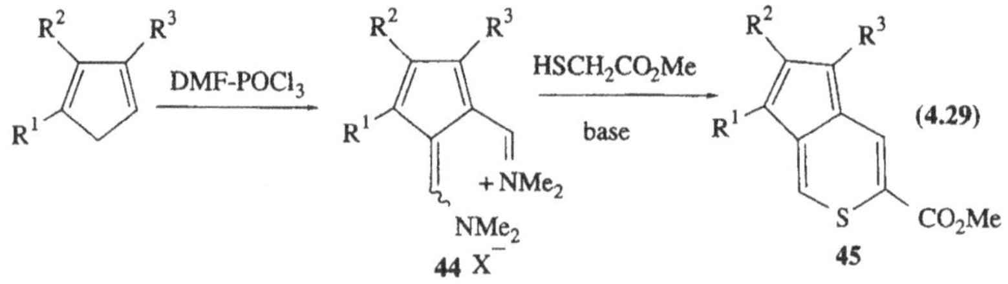

$$(4.29)$$

4.3.3 Quinolinones and Isoquinolinones

Fused pyrid-2-ones can be formed in low yields by the cyclization of enamides (scheme 4.30).[40]

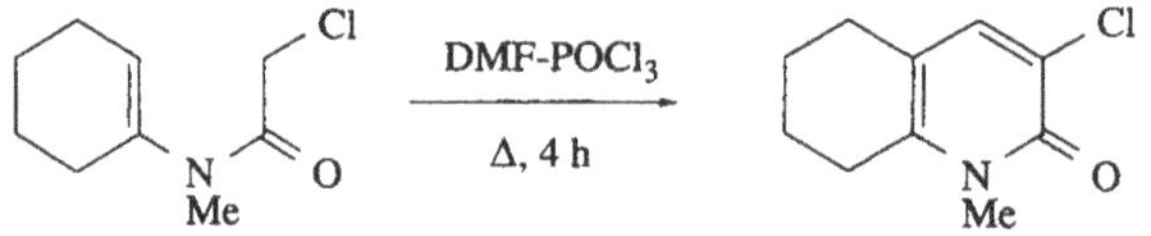

$$(4.30)$$

Although quinolines are the usual products of the action of Vilsmeier reagents on acrylamides, in cases where cyclization is slow carbonyl compounds such as **46** have been isolated (scheme 4.31).[41]

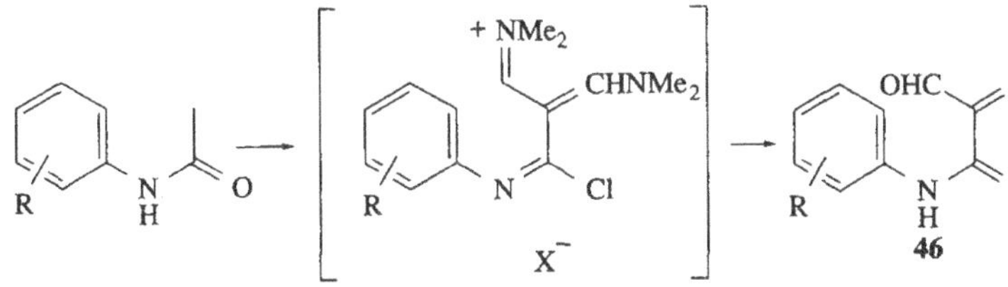

(4.31)

Carbostyrils **47** generally result from the action of the Vilsmeier reagent on α-substituted acylanilides (scheme 4.32). Electron-donating groups placed *meta* or *para* assist the cyclization whereas the same groups placed *ortho* hinder the reaction. A *p*-chloro group deactivates the ring sufficiently to prevent cyclization; the acetanilide **48** is converted by the Vilsmeier reagent into the anilide **49**.[40]

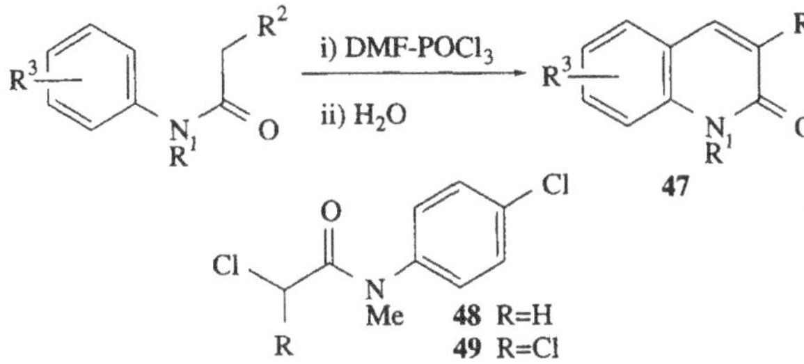

(4.32)

The formylation of *N*-phenylacetanilides **50** was initially misinterpreted,[42] but has been shown to provide an excellent route to 1-phenyl-2-quinolones such as **51**.[43,44]

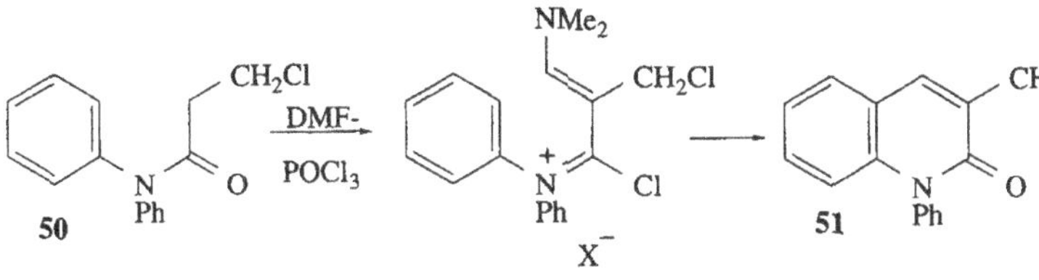

(4.33)

An unusual example of a ring-forming reaction involving iminoalkylation of an active methylene site is the formation of the isoquinolinone **52** by the Vilsmeier reaction of 3,5-dinitro-*o*-toluic acid (scheme 4.34).

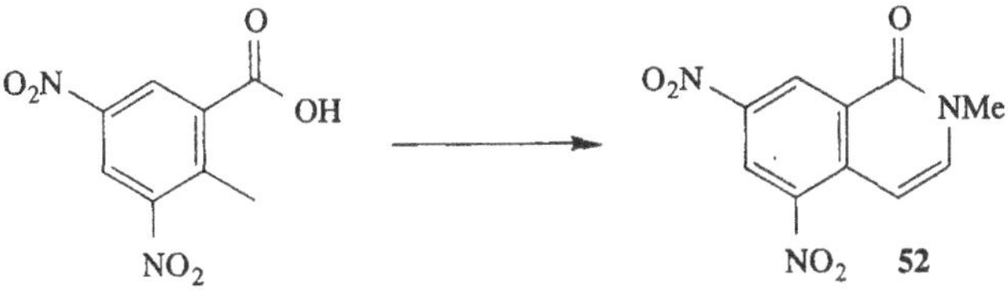

(4.34)

The carboxyisoquinolinone **54** is formed when the isochroman-1,3-dione **53** undergoes ring-opening and ring-closure with acid or POCl₃; compound

53 can itself be formed by reacting homophthalic acid with DMF-POCl$_3$ (section 4.3.5).[45]

$$(4.35)$$

53   **54**

4.3.4 Quinolizinones

Although 2-picoline (**55**, R=H) does not give an isolable product on reaction with Vilsmeier reagents,[46] pyridyl-2-ethyl acetate **55a** and pyridine-2-acetonitrile **55b** react to give the expected dimethylaminovinyl compounds **56a** and **56b**, respectively. Those react with ketene to give quinolizinones **57** and **58**.[47]

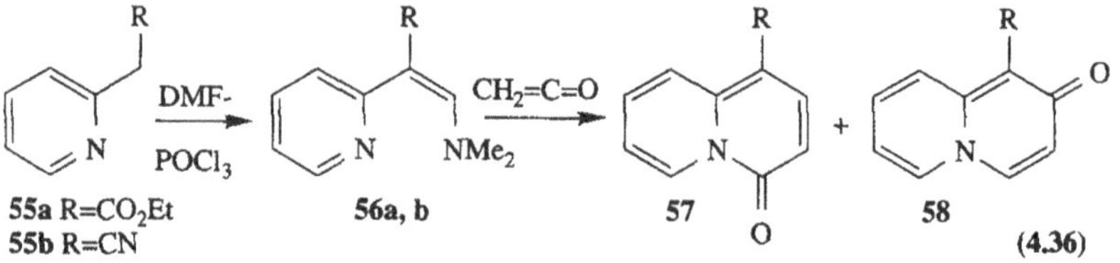

55a R=CO$_2$Et   **56a, b**   **57**   **58**
55b R=CN $$(4.36)$$

4.3.5 Pyrrolines

β-Chlorovinylaldehydes having a α–2-chloroethyl group, *e.g.* **59**, undergo condensation with primary arylamines to give substituted pyrrolines such as **60** and **61** (scheme 4.37).[48-50]

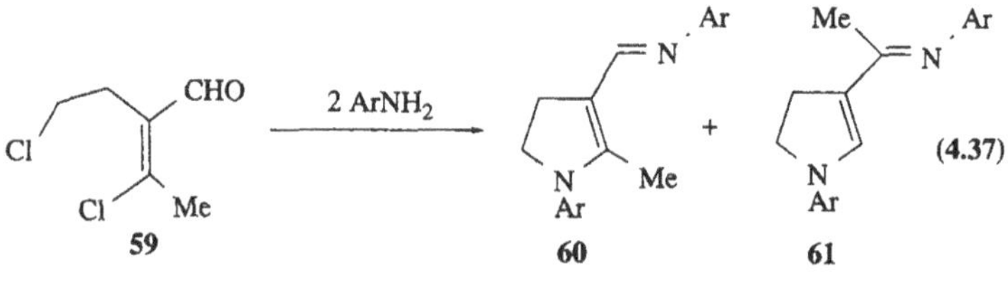

59   **60**   **61** $$(4.37)$$

4.3.6 Miscellaneous Ring Systems

The heterocycle **62** is formed by the action of NFM-POCl$_3$ on hippuric acid, in a process related to the Erlenmeyer synthesis.[51]

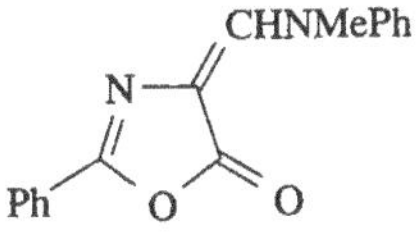

(4.38)

Benzimidazole-2-propanoic acid **63** is converted into the enamino-ketone **64** by the Vilsmeier reagent at room temperature.[52]

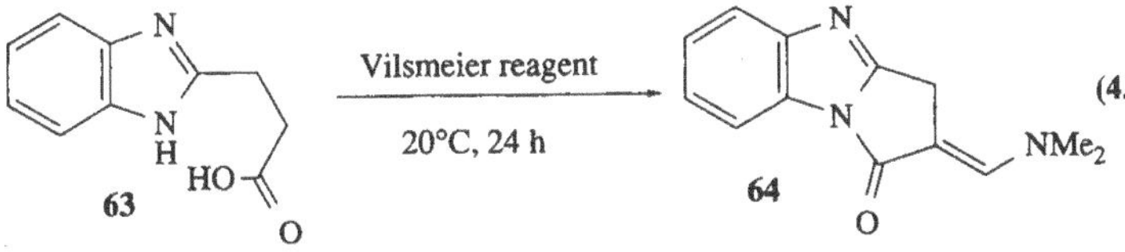

(4.39)

Homophthalic acid **65a**, its methyl ester **65b**, and homophthalic anhydride **66** all reacted with the Vilsmeier reagent to give the isochroman-1,3-dione **67** in excellent yield. Upon treatment of the Vilsmeier product **67** with HCl, isocoumarin-4-carboxylic acid **68** is obtained.[53]

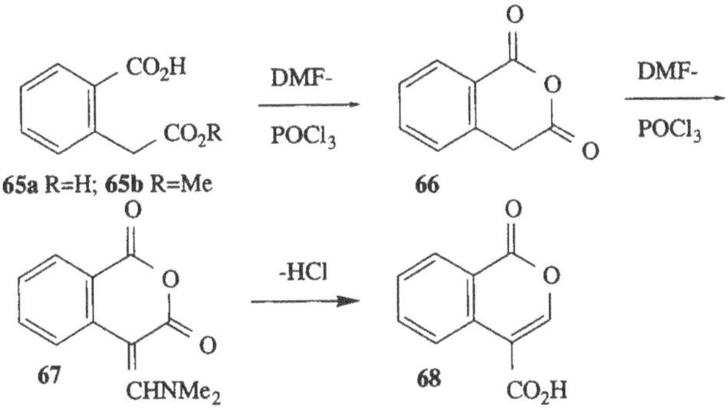

(4.40)

The inertness of an ester group towards Vilsmeier reagents is crucial to a two-step route to diazepinoindoles (scheme 4.41).[54]

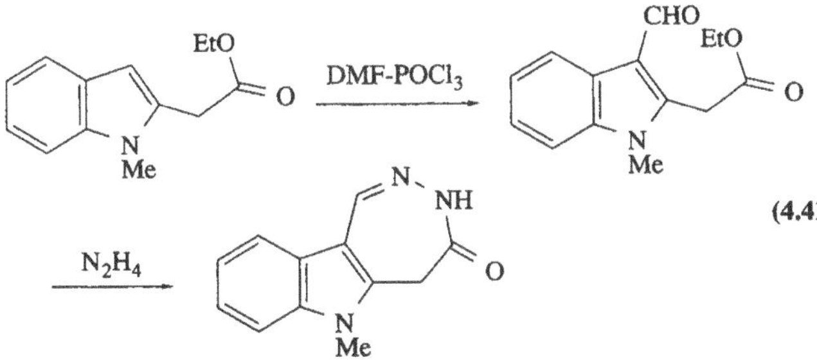

(4.41)

Iminium salt chemistry does not always predominate with (hetero) arylamides. Thus, part of the 'halogenating' agent can become incorporated as a new ring.[55-58]

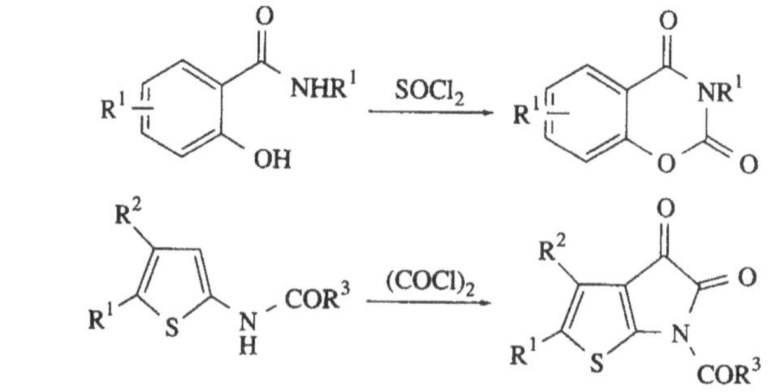

(4.42)

2-Acetylbenzamides react with DMF-COCl$_2$ *via* their cyclic tautomers that undergo dehydration and enaminic formylation to give excellent yields of the amidoaldehydes **69**.[59]

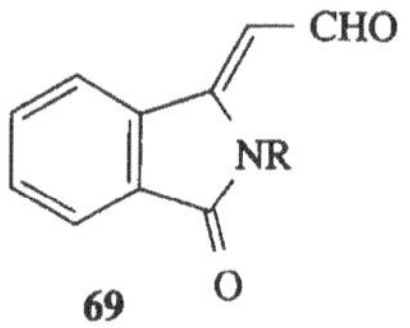

(4.43)

At 30°C, DMF-SOCl$_2$ acts upon the aminopyrrole **70** to give the amidine **72**, in which amide-to-nitrile dehydration occurred, but at higher temperatures iminoalkylation affords the fused pyrrole **71**.[60]

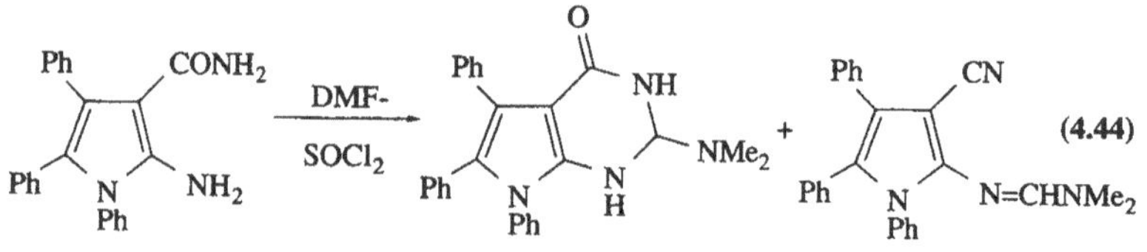

(4.44)

Two oxazole rings have been introduced into a *N*-aryl-3,5-dimethyl-1,1-dioxo-1,2-thiazine by diaminoalkylation of the methyl groups followed by reaction with excess hydroxylamine to give the bisoxazole.[61]

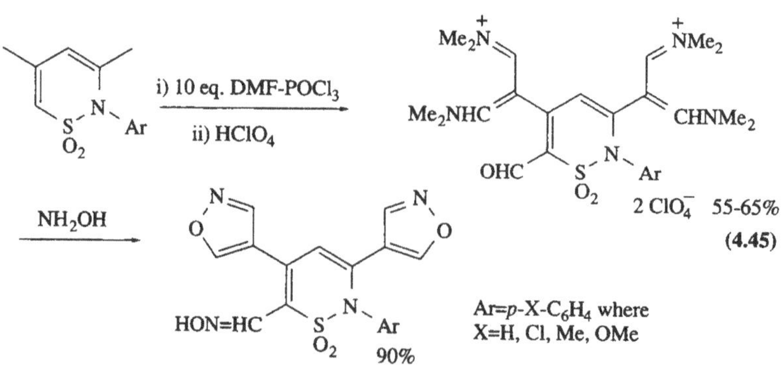

(4.45)

Ar=*p*-X-C$_6$H$_4$ where
X=H, Cl, Me, OMe

Oxadiazinones are formed by the reaction of acylguanidines with phosgene.[62]

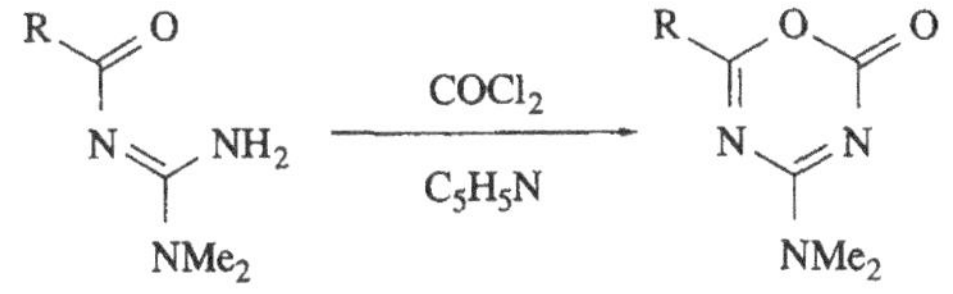

$$(4.46)$$

4.4 Synthesis of Heteroaromatic Compounds

4.4.1 Compounds Containing a Five-Membered Ring

4.4.1.1 Furans and Benzofurans

Highly substituted furans have been synthesized from aryl-1,2-diketones.[63]

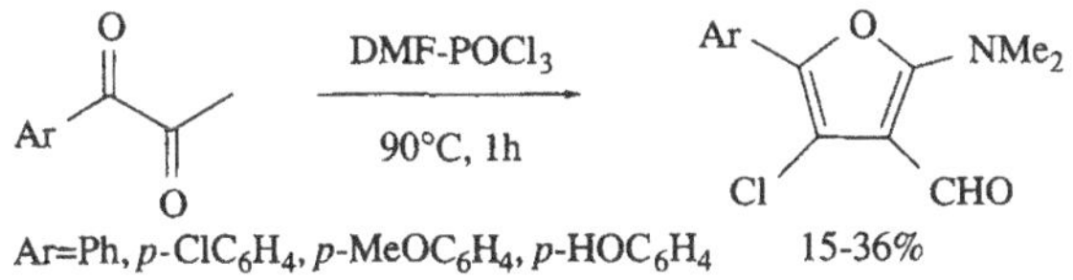

$$(4.47)$$

Ar=Ph, p-ClC$_6$H$_4$, p-MeOC$_6$H$_4$, p-HOC$_6$H$_4$ 15-36%

The reaction of Vilsmeier reagents with aryloxyacetophenones such as **74** provides a versatile synthesis of benzo[b]furans **73**.[64]

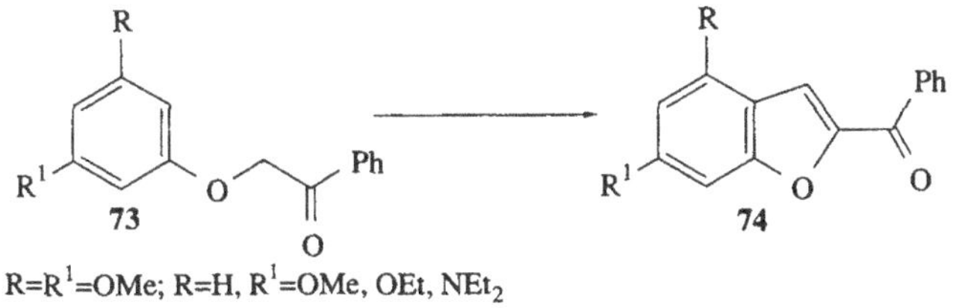

$$(4.48)$$

R=R^1=OMe; R=H, R^1=OMe, OEt, NEt$_2$

Several benzo[b]furans have been prepared from phenoxyacetonitriles. The selectivity of the reaction between cyclization and formylation is poor, but the yield of cyclized product can be improved by placing activating groups on the phenyl ring and altering the stoichiometry of the reagents (scheme 4.49).[65] It is interesting to note that phenylacetonitriles afford isoquinolines under the same reaction conditions (section 4.4.2.8).

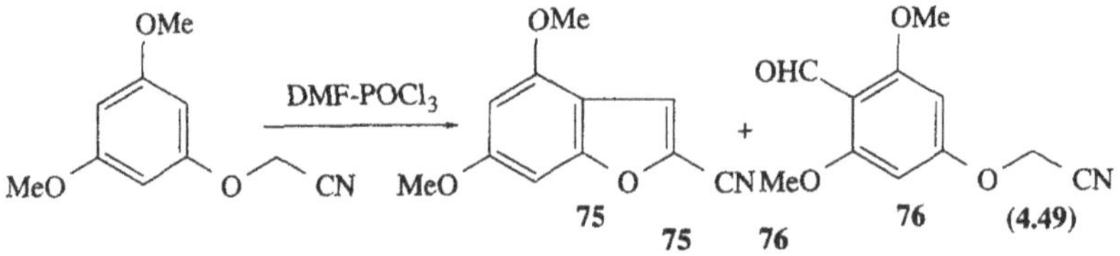

	75	76
1 eq DMF, 2 eq. POCl₃	52%	13%
1.2 eq DMF, 3.6 eq. POCl₃	61%	8.5%

Replacement of the cyano group by a diethylacetal function enabled the synthesis of benzofuran-2-carboxaldehydes in 15 to 58% yield.[66]

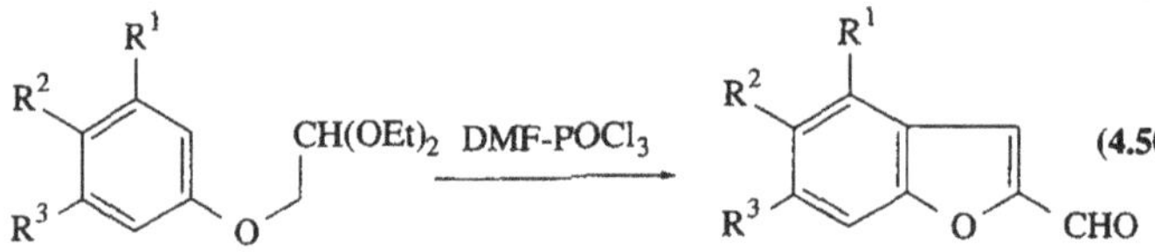

The benzo[*b*]furans **78** (R=Ph, 93%) and **78** (R=2-thienyl, 27%) result from a Fischer-type cyclization with formylation of the ketoxime ethers **77**.[67]

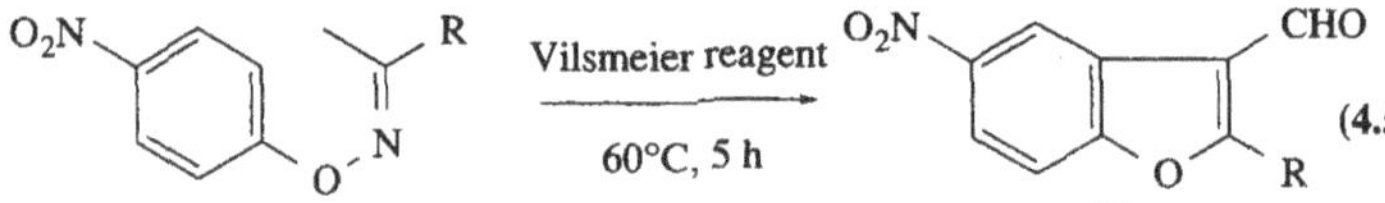

4.4.1.2 Pyrroles

A general route to pyrroles involves the reaction of β-chlorovinyl-aldehydes with amines. Hauptmann so synthesized a number of pyrrole derivatives **79**.[68-70]

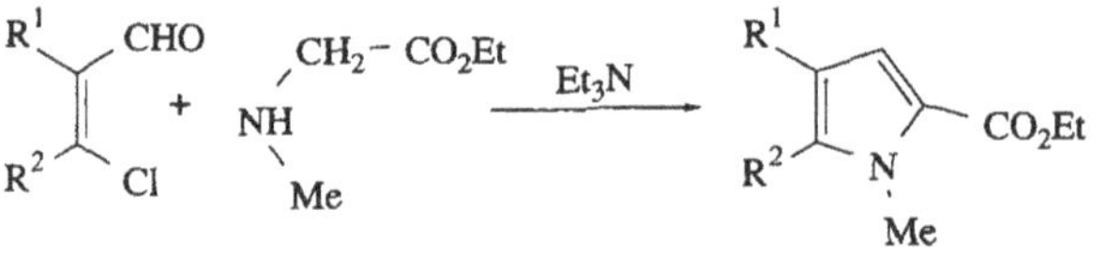

Arnold and Holy[71] showed that a 2-phenylvinamidinium salt undergoes cyclization in DMF in the presence of NaH to give *N*-methyl-3-phenyl-pyrrole. The reaction has now been extended by using the esters of α-amino acids to give a variety of 1,2,4-trisubstituted pyrroles (scheme 4.53); an azomethine ylid, as proposed by Arnold and Holy, is considered to be involved.[72]

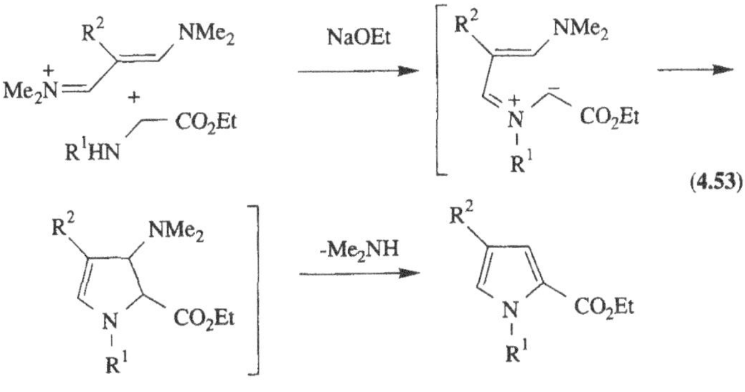

(4.53)

A number of pyrroloisoquinolines **83** has been synthesized under Vilsmeier conditions from 1-methyl-3,4-dihydroisoquinolines **80**. The aldehyde **81a** and malonaldehyde **82a** were also obtained. The yield of **83a** was increased to 45% by using 5 eq DMF-POCl$_3$ and heating the reaction mixture at 80-90°C. The structure of **84a** was confirmed by X-ray crystallography.[73]

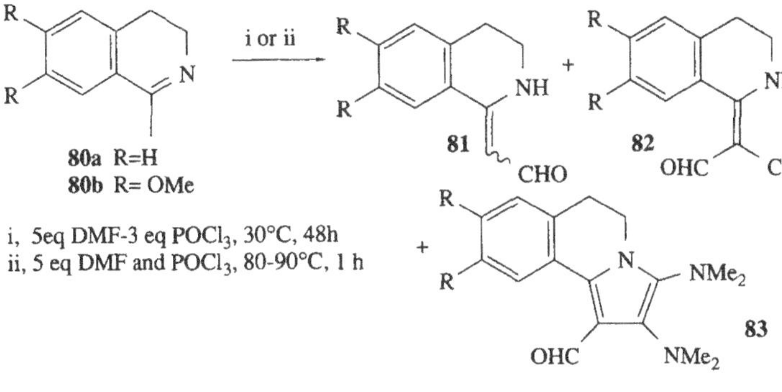

80a R=H
80b R= OMe

i, 5eq DMF-3 eq POCl$_3$, 30°C, 48h
ii, 5 eq DMF and POCl$_3$, 80-90°C, 1 h

(4.54)

Benzothiazolium salts have been converted into pyrrolobenzothiazoles by treatment with Vilsmeier reagents (scheme 4.55).[74]

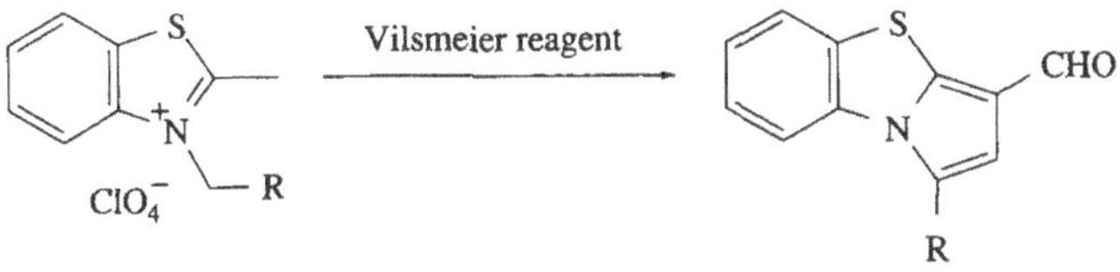

(4.55)

4.4.1.3 Indoles

A Fischer-indole type ring-closure with formylation has been reported, although the actual structure is likely to be 1,2-diphenylindole-3-carbox-aldehyde, and not the 4-formyl derivative as claimed.[75]

4.4.1.4 Thiophenes and Benzothiophenes

A general route to substituted thiophenes involves the reaction of thioglycolates with β-chlorovinylaldehydes.[69,76] Fiesselmann,[76] and Hauptmann and co-workers[69,77,78] synthesized a series of 1,2,4-trisubstituted thiophenes **84** by this method (scheme 4.57). Decarboxylation at the 2-position occurs on heating the thiophene. The principle was extended by Kvitko and co-workers,[79,80] who annealed a thiophene ring to an existing heterocyclic system (scheme 4.56).

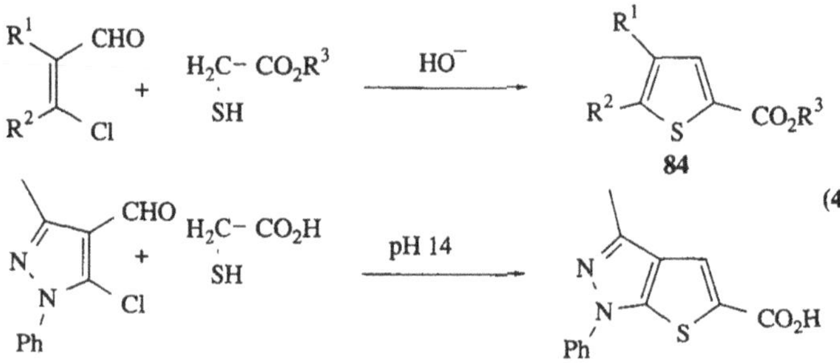

$$(4.56)$$

The synthesis of a substituted thiophene from a β-chlorovinylaldehyde and phenacylmercaptan under basic conditions, followed by acid-catalyzed ring closure has been reported.[81]

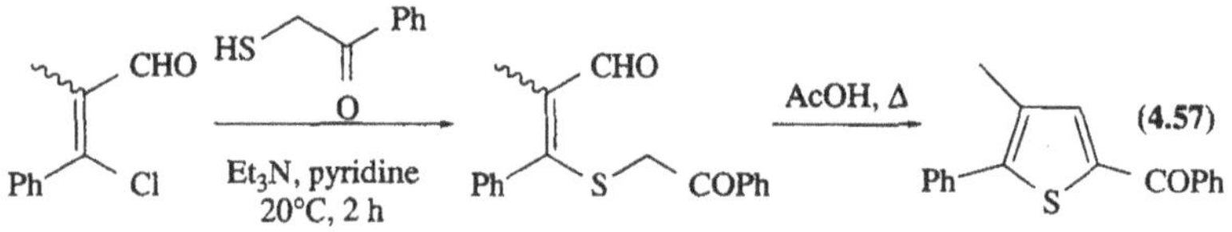

$$(4.57)$$

Caignant[82,83] prepared a series of substituted thiophenes by the action of sodium sulfide and a suitable halogenated precursor, *e.g.* ethyl bromoacetate, on a β-chlorovinylaldehyde (scheme 4.58).

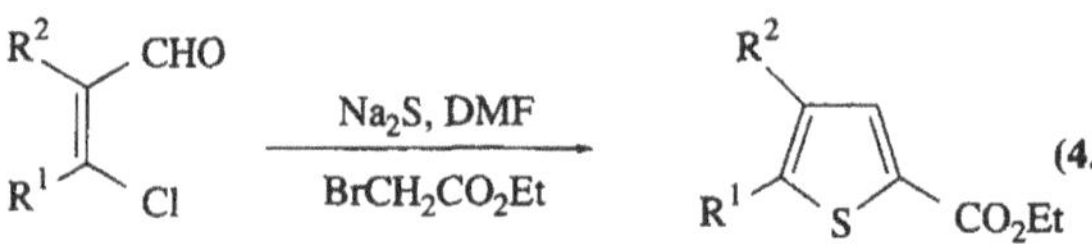

$$(4.58)$$

A series of selenophene analogues was prepared by the same authors, using sodium selenide in place of sodium sufide.[84] Tellurophene derivatives were prepared in a similiar fashion.[85]

A variety of benzo[*b*]thiophenes can be formed from activated arenes (scheme 4.59); formylation at either the 5- or the 7-position is observed in various cases. R^1 must be an electron-withdrawing group such as cyano, ester or keto.[86]

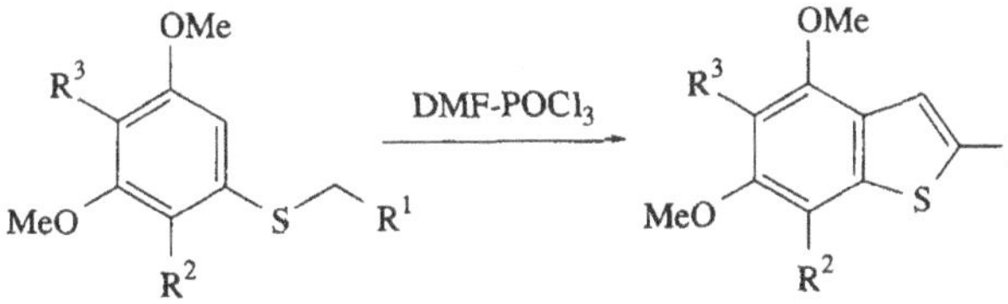

$$R^3 \quad OMe \qquad \xrightarrow{\text{DMF-POCl}_3} \qquad R^3 \quad OMe \qquad R^1 \quad (4.59)$$

MeO — with S—R^1 substituent and R^2 → benzothiophene product with MeO, R^2, S, R^1

4.4.1.5 Oxazoles and Benzoxzoles

The amide **85** reacts with POCl$_3$ to give oxazole **86** *via* a presumed iminium species.[87]

$$\text{Me} \quad \cdots \quad \text{N} \qquad \xrightarrow{\text{POCl}_3} \qquad \text{Me} \quad \cdots \quad \text{N} \qquad (4.60)$$

85   Ph   **86**

o-Aminophenol reacts with DMF-POCl$_3$ to give benzoxazole; amidinium salt intermediates have been proposed (scheme 4.61).[88]

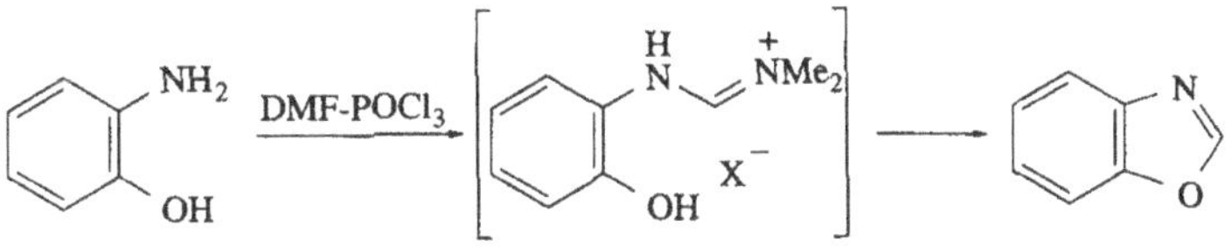

$$\text{[NH}_2, \text{OH ring]} \xrightarrow{\text{DMF-POCl}_3} \left[\text{amidinium salt intermediate } X^- \right] \longrightarrow \text{[benzoxazole]} \quad (4.61)$$

2-Substituted benzoxazoles, very useful for subsequent annelation, are formed by the reaction of Vilsmeier reagents with either 2-hydroxy-acetophenone oximes or 2-hydroxyacetanilides.[89]

4.4.1.6 Isoxazoles

The reaction of β-chlorovinylaldehydes with hydroxylamine hydro-chloride gives the corresponding oximes **87**,[37,90,91] which may be readily cyclized to give isoxazoles **89**,[91,92,93] or dehydrated to give β-chloroacrylo-nitriles **88**.[94]

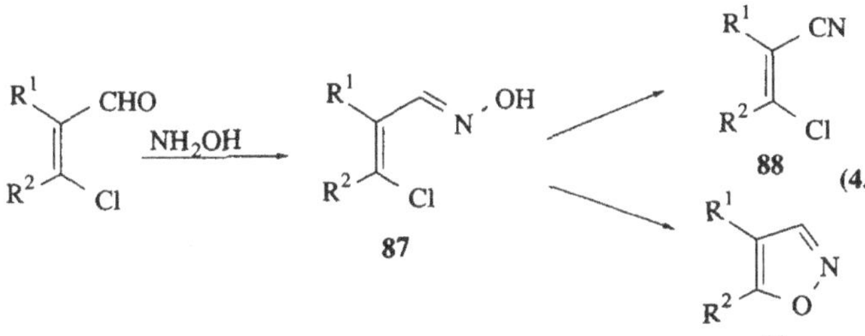

$$R^1 \quad CHO \qquad \xrightarrow{\text{NH}_2\text{OH}} \qquad R^1 \quad N\text{-OH} \qquad \nearrow \quad \begin{array}{c} R^1 \quad CN \\ R^2 \quad Cl \\ \textbf{88} \end{array} \quad (4.62)$$

$$R^2 \quad Cl \qquad \qquad R^2 \quad Cl \qquad \searrow \quad \begin{array}{c} R^1 \\ R^2 \quad O\text{-}N \\ \textbf{89} \end{array}$$

87

Iminium salts **90** react with hydroxylamine to give the isoxazoles **91**.[95]

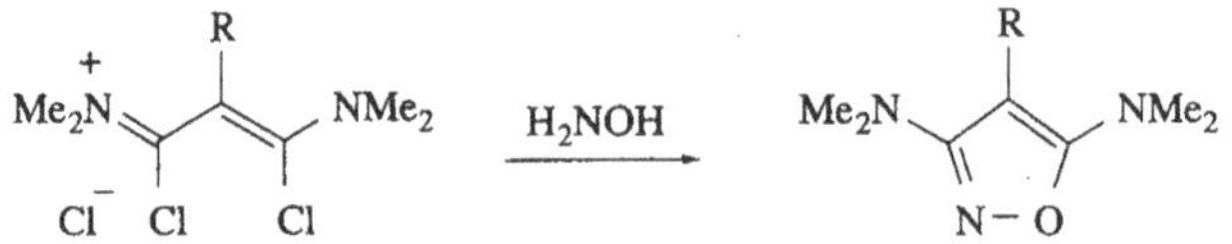

(4.63)

The 3-acylisoxazole-4-aldiminium salts **93** can be obtained from the oxime ethers **92**, but attempts to isolate the corresponding isoxazole-4-aldehydes led to cinnamaldehydes in good yields.[67]

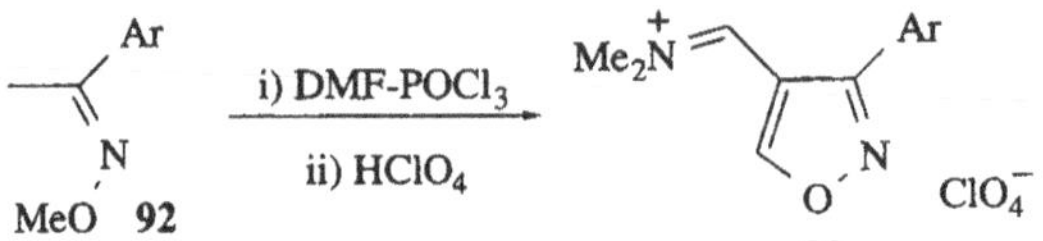

(4.64)

4.4.1.7 Pyrazoles

The reaction of β-chlorovinylaldehydes with hydrazine or substituted hydrazines gives the expected hydrazones[96] which can then undergo base-catalyzed ring-closure to give pyrazoles **94** (scheme 4.65).[92a,97,98] This provides a useful and general route to pyrazoles. Several fused heterocyclic systems, *e.g.* pyrazole3,4[*b*]pyridines[99] and 1-(*p*-chlorophenyl)-pyrazolo-[4,3-*b*]benzothiazines[100] were so synthesized.

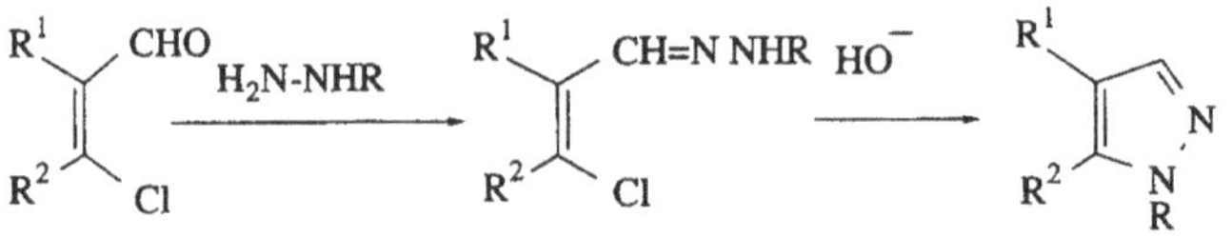

(4.65)

The action of DMF-POCl$_3$ on the phenylhydrazones and semicarbazones of 3-acetylcoumarin derivatives led to a simple synthesis of 3-(4-formylpyrazo-3-acetyl) coumarins.[101]

In a manner somewhat resembling the *O*- and *C*-alkylation of ketones by Vilsmeier reagents, hydrazines of the type **95** react at both carbon and nitrogen termini to give the iminium salts **96** which cyclize under the reaction conditions (70-80°C) to give the salts **97**; the latter are hydrolyzed by alkali to give the pyrazoles **98** in yields that range from 72-96%.[102-104] Several examples of R^1=H and Ph are given.

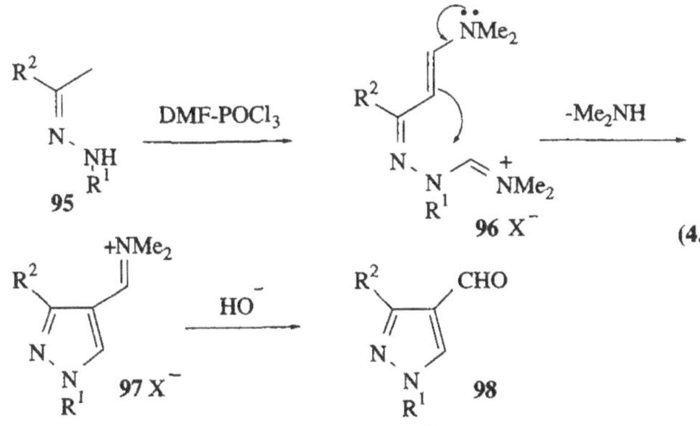

$$(4.66)$$

Hydrazines **99** afforded pyrazoles with Vilsmeier reagents.[105]

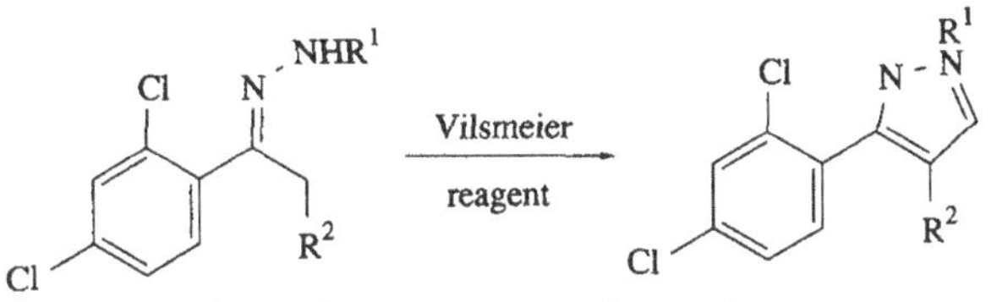

$$(4.67)$$

99 R^1=Ph, R^2=Me, Et, Pr, Bu; R^1=Me, R^2=CONH$_2$

Awad reported the synthesis of pyrazolo[3,4-*c*]pyrazoles from pyrazoline-5-ones and pyrazoline-5-thiones in good yields (scheme 4.68).[106]

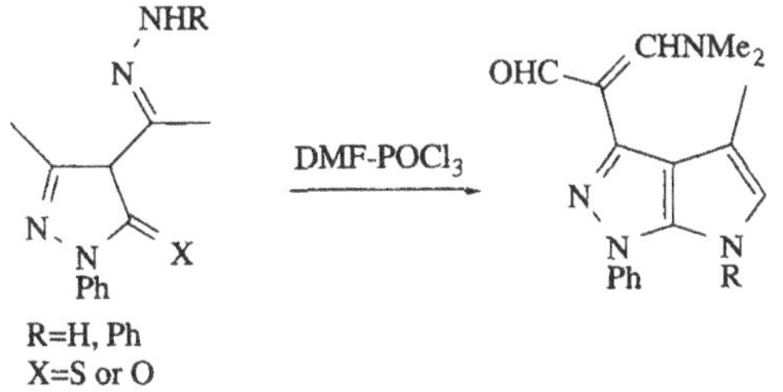

$$(4.68)$$

R=H, Ph
X=S or O

The formation of pyrazoles from β-chlorovinylaldehydes and hydrazine can provide a useful confirmation of structure. Thus, 19-nortestosterone acetate **100** reacted with DMF-POCl$_3$ to give the compounds **101a**, **101b**, and **101c**. Aldehyde **91c** was indentified by its reaction with hydrazine to form a pyrazole ring **102**.[107]

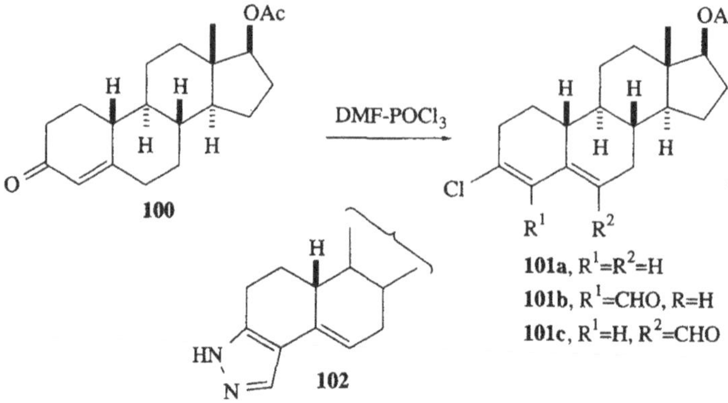

100

102

101a, R1=R2=H
101b, R1=CHO, R=H
101c, R1=H, R2=CHO

(4.69)

Similarly, the steroidal 4,6-dien-3-ones **103** reacted with DMF-$POCl_3$ to give compounds **104a**, **104b**, and **105**.[108,109] The constitution of **104a** was confirmed by its reaction to give the fused pyrazole **106**.[108]

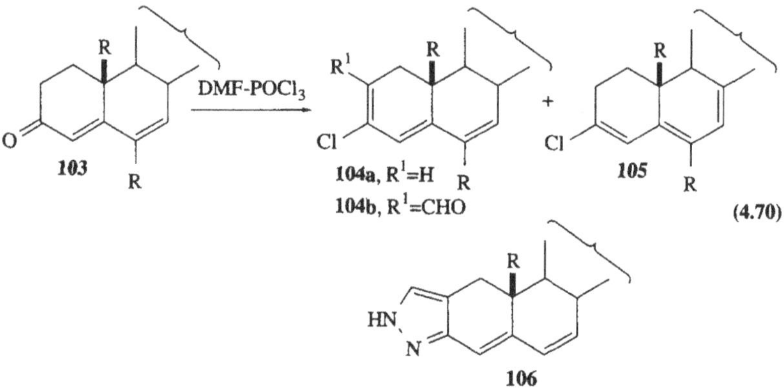

103

104a, R1=H
104b, R1=CHO

105

106

(4.70)

4.4.1.8 Benzimidazoles

Vilsmeier reagents convert *N*-carboethoxy-*o*-phenylenediamine into benzimidazole,[110] and *N,N′*-diacetyl-*o*-phenylenediamine into 1-(1-chlorovinyl)-2-methylbenzimidazole.[41] 1-Arylbenzimidazoles have been generated by the action of Vilsmeier reagents upon dibenzodiazepines.[111]

4.4.1.9 Isothiazoles

1,2-Thiazoles (isothiazoles) have been synthesized by the reaction of β-chlorovinylaldehydes with ammonium thiocyanate at temperatures higher than 40°C; below this temperature only the displacement of chloride by thiocyanate occurs.[112,113] However, heating the β-thiocyanovinylaldehyde with ammonium thiocyanate also affords the 1,2-thiazole.

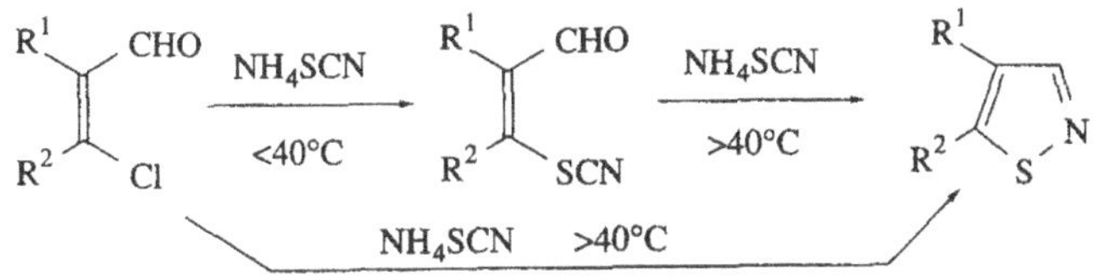

$$(4.71)$$

4.4.1.10 Pyrrolopyrazines, Pyrrolopyridines, and Pyrroloquinolines

Methyl groups that are activated by being placed *alpha* or *gamma* to an annular nitrogen atom in a heteroaromatic system are readily diformylated. Thus, an amino group placed *ortho* to a methyl group is converted into a fused pyrrole.[114,115] The pyrazines **107** react with DMF-POCl$_3$ to give the pyrrolo[2,3-*b*]pyrazine-3-carboxaldehyde **111**. One explanation advanced assumes a dialkylation to give **112** which could then undergo ring-closure to

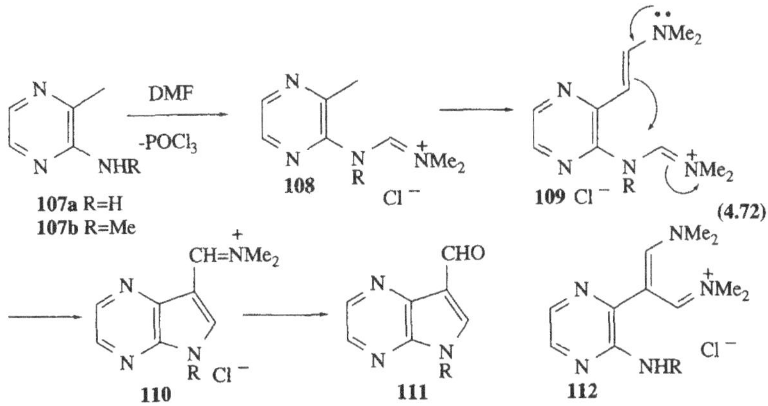

$$(4.72)$$

110. However, the isolation of the amidinium salt **108** under reaction temperatures below 50°C suggests that the pathway involves alkylation of **108** to give **109** which then ring closes to give **110**.[114]

The above formation of a fused pyrrole ring has been extended to other heterocycles that have an active methyl group and an adjacent amino group. For example, pyrrolo[2,3-*c*]pyridine-3-carboxaldehyde and pyrrolo[3,2-*b*] quinoline-3-carboxaldehyde have been so prepared (scheme 4.73).[115]

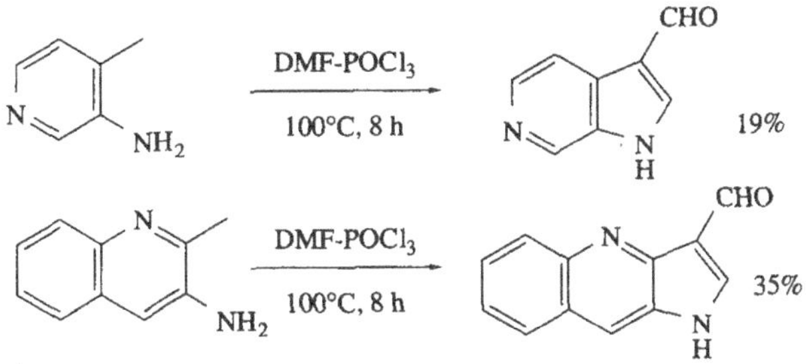

$$(4.73)$$

4.4.1.11 Miscellaneous Systems

1,3,4-Thiadiazoles can be prepared by reaction of *N,N*-diformylhydrazine with a Vilsmeier reagent to give the iminium salt **113** which undergoes ring-closure with hydrogen sulfide in a basic medium.[116]

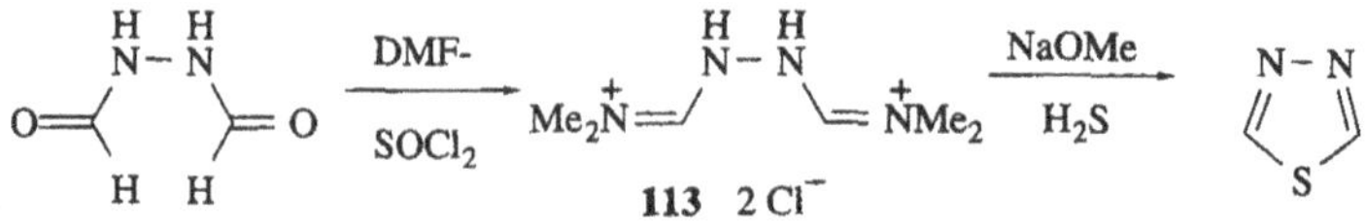

(4.74)

Thiazolopyrimidines have been formed by the treatment of amino-uracils with DMF-SOCl$_2$; however, amidines are also formed as by-products.[117]

The amide **114** was prepared by the Vilsmeier reaction of 2-ethoxy-carbonylmethylene-3-oxo-1,2,3,4-tetrahydroquinoxaline.[118]

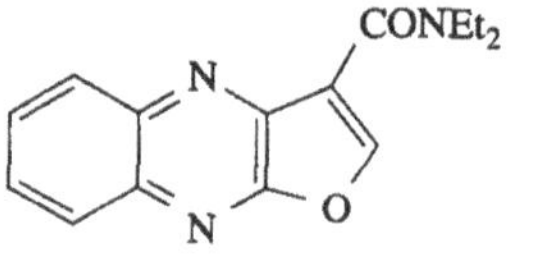

(4.75)

114

1,2,4-Oxadiazoles are formed by the reaction of amide oximes with Vilsmeier reagents.[119]

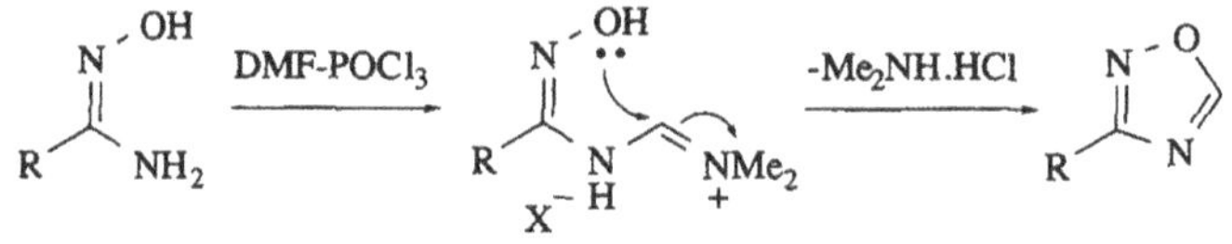

(4.76)

Dipyrazolo[3,4-*b*:4′,3′-*f*]azepines and dithieno[2,3-*b*:3′,2′-*f*]azepines have been prepared by the respective reaction of phenylhydrazine and ethyl 2-mercaptoacetate with the di-β-chlorovinylaldehyde obtained by Vilsmeier formylation of an azepine.[120]

A Vilsmeier reagent has been used to convert a sulfoxide into a fused five-membered heterocycle containing sulfur, in a reaction that proceeds initially as for a Pummerer rearrangement; *N*-(2-ethylsulfinylphenyl) pyrrole reacts with DMF-POCl$_3$ to give 1-pyrrolo[2,1-*b*]benzothiazole carbox-aldehyde (73%).[121]

An efficent route to 1,3-benzoselenazole involves ring-closure by a Vilsmeier reagent of the zinc salt of *o*-aminoselenophenol.[122]

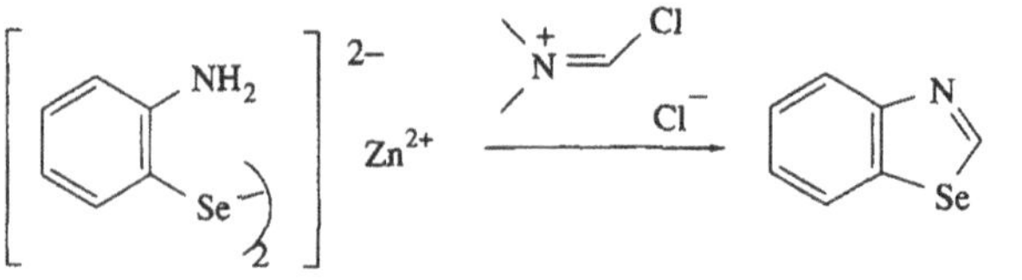

(4.77)

Various purines have been prepared by the action of Vilsmeier reagents on diaminopyrimidines.[123] The carbon atom of the 8-position in purines has been inserted into the corresponding 5-nitrosopyrimidines by Vilsmeier reagents; 8-amino-8-(*N*-methyl)amino- and 8-(*N,N*-dimethyl)amino-9-gluco-pyranosylpurines were so prepared.[123e]

The aldehyde **115**, obtained by the action of DMF-POCl₃ on the corresponding 1,4-dihydro-2,6-dimethyl derivative, was itself reacted with DMF-POCl₃ which led to cyclodehydration and also to formylation at the α-position of the pyrrole ring. This aldehyde was dehydrated upon treatment with sodium methoxide, giving the unstable diazacyclopenta[*c,d*]phenalene derivative **116**.[124]

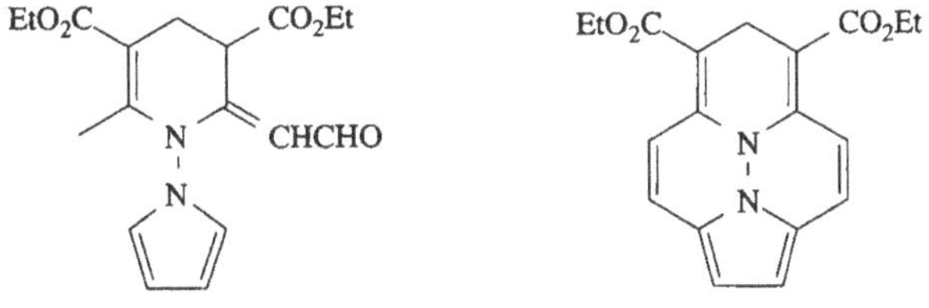

(4.78)

115   116

1,5-Dihydro-1,5-dimethyl-4*H*-pyrazolo[3,4-*d*]pyridin-4-one **118** (R = Me) was prepared from **117** (R=Me) in 88% yield by treatment with DMF-POCl₃. Similar reactions were noted for **118**, where R=H, phenyl, or benzyl in 59, 88, and 71% respectively. However, the 2-acetyl-1-methylhydrazino derivative **119** afforded 2-(1-chlorovinyl)-2,5-dihydro-5-methyl-4*H*-pyrazolo[3,4-*d*]pyridazino-4-ones **120** in 18-24% yield (various R) as well as the heterocycles **118**, as the major products.[125]

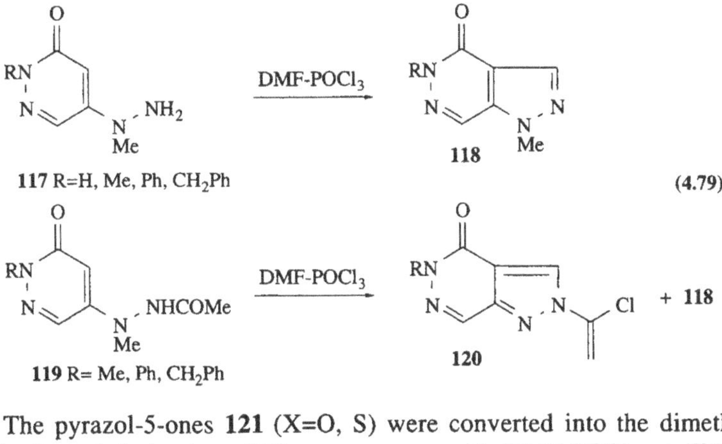

117 R=H, Me, Ph, CH₂Ph   118

119 R= Me, Ph, CH₂Ph   120

(4.79)

The pyrazol-5-ones **121** (X=O, S) were converted into the dimethyl-aminoformyl derivatives **122** on treatment with DMF-POCl₃ at 5°C; at higher temperatures, the thieno[2,3-*c*:5,4-*c′*]dipyrazines **123** were isolated.[126,127] Awad further demonstrated that heating compound **122** at

70°C afforded **123** in high yields. Hydrolysis of the dimethylaminoformyl group afforded malonaldehyde derivatives.[126,127]

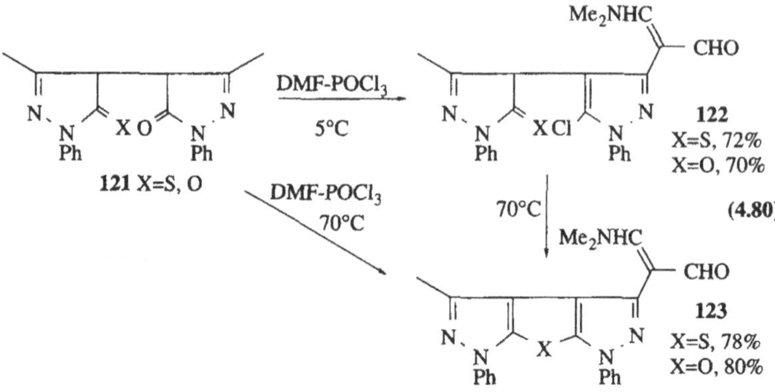

(4.80)

Awad has also reported the conversion of activated methyl groups into dimethylaminoformyl groups in pyrazolin-5-thione systems,[128] and also in 3-methyl-2(1H)-quinoxalinethiones[129] used for the synthesis of dyes.

Chuiguk and co-workers[130-134] reacted a series of α–amino-N-hetero-cycles with β-chlorovinylaldehydes to give a series of azolium salts. Of particular elegance is the condensation of 5-phenyl-2-amino-1,3,4-selenodiazole with β-methyl-β-chloroacrylaldehyde to give the substituted 1,3,4-selenodiazolo-3,2-[a]-pyrimidinium perchlorate **124** (scheme 4.81).[132]

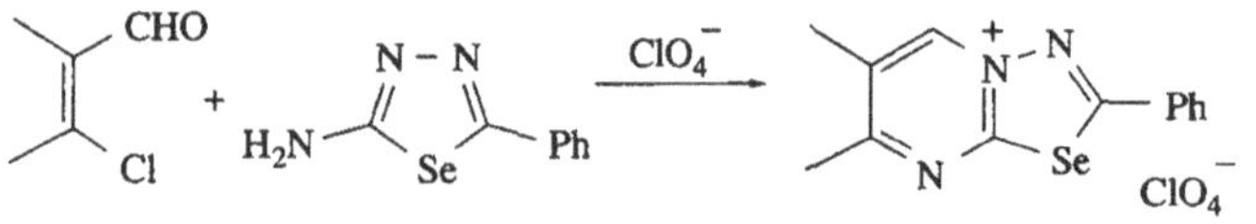

(4.81)

Cyclic β-thiocyanatovinylaldehydes when condensed with hydrazones, afford N-aroyl isothiazole-2-imines **125** in high yields.[135]

(4.82)

n=3,4,5; R=PhCO- or R=p-Tosyl

Reaction of β-chlorovinylaldehyde **126** with ethyl-2-mercaptoacetate affords ethyl thieno[2,3-d]thiazole-5-carboxylate **127**.[136]

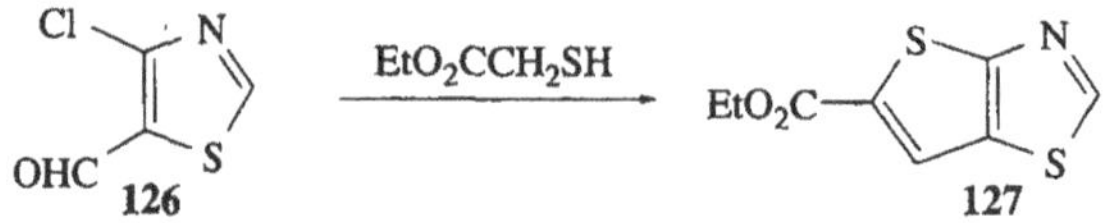

(4.83)

The temperature of Vilsmeier reaction is often crucial to the outcome of the reaction. For instance, the fused pyrimidine **129** was the major produc from **128** at 0-5°C, along with traces of **130**. At 70°C, the 7-formamido-[1,2,4]triazole[1,5-*c*]pyrimidin-5(6*H*)-4-one **131** was the exclusive product. At intermediate temperatures, a mixture of the **130** and **131** was isolated.[137] With dimethylacetamide-$POCl_3$ and diphenylacetamide-$POCl_3$ the products **130** and **131** (R=Me) and **130** and **131** (R=Ph) were isolated, respectively.

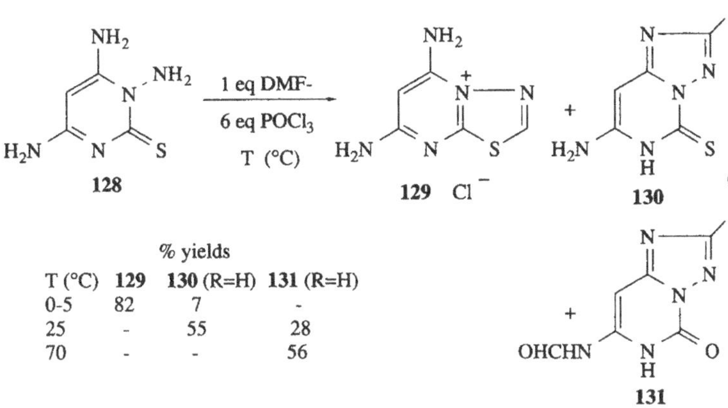

(4.84)

	% yields		
T (°C)	**129**	**130** (R=H)	**131** (R=H)
0-5	82	7	-
25	-	55	28
70	-	-	56

4.4.2 Compounds Containing a Six-Membered Ring

4.4.2.1 Pyrylium Salts and Thiopyrylium Salts

Condensation of β-chlorovinylaldehydes with ketones under acidic conditions affords a general and very useful synthesis of pyrylium salts.[138,139]

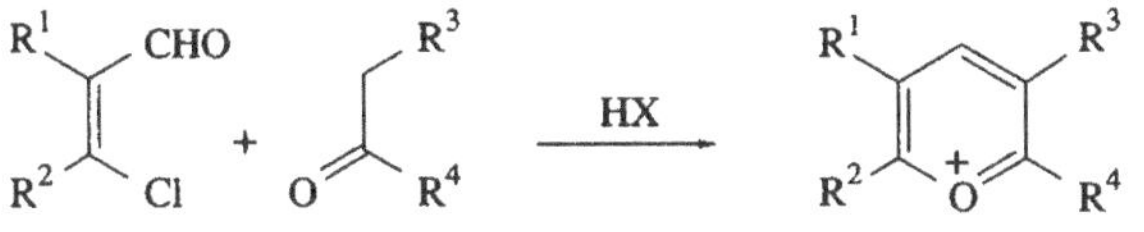

(4.85)

Using Vilsmeier methodology, the diketone **132** was converted into the dialdehyde **133** which afforded a convergent and unambiguous route to the intricate macrocyclic pyrylium salt **134** (scheme 4.86).[140]

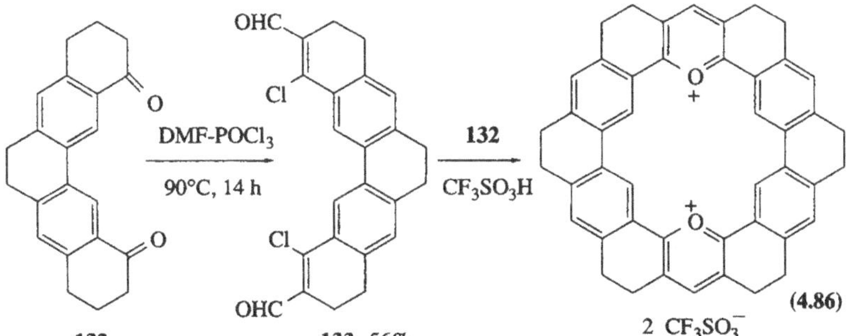

(4.86)

132 **133** 56% 2 $CF_3SO_3^-$ **134** 86%

Benzonitriles and phenylacetonitriles react with DMF-POCl$_3$ in the presence of HCl to give the amidinium salts 135 which as their perchlorates have been used to prepare pyrylium salts.[141]

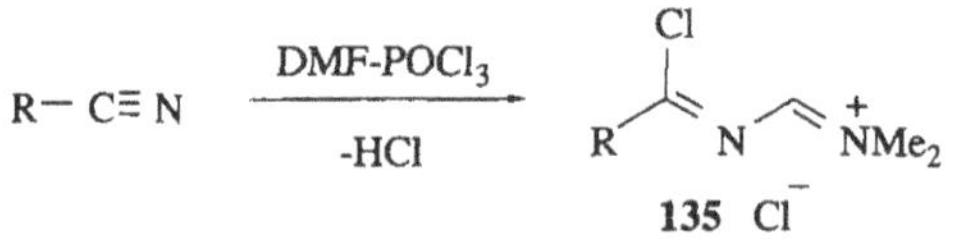

$$R-C\equiv N \xrightarrow[\text{-HCl}]{\text{DMF-POCl}_3} \quad \text{135 Cl}^- \qquad (4.87)$$

The pyrylium salt 137, described as the tetraperchlorate, was obtained by reaction of perchloric acid and acetic acid with the key β-chlorovinyl-aldehyde 136, obtained by chloroformylation (84%) of the corresponding symmetrical diketone with DMF-POCl$_3$.[142] Reaction of the salt 137 with ammonia almost certainly afforded the dodecahydrohexaazakekulene 138.

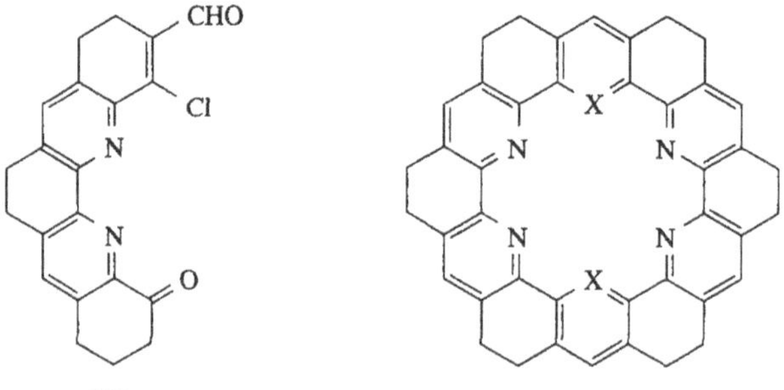

$$(4.88)$$

136

137 X=O$^+$, ClO$_4^-$
138 X=N

4H-Pyrans can be formylated under Vilsmeier conditions. Subsequent hydride abstraction with trityl perchlorate affords the corresponding 3-formylpyrylium salt (scheme 4.89).[33]

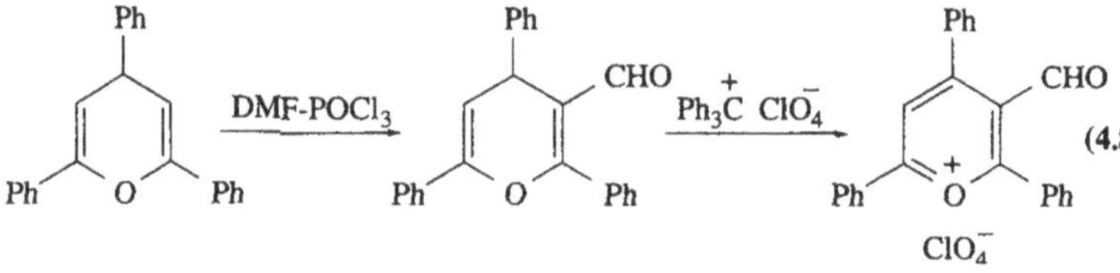

$$(4.89)$$

Chromones can be formylated in the 3-position, if the 2-position is blocked. Treatment of aldehyde 139 with DMF-POCl$_3$ gave the benzo[b]indeno[2,1-e]pyrylium salt 140.[143]

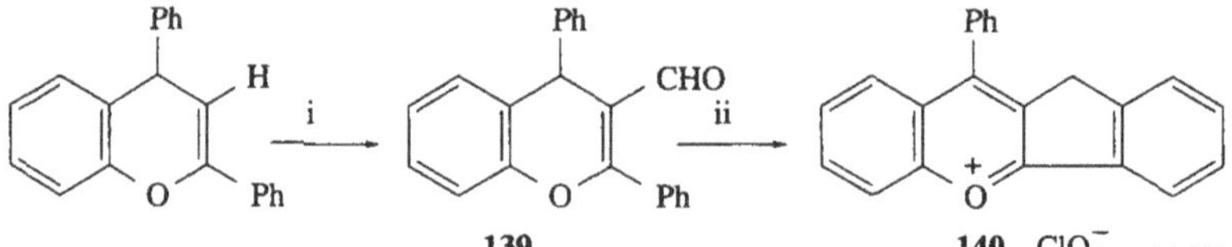

139   140 ClO$_4^-$ (4.90)

i, DMF-POCl$_3$; ii, Ac$_2$O, HClO$_4$

Thiopyrylium salts have been synthesized by condensing iminium species such as **141** with thioacetamides.[144]

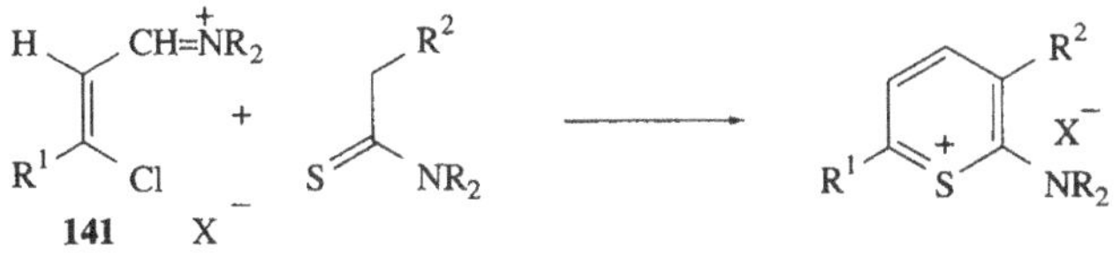

$$(4.91)$$

4.4.2.2 Benzopyrones and Naphthopyrones

o-Hydroxyacetophenones cyclize in good yields to give the valuable intermediates 3-formylchromones.[145-147] Related cyclization of acetophenone derivatives to give 3-phenoxychromones are catalyzed by $BF_3.OEt_2$.[148] Cyclization of 2-hydroxy-α-phenoxyacetophenone derivatives by Vilsmeier reagents, catalyzed by $BF_3.OEt_2$, to give 3-phenoxychromone derivatives has been reported.[148,149] In these cases, initial attack on the hydroxyl group has been considered more likely than double alkylation to give cation **142** (followed by ring-closure).

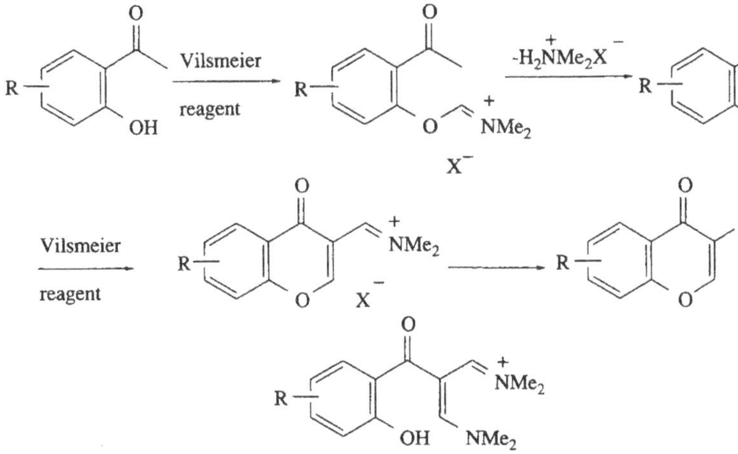

$$(4.92)$$

142

Appropriately substituted naphthalenes and coumarins react similarly.[145-147] A variation involves conversion of the 1-acetyl-2-hydroxynaphthalene into a difluoro-1,3,2-dioxaborin with $BF_3.OEt_2$, and subsequent formylation, by which the phenalenone **143** can be generated, as well as the expected chromone **144** (scheme 4.93).[150-152]

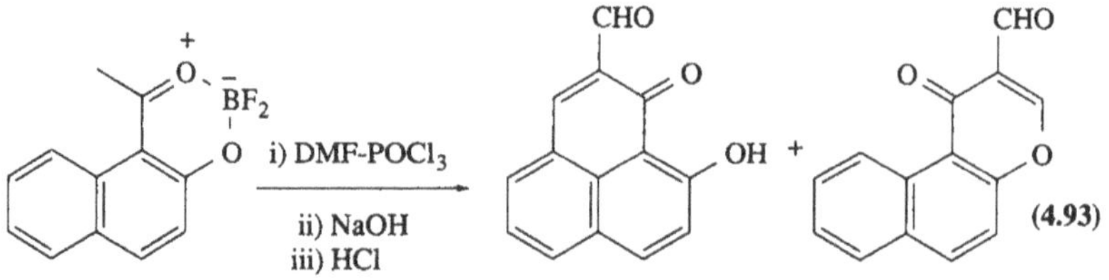

143   62%   **144**   20%   (4.93)

In scheme 4.94, initial attack on oxygen cannot be excluded, although attack by the Vilsmeier reagent on the enolized methylene group to give intermediate **145** is presumed to occur, prior to cyclization with loss of methylamine.

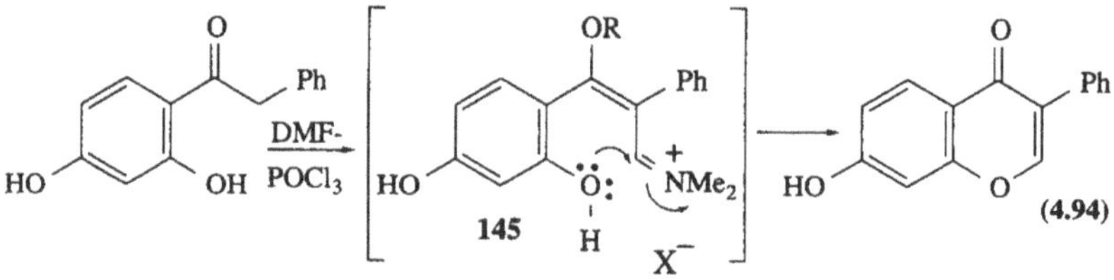

(4.94)

145

β-Naphthol reacts with the amides **146** to give the 1-oxo-1*H*-naphtho-[2,1-*b*]pyrans **147**.[153] In the case of β-naphthols, the ketene-*O,N*-acetals **148** have been isolated, and these can undergo cyclization reactions.

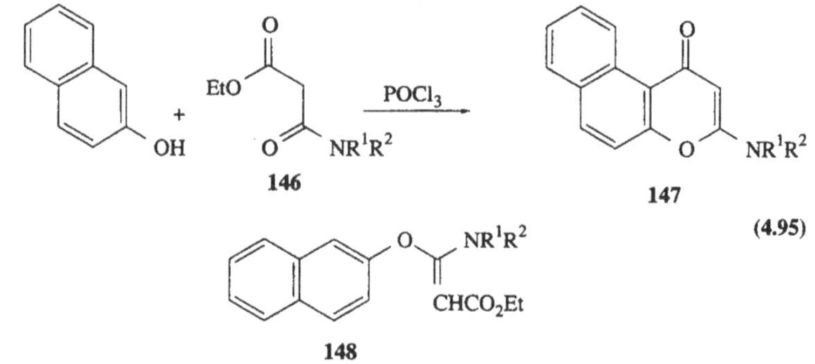

146   **147**

(4.95)

148

Substituted coumarins can be prepared by the condensation of 2-hydroxyphenyl aldehydes and ketones with carboxylic acids using DMF-POCl$_3$ (scheme 4.96).[154] A mixed phosphoric carboxylic acid anhydride has been invoked as the activated species.

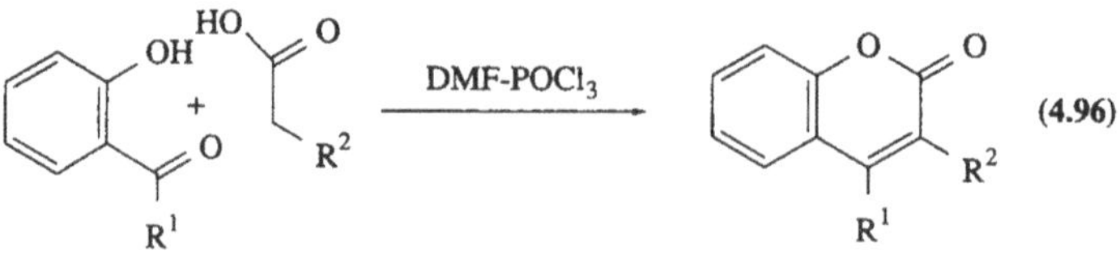

(4.96)

4.4.2.3 Pyridines

A variety of substituted pyridines can be readily prepared from the appropriate pentamethinium salts (section 2.29.4) (or the corresponding 1,5-dialdehydes which can sometimes be isolated by hydrolysis of those salts).

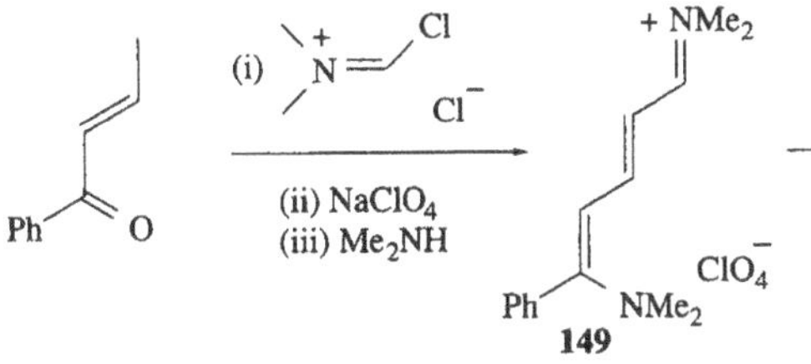

(4.97)

The formation of the pentamethinium salt **149** from crotonophenone has been discussed in section 2.29.4; it reacts with ammonium chloride by displacement of dimethylamine and subsequent ring-closure (either by nucleophilic displacement or by an electrocyclic mechanism.[155] The overall yield of 2-phenylpyridine from crotonophenone is about 20% (scheme 4.98).

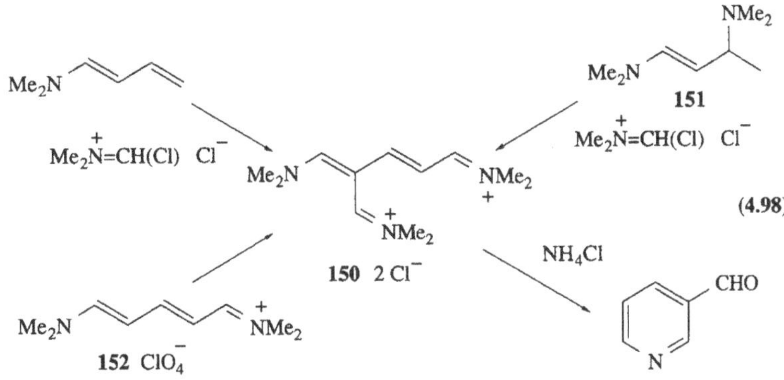

(4.98)

Pyridine-3-carboxaldehyde is formed in poor yield from the pentamethinium salt **150** and ammonium chloride.[156] The salt **150** is formed from 1-dimethylamino-1,3-butadiene and $[Me_2N=CHCl]^+$ Cl^-, but a better yield is obtained from the enamine **151**, itself readily prepared from crotonaldehyde and dimethylamine; the overall yield of pyridine-3-carboxaldehyde from crotonaldehyde is 40%. A pentamethinium salt derived from the reaction of acetone with a Vilsmeier reagent, upon treatment with potassium carbonate followed by ammonium chloride, is converted into 4-dimethylamino-pyridine-3-carboxaldehyde.[157]

Knoevenagel condensation of malononitrile with various aldehydes and ketones gave a series of alkylidene malononitriles which with DMF-POCl$_3$ afforded a wide variety of variously substituted 2-chloro-3-cyanopyridines (scheme 4.99).[158]

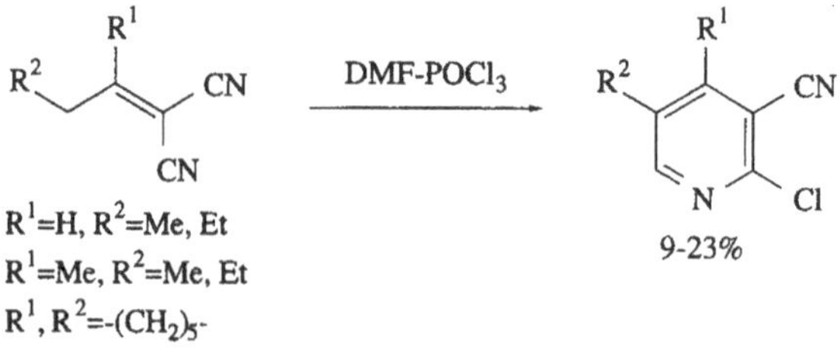

$$R^1=H, R^2=Me, Et$$
$$R^1=Me, R^2=Me, Et$$
$$R^1, R^2=-(CH_2)_5-$$

(4.99)

9-23%

Fused pyridine rings have been generated from pentamethinium salts such as **153** (scheme 4.100), and also from intermediate 1,5-dialdehydes such as **154** (scheme 4.101).

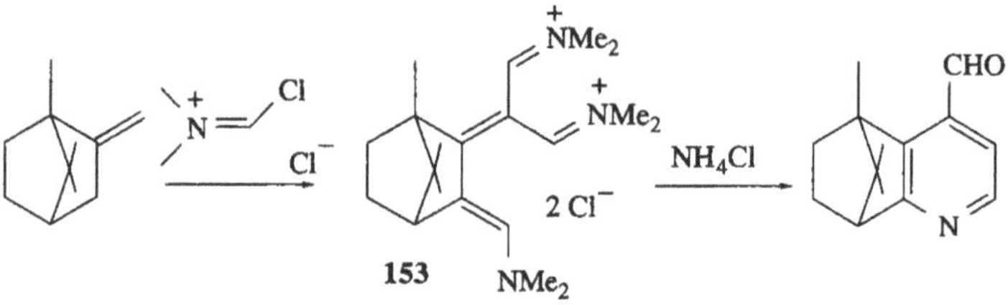

(4.100)

The formation of iminium salts such as **153** has been discussed in section 2.29). Conversion of salt **153** into the corresponding fused pyridine (93%) by ammonium chloride was used to confirm the structure of the salt **153**.[159]

Dialdehyde **154** was obtained by Vilsmeier formylation of 3-*N*-pyrrolo-dino-3,5-androstadien-17β-ol with DMF-POCl$_3$ in trichloroethene (50°C, 2h).[160]

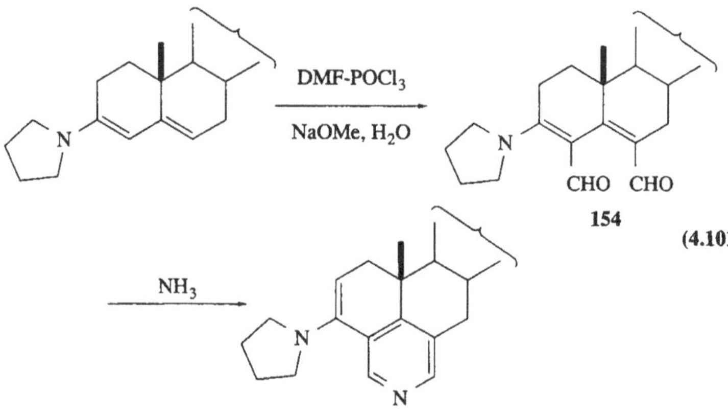

(4.101)

The action of Vilsmeier reagents upon α-monosubstituted carboxylic acids affords amidinium salts (section 2.29.2) such as **155** which when condensed with ketones afford substituted pyridines. The pyridine **156** possesses a smectic C liquid crystalline phase.[161]

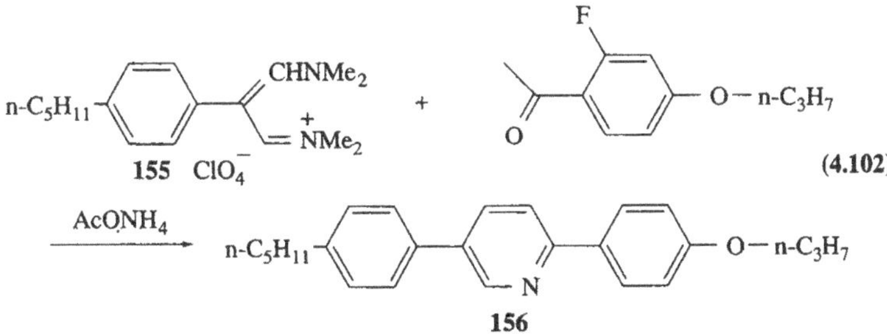

155 156 (4.102)

Several syntheses of the pyridine ring involve the action of Vilsmeier reagents upon alkenes, enamines, vinyl ethers, or ketones, followed by a separate cyclization of the iminium salts with ammonium chloride or acetate.[156,157,159,162-164] Thus, two different amides such as **157** and **158** can be condensed.[165] If two moles of $NCCH_2CONR^1R^2$ are used then good yields of the pyridines **159** are obtained.[165,166]

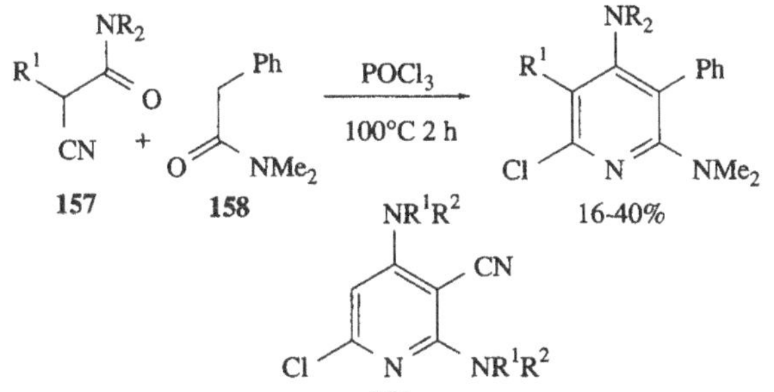

157 158 159 (4.103)

Malononitrile reacts with DMF-$POCl_3$ to give the 2-aminopyridine **160**.[167]

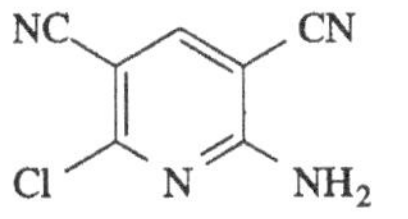

160 (4.104)

Reaction of 8-chlorobenzodiazepinone **161** with DMF-$POCl_3$ gave the formylpyridobenzodiazepinone **162**.[168]

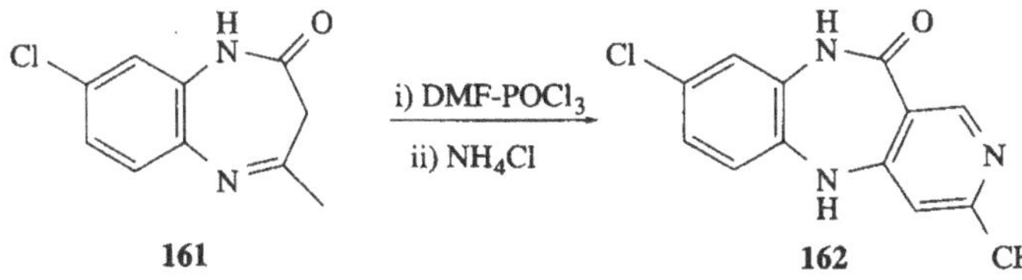

161 162 (4.105)

In the case of amides **163** and **164** whether or not the newly formed pyridine ring is formylated can be controlled by the quantities and ratio of DMF and $POCl_3$ used.[169]

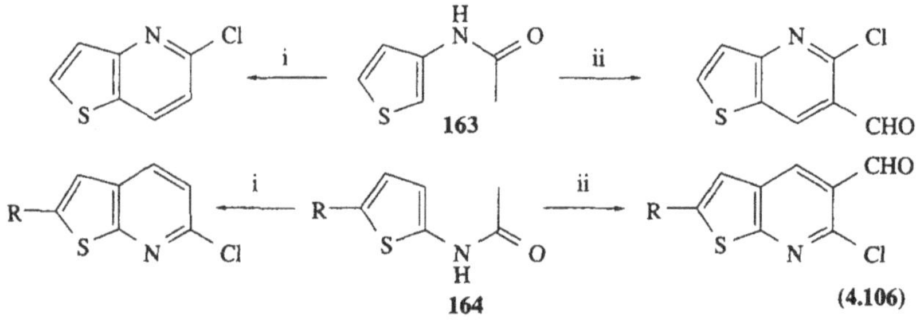

i, DMF-POCl$_3$ in ClCH$_2$CH$_2$Cl, reflux; ii, DMF-POCl$_3$, reflux

A thienopyridine **165** has been prepared in 34% yield by reaction of 3-thienylacetonitrile with a Vilsmeier reagent.[170]

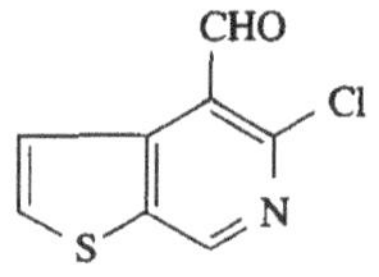

A range of 4-alkyl- and 4-aryl-aminocoumarins **166** were reacted with DMF-POCl$_3$ at 90°C. The 4-arylcoumarins afforded 6*H*-benzopyrano[4,3-*b*]-quinoline-6-one derivatives **167** in high yields, except where the *ortho*-position of the phenyl group was blocked, in which case the 3-formyl derivative was isolated. The 4-alkyl derivatives all gave 3-formylcoumarins (scheme 4.108).[171]

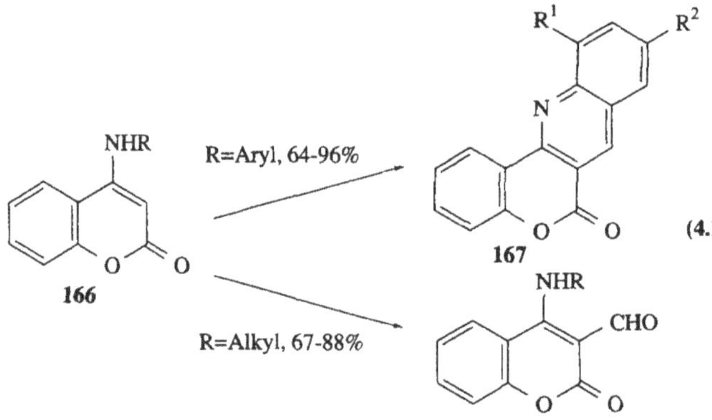

The synthesis of a pyrylium ring (*e.g.* by the condensation of a β-chloro-vinylaldehydes with a ketone in an acidic medium, section 4.4.2.1) provides a useful route to certain pyridines by means of a subsequent reaction with ammonia. Both of these steps were crucial to a synthesis of dodecahydro-hexazakekulene (section 4.4.2.1).[142]

4.4.2.4 Pyridinium Salts

Two moles of $NCCH_2CONHR$ react with $POCl_3$ to give the pyridinium salts **168**.[166b,172]

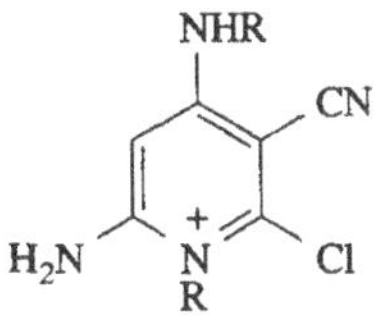

(4.109)

4.4.2.5 Quinolines

Since the pioneering work of Fischer, Mueller and Vilsmeier,[173] numerous quinolines have been prepared by Vilsmeier reactions of acylanilides (scheme 4.110). These include 2-chloro-3-cyanoquinolines, prepared by i) the action of hydroxylamine hydrochloride on the Vilsmeier reaction mixture,[174] ii) directly from the ketone oximes,[175] or iii) from cyanoacetamides.[176] Other 2-chloro-3-substituted quinolines have also been prepared.[177-180] In certain cases, the uncyclized intermediate was identified as the salt **169** (scheme 4.110).[41]

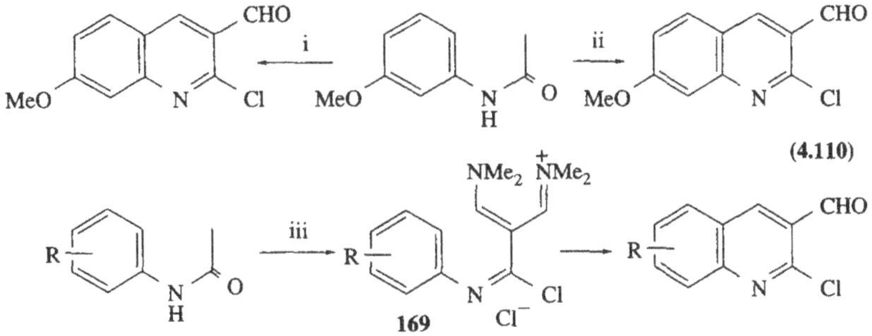

(4.110)

i, DMF-3 $POCl_3$ in $ClCH_2CH_2Cl$, reflux; ii, 3 DMF- 7 $POCl_3$, reflux; iii, DMF-$POCl_3$

Adams[181] reported the synthesis of 3-carboalkoxyquinolines from the unsaturated esters **170** under Vilsmeier conditions.

(4.111)

Quinolines have been prepared from β-chlorovinylaldehydes and anilines;[182-184] this process also allows the annelation of a quinoline ring onto an existing system (scheme 4.112).[185]

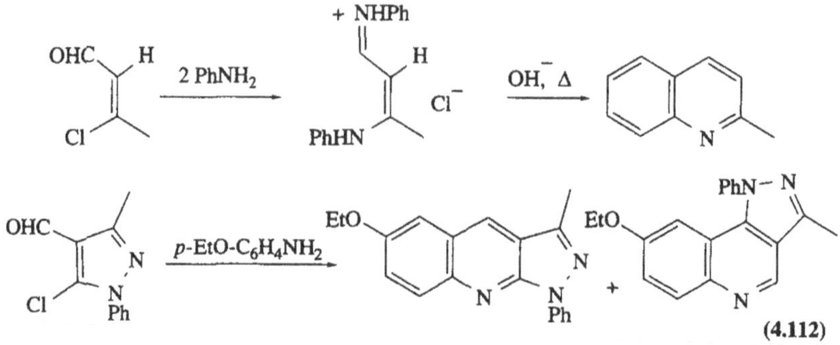

(4.112)

4.4.2.6 Quinolinium Salts

Certain N-arylamides react with phosgene, *via* iminium intermediates, to give quinolinium salts **171** (scheme 4.113).[186] Meth-Cohn has reported the condensation of a mixture of N-methylacetanilide (2 eq.) and POCl₃ to give the 4-chloroquinolinium salt **172**.[187]

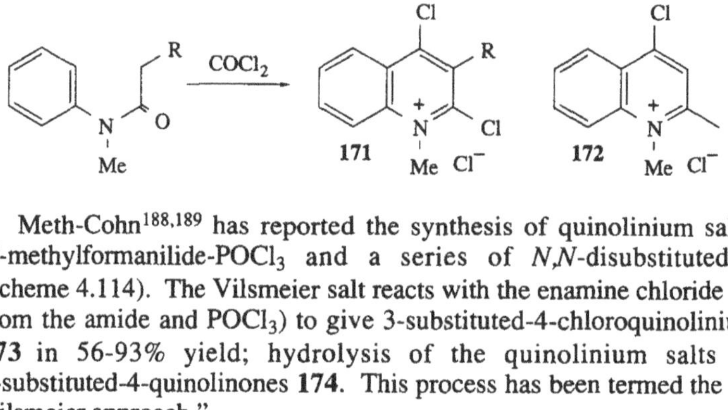

(4.113)

Meth-Cohn[188,189] has reported the synthesis of quinolinium salts from N-methylformanilide-POCl₃ and a series of N,N-disubstituted amides (scheme 4.114). The Vilsmeier salt reacts with the enamine chloride (formed from the amide and POCl₃) to give 3-substituted-4-chloroquinolinium salts **173** in 56-93% yield; hydrolysis of the quinolinium salts affords 3-substituted-4-quinolinones **174**. This process has been termed the "reverse Vilsmeier approach."

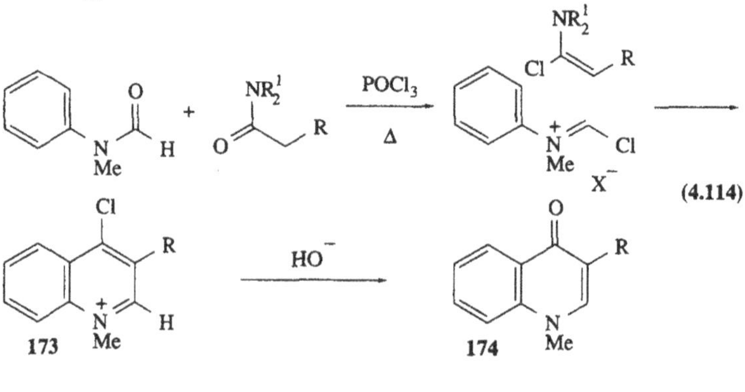

(4.114)

4.4.2.7 Quinolizinium and Fused Quinolizinium Rings

The fused quinolinium rings 176[190] and 178[190] result from the reaction of methylpapaverines 175, and their isomers 177, with DMF-POCl$_3$. It is possible that the Vilsmeier reagent first attacks the benzene ring, giving an aldiminium group which subsequently attacks the isoquinoline nitrogen atom, prior to elimination of dimethylamine.

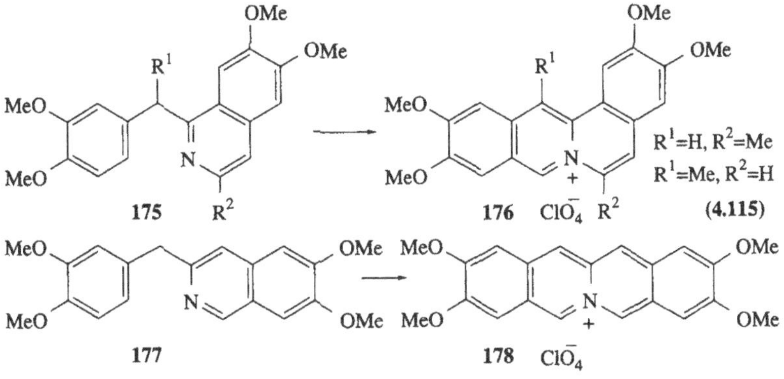

(4.115)

4.4.2.8 Isoquinolines

Among other products, 6-methoxy-3-methylisoquinoline is formed by reacting 3-methoxyphenylacetone with Vilsmeier reagents.[191]   Hirota reported that substituted phenylacetonitriles afford chloroisoquinolines on reaction with DMF-POCl$_3$. The reaction proved to be general, but less activated rings afforded very poor yields of chloroformyl isoquinolines (3-8%).[192] Conducting the reaction at 65-70°C afforded a complex mixture of products, none containing the chloroisoquinoline structure.[193]

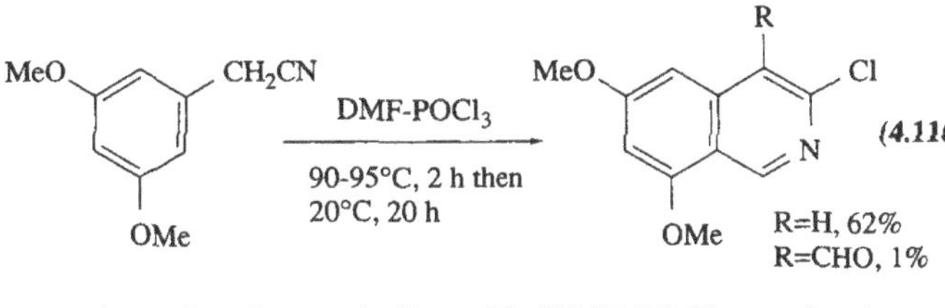

(4.116)

R=H, 62%
R=CHO, 1%

The reaction of arylacetonitriles with DMF-POCl$_3$ to give isoquinolines (scheme 4.117) is considered to proceed by elimination of dimethylamine from intermediates, which in principle could be formed either from the amidinium salts 179 or from the aldiminium salts 180. Which of these two intermediates is involved has not been established.[194]

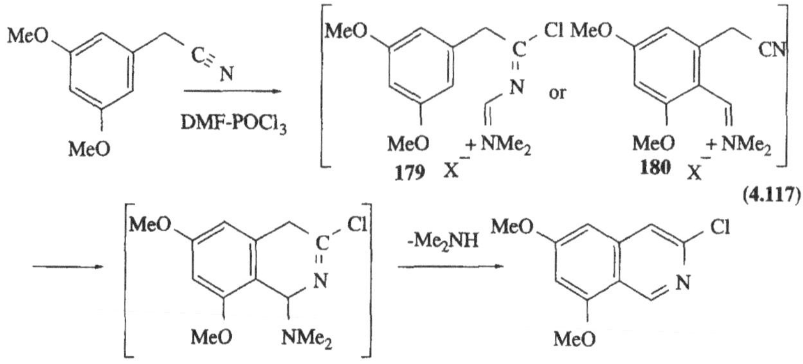

(4.117)

In 1968, Koyama and co-workers reported a synthesis of isoquinolines **182** from α-acyl-α-arylacetonitriles **181** and formamide-POCl₃ (modified Vilsmeier reaction; scheme 4.118).[195] The formation of isoquinolines *versus* pyrimidines was further studied[196] revealing that MeCONH₂-POCl₃ can afford 4-(3*H*)-pyrimidinones. However, the same Vilsmeier reagent was shown to convert certain alkoxy-substituted acetophenones into either isoquinolines or naphthalenes, depending upon the substitution.[196]

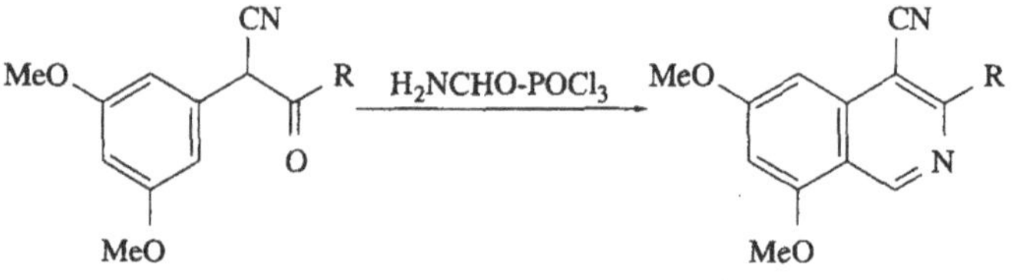

(4.118)

181   **182** R=H, Me, Et, Ph

Reaction of benzylacetamide-POCl₃ in the presence of alkyl nitriles afforded the corresponding isoquinolines. However, acrylonitrile afforded 1-(2-chlorovinyl)-isoquinoline, presumably *via* addition of HCl.[197]

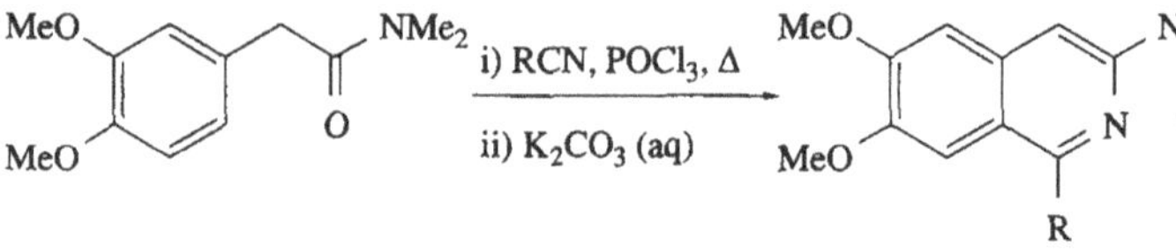

(4.119)

4.4.2.9 Pyrimidines

4-Phenylpyrimidine can be prepared by the chloroformylation of phenylethyne with DMF-POCl₃, followed by condensation of the β-chloro-vinylaldehyde with formamide (scheme 4.120).[198-200]

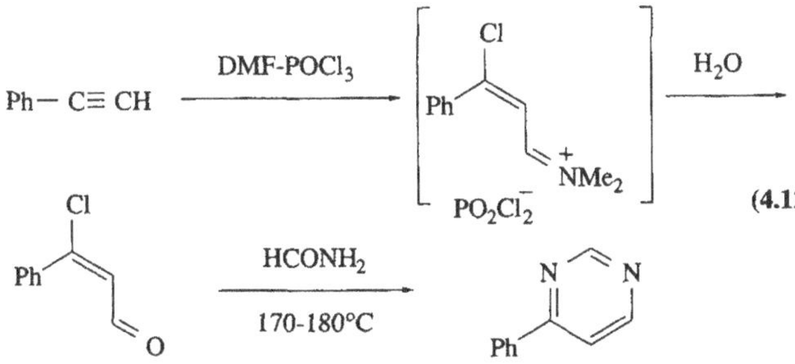

$$(4.120)$$

Phenylacetone reacts with Vilsmeier reagents followed by formamide to give a mixture of substituted pyrimidines (derived from chloromethylene-iminium units) as well as a tetrasubstituted benzene.[26] 3-Methoxyphenyl-acetone affords a mixture of these compounds containing a pyrimidine ring, and 6-methoxy-3-methylisoquinoline.[191]

A remarkable set of syntheses of aminopyrimidines can be achieved by the action of formamide and $POCl_3$ upon acid amides (scheme 4.121).[202-205] Formamide and $POCl_3$ alone form adenine **183**. Aliphatic acid amides react to give the aminopyrimidines **184**; lactams may also be used, in which case fused pyrimidines **185** are formed. α,ω-Diamides have been shown to condense in an intramolecular fashion, giving fused pyrimidines such as **186**.

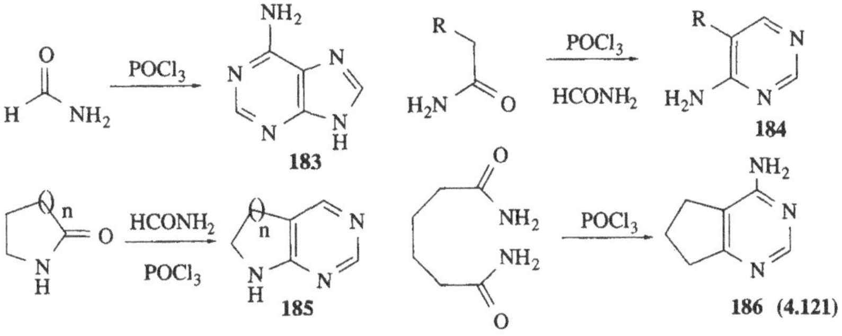

183

184

185

186 (4.121)

A reaction with a mechanism consistent with the formation of pyrimidines described above is the reaction of *o*-substituted anilines **187** with chloromethyleneiminium chlorides to give the isolable salts **188** which undergo condensation with primary amines to give fused pyrimidine derivatives (scheme 4.122).[206]

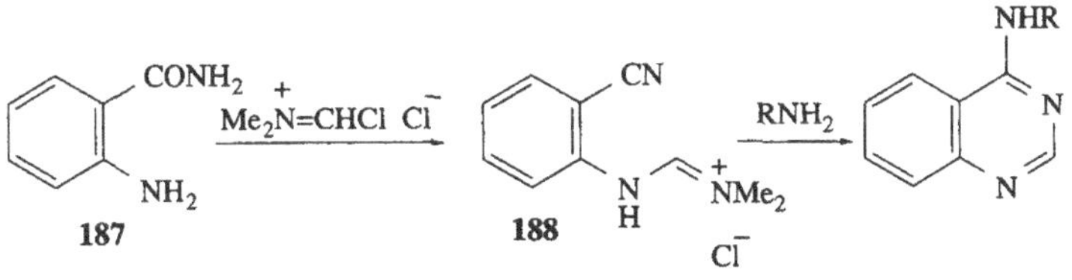

187

188

$$(4.122)$$

Pyrimidines may be synthesized by the condensation of amides with β-chlorovinylaldehydes. A number of 2-substituted-[94,207] and 2,3-

disubstituted pyrimidines[92,208] have been made in this manner. Julia[209] reported the synthesis of 2-amino-4-methylpyrimidine from guanidine and 3-chloro-3-methylacrylaldehyde. Some heterocyclic β-chlorovinylaldehydes also react in this way.[210]

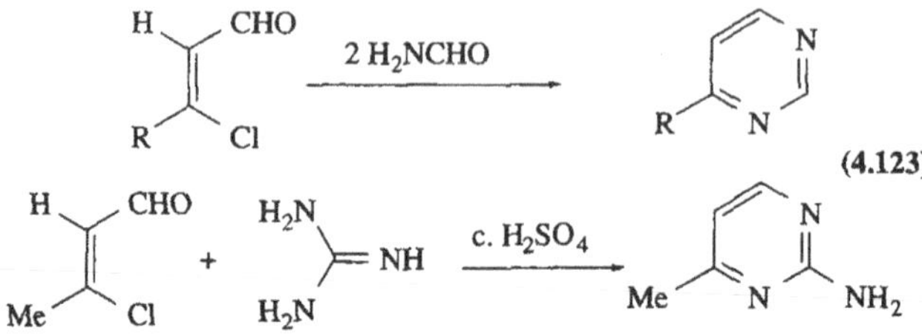

(4.123)

Other routes to substituted pyrimidines **189** include the condensation of 3-aminoacrylonitriles with arylamides (scheme 4.124).[211,212] The cross-condensation of different amides[213] is also possible; the pyrimidines **190** or the pyrimidinones **191** (scheme 4.125) are formed, the former being favored if R^2 is a secondary alkyl group, which undergoes dealkylation readily.

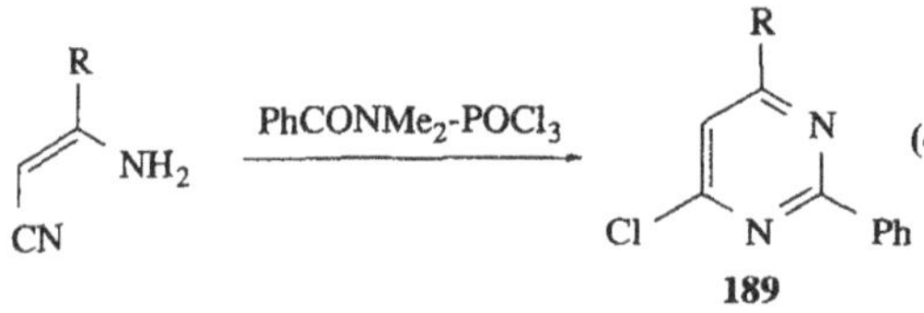

(4.124)

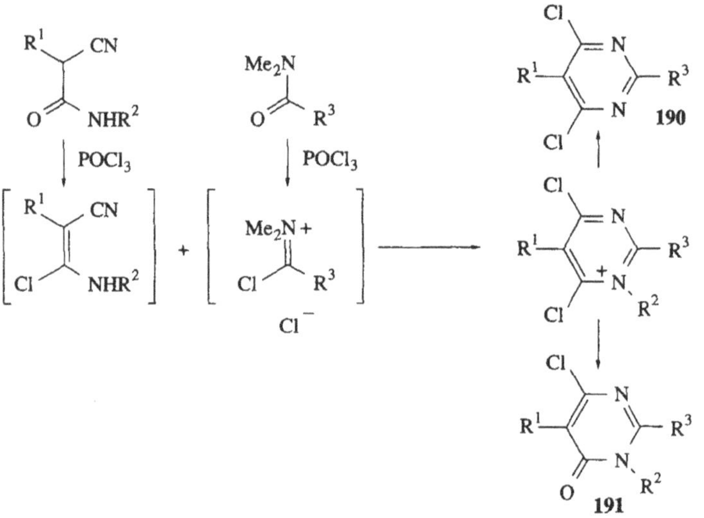

189

(4.125)

190

191

4.4.2.10 1,3-Thiazinium Salts

1,3-Thiazinium salts have been prepared from the amidinium salts **135**.[141]

4.4.2.11 Adenine
Adenine is formed by the reaction of formamide with $POCl_3$ (section 4.4.2.9).

4.4.2.12 Naphthyridines
Reaction of 2-methylpropene with $[Me_2N=CHCl]^+$ Cl^- leads by repeated iminoalkylation to the cross-conjugated salts **192**, of which the triperchlorate has been isolated; double ring-closure with aqueous ammonium chloride affords 2,7-naphthyridine-4-carboxaldehyde.[162,163]

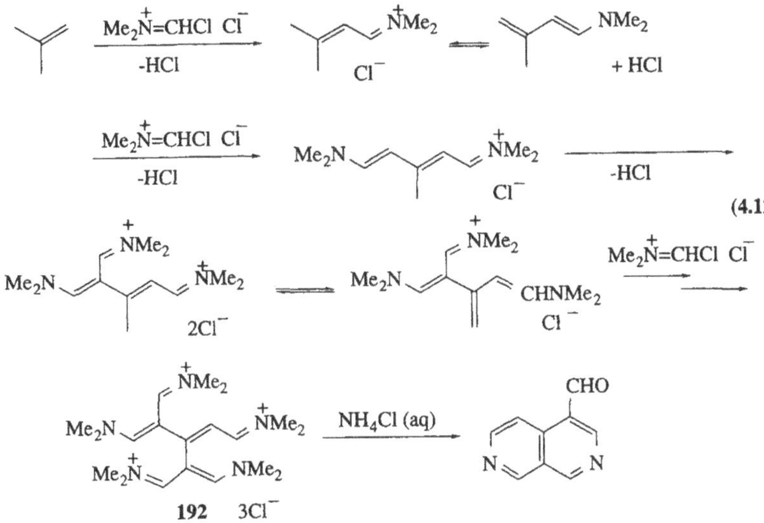

(**4.126**)

4.4.2.13 Pyrimidoindoles
An efficient route to the pyrimidoindoles **194** involves annulation of the chloroaldehyde **193**, obtained by the reaction of DMF-$POCl_3$ upon oxindole.[210]

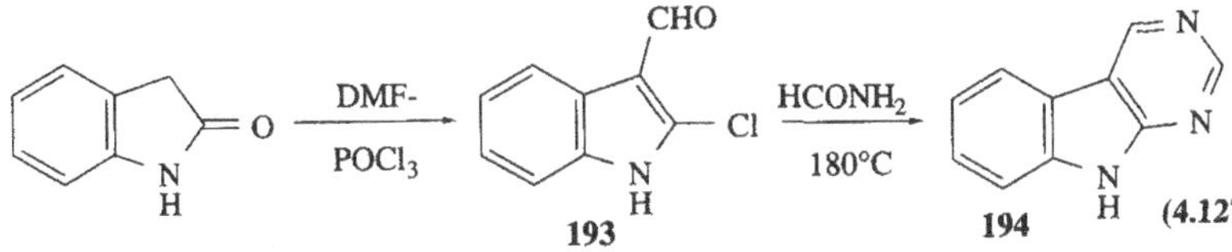

(**4.127**)

4.4.2.14 Pyrimidobenzodiazepines
The pyrimidobenzodiazepines **196** are formed in an annulation reaction from the 1,4-benzodiazepines **195** and the formamide-$POCl_3$ adduct.[214]

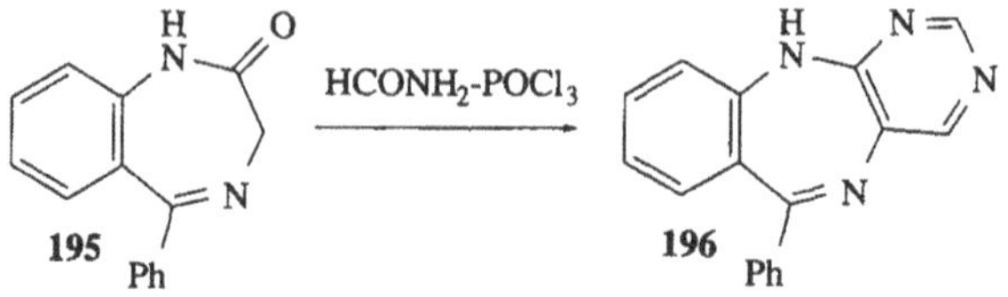

(4.128)

195   196

4.4.2.15 Miscellaneous Ring Systems

Tricyclic fused pyridine systems result from the reaction of the corresponding acetanilide derivatives with Vilsmeier reagents, presumably by electrocyclic ring-closure of the bis(trimethinium) dication (scheme 4.129).[177]

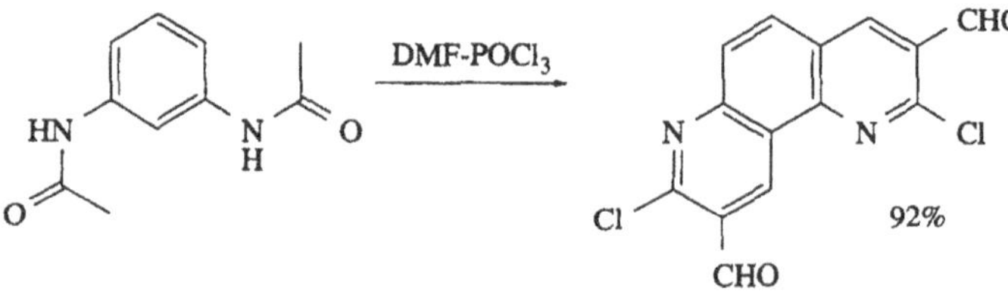

(4.129)

92%

4-Oxo-3,4-dihydro-5H-pyridazino[4,5-b]indoles **197** have been prepared in excellent yield by a one-pot procedure from 2-indolecarbohydrazides.[215]

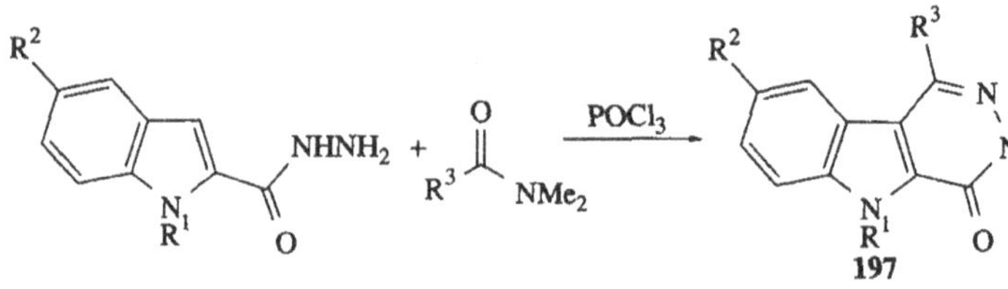

(4.130)

197

5-Chlorothieno[3,2-b]pyridine-6-carbonitrile is formed by reacting 3-acetamidothiophene with DMF-POCl$_3$ followed by addition of hydroxylamine hydrochloride.[216]

Benzo[1,2-b;5,4-b']dithiophene (33%) is formed in a ring-closure by the reaction of DMF-POCl$_3$ on 3,3'-dithienylmethane.[217]

A pyrrolo[1,2-b]cinnoline derivative has been prepared by reaction of a 1,4-dihydropyridine with DMF-POCl$_3$; one of the 2,5-dimethyl groups undergoes iminoalkylation, and the electrophilic carbon atom so added is then attacked by the 1-pyrrolyl substituent with the formation of a central six-membered heterocyclic ring.[218] The reaction pathway was probed using carbon-13 enriched DMF.[124]

Pyrrolo[1,2-b][1,2,4]triazinium salts are formed by the action of the isolated Vilsmeier reagent form DMF-(COCl)$_2$ on tetramethylhydrazones of arylaldehydes; after iminoalkylation (section 2.8.2.3), a formal 1,5-H shift occurs, followed by intramolecular ring-closure.[219]

Condensation of complex heterocyclic β-chlorovinylaldehydes with a suitable amine allows the synthesis of cyanine dyes. Harnish reacted

4-chloro-7-dimethylamino-1-methylquinolin-2-one-3-carboxaldehyde **198** with *m- N,N*-dimethylaminophenol, to give the violet-blue dye **199** (scheme 4.131).[220]

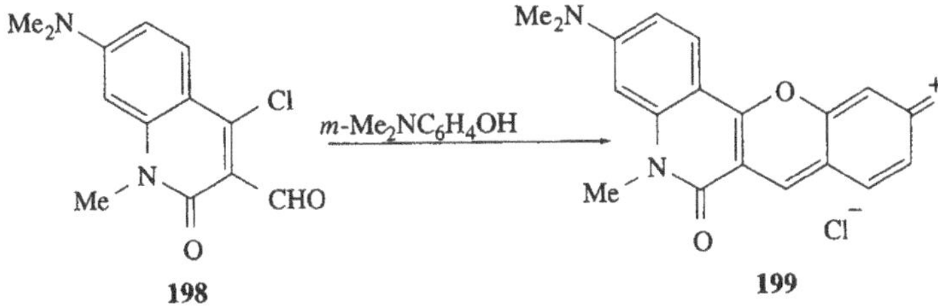

(4.131)

198   199

Reaction of tetrahydroisoquinolinone **200** with 6 eq of POCl$_3$ in excess DMF at 90°C, gave a mixture of **201, 202**, and **203**. Chloroaldehyde **201** could be formed by a decyanation, followed by chlorination as reported by Kashiri.[221] The first step of the formation of **202** and **203** is likely to be attack of DMF-POCl$_3$ at the benzylic C-5 position, followed by attack at the cyano group and subsequent cyclization.[222]

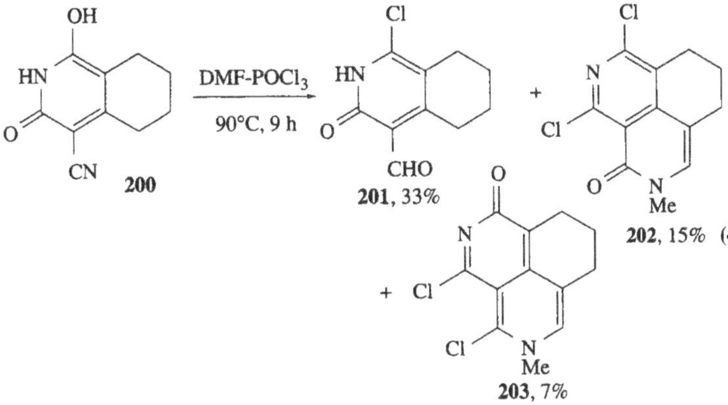

(4.132)

Fervenulins, some of which are antibiotics, have been prepared by ring closures effected by Vilsmeier reagents.[223,224] The use of an aza-Vilsmeier reagent is notable (scheme 4.133).[224]

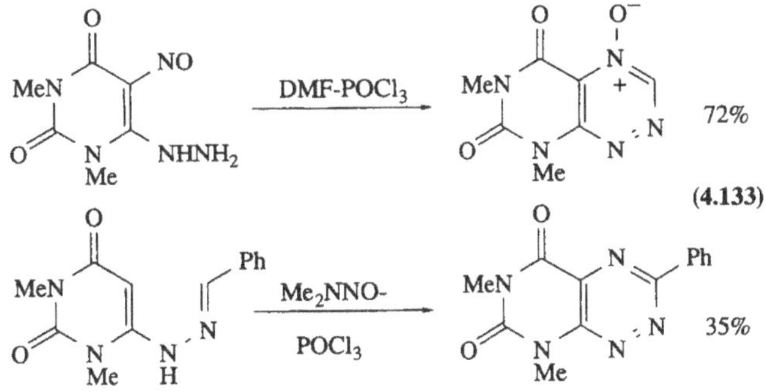

(4.133)

4.5 Formation of Seven-Membered Rings

4.5.1 2-Azaazulenes

Formylation of 2,5-dimethylpyrrole and of 1,2,5-trimethylpyrrole with
DMF-POCl$_3$ leads to the corresponding pyrrole-3,4-dialdehydes which are
precursors in a new route to 2-azaazulenes.[225]

4.5.2 Benzodiazepines

o-Phenylenediamine and β-chlorovinylaldehydes condense to give
substituted benzo[b]diazepine salts 204.[226,227]

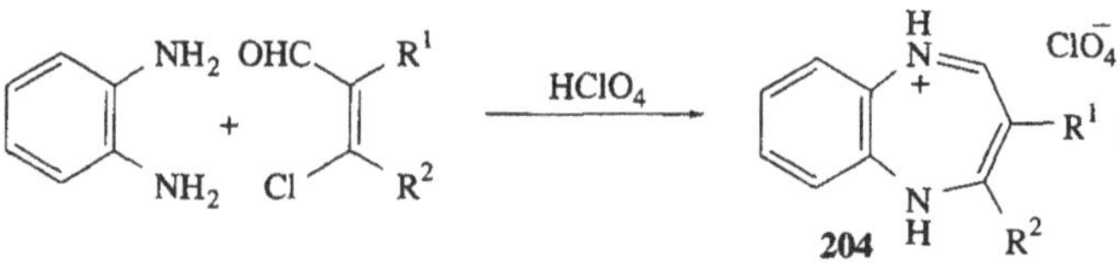

4.5.3 Benzoxazolodiazepines

Benzoxazolodiazepines have been prepared from 6-chlorobenzoxazol-
2-malondialdehyde, itself obtained by reaction of 4-chloro-2-hydroxyaceto-
phenone oxime with DMF-POCl$_3$.[228]

4..5.4 Thiazepines and Benzothiazepines

The condensation of β-chlorovinylaldehyde 205a with 2-aminoethane-
thiol in the presence of NaH affords thiazepine 206 in quantitative yield.[229]
Similarly, reaction of 2-aminothiophenol with 205a gave benzothiazepine
207a.[229] Both of these reactions proceed via an initial 1,4-addition of sulfur
followed by a subsequent 1,2-addition of nitrogen to the aldehyde.

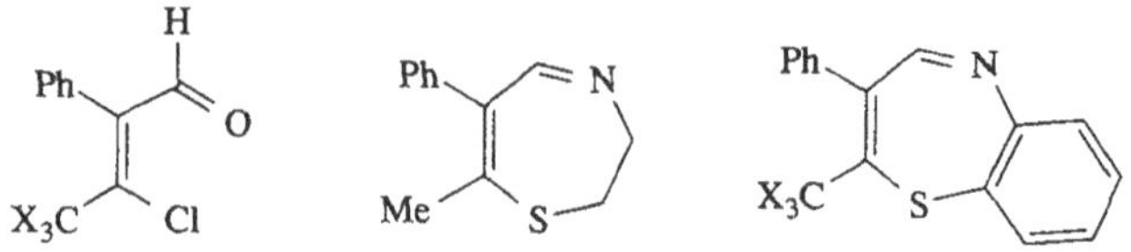

205a, X=H
205b, X=F

206

207a, X=H
207b, X=F

(4.135)

However, the trifluoromethyl analog of 205a reacted in a different mode.
The strongly electron-withdrawing effect of the trifluoromethyl group
reverses the reactivity of the conjugated aldehyde. The reaction of 205b with
2-aminoethanethiol afforded the thiazolidine 208 by a subsequent
1,4-addition of the nitrogen group. Condensation of 2-aminothiophenol with
205b affords the benzothiazole 209 by 1,4-addition of the nitrogen group,
followed by a retro-aldol reaction. The unstable benzothiazepine 207b was
also formed, but the electron-withdrawing effect of the trifluoromethyl group

encourages the elimination of sulfur *via* a thiirane intermediate, to yield the quinoline **210**.[229]

$$(4.136)$$

References

1. J. Zemlicka and Z. Arnold, *Collect. Czech. Chem. Commun.*, 1961, **26**, 2852.
2. C. Jutz and R. Heinicke, *Chem. Ber.*, 1969, **102**, 623.
3. K. Hafner and J. Schneider, *Liebigs Ann. Chem.*, 1959, **624**, 37.
4. W. Flitsch and E. R. Gesling, *Chem. Ber.*, 1983, **116**, 1174.
5. M. S. C. Rao and G. S. K. Rao, *Synthesis*, 1987, 231.
6. C. Jutz, R. Kirchlechner, and H. J. Seidel, *Chem. Ber.*, 1969, **102**, 2301.
7. C. Jutz and E. Schweiger, *Chem. Ber.*, 1974, **107**, 2383.
8. C. Jutz and M. Wagner, *Angew. Chem.*, 1972, **84**, 299.
9. C. Jutz, *Angew. Chem.*, 1974, **86**, 781.
10. A. Holy and Z. Arnold, *Collect. Czech. Chem. Commun.*, 1965, **30**, 53.
11. M. Sreenivasulu and G. S. K. Rao, *Indian J. Chem., Sect. B.*, 1989, **28B**, 494.
12. A. R. Katritzky and C. M. Marson, *J. Org. Chem.*, 1987, **82**, 2726.
13. M. Weissenfels, M. Pulst, M. Haase, U. Pawlowski, and H.-F. Uhlig, *Z. Chem.* 1977, **17**, 56.
14. R. E. Mewshaw, *Tetrahedron Lett.*, 1989, **30**, 3753.
15. T. Schirado, E. Gennari, R. Merello, A. Deciniti, and S. J. Bunel, *Inorg. Nucl. Chem.* 1971, **33**, 3417.
16. B. Raju and G. S. K. Rao, *Indian J. Chem., Sect. B*, 1987, **26B**, 175.
17. J. Bergman and B. Pelcman, *Tetrahedron Lett.*, 1985, **26**, 6389.
18. L. D. Napoli, L. Mayol, G. Piccialli, C. Santcroce, *Tetrahedron*, 1988, **44**, 215.
19. V. Dressler and K. Bodendorf, *Arch Pharm (Weinheim, Ger.)*, 1970, **303**, 481.
20. K. Bodendorf and R. Mayer, *Chem. Ber.*, 1965, **98**, 3565.
21. I. W. Elliot, S. L. Evans, L. T. Kennedy, and A. E. Parrish, *Org. Prep. Proc. Int.*, 1989, **21**, 368.
22. Z. Arnold, *Collect. Czech. Chem. Commun.*, 1965, **30**, 2783.
23. M. Venngopal and P. T. Perumal, *Syn. Commun.*, 1991, **21**, 515.

24.   (a) N. Narasimhan and T. Mukhopadhyay, *Tetrahedron Lett.*, 1979,
      1341; (b) N. Narasimhan, T. Mukhopadhyay, and S. S. Kusarhar,
      Indian. J. Chem., Sect. B, 1981, **20B**, 546.
25.   (a) M. Ahmed, J. Ashby, and O. Meth-Cohn, *J. Chem. Soc. (D)*, 1970,
      1094; (b) M. Ahmed, PhD thesis, University of Salford, 1969.
26.   G. W. Fischer and W. Schroth, *Chem. Ber.*, 1969, **102**, 590.
27.   A. R. Katritzky, Z. Wang, C. M. Marson, R. J. Offermann, A. E.
      Koziol, and G. J. Palenik, *Chem. Ber.*, 1988, **121**, 999.
28.   P. C. Traas, H. Boelens, and H. J. Takken, *Recl. Trav. Chim. Pay-Bas*,
      1976, **95**, 308.
29.   B. Hirsch and N. Hoefgen, Ger. (East) DD, 226892 (1985); *Chem.
      Abstr.*, 1986, **105**, 42647.
30.   D. Mohlo and M. Giraud, *Bull. Chim. Soc. Fr.*, 1968, 2603.
31.   M. Weissenfels, P. Schneider, and D. Schmiedl, *Z. Chem.*, 1972, **12**,
      263.
32.   M. Vengopal, R. Umarani, P. T. Perumal, and S. Rajadurai,
      Tetrahedron Lett., 1991, **32**, 3235.
33.   A. V. Koblik and K. F. Suzdalev, *Zh. Org. Khim.*, 1989, **25**, 1342;
      Chem. Abstr., 1990, **112**, 76874q.
34.   M. Weissenfels and S. Illing, *Z. Chem.*, 1973, **13**, 130.
35.   M. Weissenfels and M. Pulst, *Tetrahedron*, 1972, **28**, 5197.
36.   M. Weissenfels and M. Pulst, *J. Prakt. Chem.*, 1973, **315**, 873.
37.   J. Schmitt, J. J. Panouse, A. Hallot, P. J. Cornu, H. Pluchet, and P.
      Comoy, *Bull. Chem. Soc. Fr.*, 1964, 2753.
38.   K. Hafner, K. H. Voepel, G. Ploss, and C. Koenig, *Liebigs Ann.
      Chem.*, 1963, **661**, 52.
39.   T. Kaempchen, G. Moddelmog, and G. Sietz, *Synthesis*, 1984, 262.
40   J. P. Chupp and S. Metz, *J. Heterocycl. Chem.* 1979, **16**, 65.
41.   O. Meth-Cohn, B. Narine, and B. Tarnowski, *J. Chem. Soc., Perkin
      Trans. 1*, 1981, 1520.
42.   K. E. Schulte and D. Bergenthal, *Arch. Pharm. (Weinheim, Ger.)*,
      1979, **312**, 265.
43.   R. Hayes, O. Meth-Cohn, and B. Tarnowski, *J. Chem. Research (S)*,
      1980, 414.
44.   K. E. Schulte and D. Bergenthal, *Arch. Pharm. (Weinheim, Ger.)*,
      1980, **313**, 890.
45.   V. H. Belgaonkar and R. N. Usgaonkar, *Tetrahedron Lett.*, 1975,
      3849.
46.   Z. Arnold, *Collect. Czech. Chem. Commun.*, 1963, **28**, 863.
47.   T. Kato and T. Chiba, *J. Pharm. Soc. Jap.*, 1969, **89**, 1464.
48.   M. A. Volidina, V. A. Kudrjasova, and A. P. Terent'ev, *Zh. Obsch.
      Khim.*, 1964, **31**, 3130; *Chem. Abstr.*, 1964, 61, 16034b.
49.   A. P. Terent'ev, M. A. Volidina, and V. A. Kudrjasova, *Dokl. Akad.
      Nauk. SSSR*, 1965, **164**, 115; *Chem. Abstr.*, 1965, **63**, 16290c.
50.   M. A. Volidina, A. P. Terent'ev, V. A.Kudrjasova, and L. N.
      Kabosina, *Khim. Geterotsikl. Soedin., Sb 1*, 1967, 5; *Chem. Abstr.*,
      1969, **70**, 77688h.
51.   J. W. Cornforth, *Heterocyclic Comp.*, 1957, **5**, 346

52. H. A. Naik, V. Purnaprajna, and S. Seshadri, *Indian J. Chem., Sect. B,* 1977, **15B**, 338.
53. V. H. Belgaonkar and R. N. Usgaonkar, *Chem. Ind. (London),* 1976, 954.
54. A. Monge, J. A. Palop, T. Goni, A. Martinez, and E. Fernandez-Alvarez, *J. Heterocycl. Chem.,* 1985, **22**, 1445.
55. M. Golfier and M. G. Guillerez, *Tetrahedron Lett.,* 1976, 267.
56. L. R. Morris and L. R. Collins, *J. Heterocycl. Chem.,* 1975, **12**, 305.
57. J. A. Deyrup and H. L. Gingrich, *J. Org. Chem.,* 1977, **42**, 1015.
58. V. I. Shuedov, and V. K. Vasil'eva, I. Z. Kharizomenova, and A. N. Grinev, *Khim. Geterotsikl. Soedin.,* 1975, 769.
59. H. R. Muller and M. Seefelder, *Liebigs Ann. Chem.,* 1969, **728**, 88.
60. M. Kim-Su, K. Eger, and H. J. Roth, *Arch. Pharm. (Weinheim, Ger.)* 1976, **309**, 729.
61. H. Hasan, R. Radeglia, E. Fanghaevel, *J. Prakt. Chem.,* 1990, **332**, 666.
62. I. T. Kay and I. T. Streeting, *Synthesis,* 1976, 38.
63. A. M. Jones, A. J. Simpson, and S. P. Stanforth, *J. Hetercycl. Chem.,* 1990, **27**, 1843.
64. T. Hirota, H. Fujita, K. Sasaki, T. Namba, and S. Hayakawa, *Heterocycles,* 1986, **24**, 771.
65. T. Hirota, H. Fujita, K. Sasaki, S. Hayakawa, and T. Namba, *J. Heterocycl. Chem.,* 1986, **23**, 1347.
66. T. Hirota, H. Fujita, K. Sasaki, and T. Namba, *J. Heterocycl. Chem.,* 1986, **23**, 1715.
67. M. A. Kira, M. O. Abdel-Rahman, and Z. M. Nofal, *Egypt. J. Chem.,* 1976, **19**, 109.
68. S. Hauptmann, M.Weissenfels, M. Scholz, E. -M. Werner, H. J. Kohler, and J. Weissflog, *Tetrahedron Lett.,* 1968, 1317.
69. S. Hauptmann, M. Schulz, H.J. Kohler, and H. J. Hoffmann, *J. Prakt. Chem.,* 1969, **311**, 614.
70. S. Hauptmann and J. Weissflog, *J. Prakt. Chem.,* 1972, **214**, 353.
71. Z. Arnold and Z. Holy, *Collect. Czech. Chem. Commun.,* 1965, **30**, 346.
72. J. T. Gupton, D. A. Krolikowski, R. H. Yu and S. W. Riesinger, *J. Org. Chem.,* 1990, **55**, 4735.
73. K. Nagarajan, P. J. Rodriguez, and M. Nethaji, *Tetrahedron Lett.,* 1992, **33**, 7229.
74. Yu. L. Briks and N. N. Romanov, *Khim. Geterosikl. Soedin.,* 1991, 549; *Chem. Abstr.,* 1992, **115**, 159044g.
75. M. A. Kira, Z. M. Nofal, and K. Z. Gadalla, *Tetrahedron Lett.,* 1970, 4215.
76. H. Feisselmann, *Angew. Chem.,* 1960, **72**, 573.
77. S. Hauptmann and E. M. Werner, *J. Prakt. Chem.,* 1972, **314**, 499.
78. N. D. Trieu and S. Hauptmann, *Z. Chem.,* 1973, **13**, 57; *Chem. Abstr.,* 1973, **78**, 147716g.
79. I. J. Kvitko and T. M. Galkina, *Zh. Org. Khim.,* 1969, 5, 1498; (b) I. J. Kvitko, *Khim. Geterotsikl. Soedin.,* 1975, 769.

80. J. N. Koselv, I. J. Kvitko, and L. S. Efros, *Zh. Org Khim.*, 1972, **8**, 1750.
81. R. T. LaLonde, R. A. Florence, B. A. Horenstein, R. C. Fritz, L. Silveira, J. Clardy, and B. S. Krishnan, *J. Org. Chem.*, 1985, **50**, 85.
82. P. Caignant and G. Kirsch, *C. R. Hebd. Seances Acad. Sci., Ser. C*, 1975, **281**, 35.
83. P. Caignant and G. Kirsch, *C. R. Hebd. Seances Acad. Sci., Ser. C*, 1975, **281**, 393.
84. (a) P. Caignant, P. Perin, G. Kirsch, and D. Caignant, *C. R. Hebd. Seances Acad. Sci., Ser. C*, 1973, **277**, 37; (b) P. Caignant, P. Perin, and G. Kirsch, *C. R. Hebd. Seances Acad. Sci., Ser. C*, 1974, **278**, 1201.
85. (a) P. Caignant, G. Kirsch, and P. Perin, *C. R. Hebd. Seances Acad. Sci., Ser. C*, 1974, **279**, 851; (b) P. Caignant, R. Close, G. Kirsch, and D. Caignant, *C. R. Hebd. Seances Acad. Sci., Ser. C*, 1975, **281**, 187.
86. T. Hirota, Y. Tashima, K. Sasaki, T. Namba and S. Hyakama, *Heterocycles*, 1987, **26**, 2717.
87. R. G. Harrison, M. R. J. Jolley, and J. C. Saunders, *Tetrahedron Lett.*, 1976, 293.
88. C. S. Davis, A. D. Kneval, and G. L. Jenkins, *J. Org. Chem.*, 1962, **27**, 1919.
89. (a) M. R. Jayanth, H. A. Naik, D. R. Tatke, and S. Seshadri, *Indian. J. Chem.*, 1973, **11**, 1112; (b) S. M. Jain and R. A. Pawar, *Indian. J. Chem.*, 1975, **13**, 304; (c) M. R. Chandramohan and S. Seshadri, *Indian. J. Chem.*, 1972, **10**, 573.
90. Z. Arnold and A. Holy, *Collect. Czech. Chem. Commun.*, 1961, **26**, 3059.
91. M. G. Lester, V. Petrov, and O. Stephenson, *Tetrahedron*, 1964, **20**, 1407.
92. (a) M. A. Volidina, A. P. Terent'ev, V. A. Kudrjasova, and V. G. Mesina, *Zh. Obsch. Khim.*, 1964, **34**, 469; (b) M. A. Volidina, A. P. Terent'ev, L. G. Rosupkina, and V. Misina, *Zh. Obsch. Khim.*, 1964, **34**, 473.
93. Hoffmann-La Roche, Dutch patent 6604628 (1967); *Chem. Abstr.*, 1967, **67**, 21903.
94. W. R. Benson and A. E. Pohland, *J. Org. Chem.*, 1965, **30**, 1126.
95. G. J. de Voghel, T. L. Eggerichs, B. Clannot, and H. G. Viehe, *Chimia*, 1976, **30**, 191.
96. M. Weissenfels, H. Schurig and G. Huesham, *Z. Chem.*, 1966, **6**, 471.
97. R. Sciaky and U. Pallini, *Tetrahedron Lett.*, 1965, 167.
98. G. Gaudiano, A. Quilico, and A. Ricca, *Atti accad. nazl. Lincei Rend., Classe Si. fis., nat. e mat.*, 1956, **21**, 253; *Chem. Abstr.*, 1957, **51**, 10500e.
99. D. Bonnetaud, G. Quiegner, and P. Pasteur, *J. Heterocycl. Chem.*, 1972, **9**, 165.
100 O. Aki and Y. Nakagowa, *Chem. Pharm. Bull. Jap.*, 1972, **20**, 1325.
101. N. K. Chodankar and S. Sequeria, *Dyes Pigm.*, 1986, **7**, 231; *Chem. Abstr.*, 1986, **105**, 62194m.

102. M. A. Kira, A. Bruckner-Wilhelm, F. Ruff, and J. Borsi, *Acta Chim. Acad. Sci. Hung.*, 1968, **56**, 189.

103. M. A. Kira, M. O. Abdel-Rahman, and K. Z. Gadalla, *Tetrahedron Lett.*, 1969, 109.

104. M. A. Kira, M. N. Aboul-Enein, and M. I. Korkor, *J. Heterocycl. Chem.*, 1970, **7**, 25.

105. R. A. Pawar and A. P. Borse, *J. Ind. Chem. Soc.*, 1989, **66**, 2305; *Chem. Abstr.*, 1989, **112**, 77026b.

106. I. M. A. Awad, *Monatsch. Chem.*, 1990, **121**, 1023.

107. R. Sciaky and F. Mancini, *Tetrahedron Lett.*, 1965, 137.

108. H. Laurent and R. Wiechart, *Chem. Ber.*, 1968, **101**, 2393.

109. F. Gonsonni, F. Mancini, U. Pallini, B. Patelly, and R. Sciaky, *Gazz. Chim. Ital.*, 1970, **100**, 244.

110. W. Schulze, P. Held, and A. Jumar, *Z. Chem.*, 1975, **15**, 184.

111. K. Nagarajan and R. K. Shah, *Indian. J. Chem., Sect. B*, 1976, **14B**, 1.

112. M. Mühlstüdt, R. Braimer, and B. Schulze, *J. Prakt. Chem.*, 1976, **318**, 507.

113. B. Schulze, G. Kirsten, S. Kirrbach, A. Rahm, and H. Heingarter, *Helv. Chem. Acta.*, 1991, **74**, 1059.

114. S. Klutchko, H. V. Hansen, and R. I. Meltzer, *J. Org. Chem.*, 1965, **30**, 3454.

115. B. A. Clark, J. Parrick, P. J. West, and A. H. Kelly, *J. Chem. Soc. C*, 1970, 498.

116. B. Foehlisch, R. Braun, and K. W. Schultze, *Angew. Chem., Int. Ed. Engl.*, 1967, **6**, 361.

117. Y. Furukawa, O. Miyashita, and S. Shima, *Chem. Pharm. Bull. Jap.*, 1976, **24**, 970.

118. Y. Kurasawa and S. Takada, *Heterocycles*, 1980, **14**, 281.

119. C. A. Ainsworth, W. E. Butin, J. Davenport, M. E. Callender, and M. C. McCowen, *J. Med. Chem.*, 1967, **10**, 208.

120. T. Aubert, M. Farnier, I. Meunier, and R. Guilard, *J. Chem. Soc., Perkin Trans. 1*, 1989, 2095.

121. D. K. Bates, B. A. Sell, and J. A. Picard, *Tetrahedron Lett.*, 1987, **28**, 3535.

122. M. R. Bryce and M. E. Fakley, *Syn. Commun.*, 1988, **18**, 181.

123. (a) J. Clark and G. R. Ramage, *J. Chem. Soc.*, 1958, 2821; (b) J. Clark, and J. H.Lister, *J. Chem. Soc.*, 1961, 5048; (c) J. H. Lister, *J. Chem. Soc.*, 1963, 2228; (d) F. Yoneda, K. Senga, and S. Nishigaki, *Chem. Pharm. Bull.*, 1973, **21**, 260; (e) M. Melguizo, M. Nogueras, A. Sanchez, and L. Quijano, *Tetrahedron Lett.*, 1989, **30**, 2669.

124. W. Flitsch, U. Lewinski, R. Mattes, and B. Wibbeling, *Liebigs Ann. Chem.*, 1990, 623.

125. K. Kaji, H. Nagashima, Y. Hirose, and H. Oda, *Chem. Pharm. Bull. Jap.*, 1985, **33**, 982.

126. I. M. A. Awad and K. H. Hassan, *Phosphorus, Sulfur, Silicon*, 1992, **72**, 81.

127. I. M. A. Awad and K. H. Hassan, *Bull. Chem. Soc. Jap.*, 1992, **65**, 1652.

128. I. M. A. Awad and K. H. Hassan, *Phosphorus, Sulfur, Silicon,* 1990, **47**, 311.

129. I. M. A. Awad, *Bull. Chem. Soc. Jap.,* 1993, **66**, 167.

130. S. J. Sulga, N. F. Fursaeva, and V. A. Cuiguk, *Khim Geterosikl. Soedin.,* 1972, 629; *Chem. Abstr.,* 1962, 77, 126559h.

131. V. A. Cuiguk and V. V. Oksanic, *Khim Geterosikl. Soedin.,* 1973, 1285; Chem. Abstr., 1973, **79**, 6745z.

132. V. A. Cuiguk, D. J. Seiko, and V. G. Glusakov, *Khim Geterosikl. Soedin.,* 1974, 1435; *Chem. Abstr.,* 1975, **82**, 43317a.

133. V. A. Cuiguk, K. V. Fedotov, J. P. Bojko, J. P. Backovskij, G. M. Golubusina, and O. M. Mostovaja, *Khim Geterosikl. Soedin.,* 1973, 1432; *Chem. Abstr.,*1974, **80**, 27191b.

134. S. J. Sulga and V. A. Cuiguk, *Ukr. Khim. Zh. (Russ. Ed.),* 1973, **39**, 66; *Chem. Abstr.,* 1973, **78**, 111247r.

135. B. Schulze, K. Mütze, D. Selle, and R. Kempe, *Tetrahedron Lett.,* 1993, **34**, 1909.

136. (a) S. Athrain and B. Iddon, *Tetrahedron,* 1992, **48**, 7689; (b) S. Athrain, M. F. Farhat, and B. Iddon, *J. Chem. Soc., Perkin Trans. 1,* 1992, 973.

137. T. Tjusi and K. Takeneda, *J Heterocycl. Chem.,* 1990, **27**, 851.

138. (a) G. N. Dorofeenko and A. I. Pyschev, *Zh. Obshch. Khim.,* 1973, **9**, 1084; *Chem. Abstr.,* 1973, **79**, 66123; (b) G. N. Dorofeenko and A. I. Pyschev, *Khim. Geterotsikl. Soedin.,* 1974, 1031; *Chem. Abstr.,* 1974, **81**, 169392n.

139. J. Andieux, J.-P. Battioni, M. Gerard, and D. Mohlo, *Bull. Chim. Soc. Fr.,* 1974, 2093.

140. A. R. Katritzky and C. M. Marson, *J. Am Chem. Soc.,* 1983, **105**, 3279.

141. J. Liebscher and H. Hartmann, *Z. Chem.,* 1975, **15**, 16.

142. J. E. Butler-Ransohoff and H. A. Staab, *Tetrahedron Lett.,* 1985, **26**, 6179.

143. A. V. Koblik and K. F. Suzdalev, *Zh Org. Khim.,* 1989, **25**, 171; *Chem Abstr.,* 1989, **111**, 194516b.

144. H. Hartmann, *J. Prakt. Chem.,* 1971, **313**, 1113.

145. H. Harnisch, *Liebigs Ann. Chem.,* 1972, **765**, 8.

146. A. Nohara, T. Umetani, and Y. Sanno, *Tetrahedron Lett.,* 1973, 1995.

147. A. Nohara, T. Umetani, and Y. Sanno, Ger. Offen, 2317899 (1973); *Chem. Abstr.* 1974, **80**, 14932u.

148. V. G. Pivovarenko, V. P. Khilya, and S. A. Vasil'ev, *Khim. Prir. Soedin.,* 1989, **5**, 639; *Chem. Abstr.* 1990, **113**, 5988q.

149. S. A. Vasil'ev, V. G. Pivovarenko, and V. P. Khilya, *Dokl. Akad. Nauk. Ukr, SSR. Ser. B: Geol., Khim. Biol. Nauki,* 1989, **4**, 34; *Chem. Abstr.,* 1990, **112**, 20813b.

150. G. A. Reynolds and J. A. Van Allan, *J. Heterocycl. Chem.,* 1969, **6**, 375.

151. (a) D. Kaminsky, S. Klutchko, and M. Von Strandtmann, U.S. Pat. 3887585 (1975); *Chem. Abstr.,* 1975, **83**, 178816x.

152. S. Klutchko, D. Kaminsky, and M. Von Strandtmann, U.S. Pat. 4008252 (1977); *Chem. Abstr.*, 1977, **87**, 5808a.
153. G. Rowa, A. Ermili, and M. Mazzei, *J. Heterocycl. Chem.*, 1975, **12**, 31.
154. A. K. Awasthi and R. S. Tewari, *Synthesis*, 1986, 1061.
155. Z. Arnold and A. Holy, *Collect. Czech. Chem. Commun.*, 1965, **30**, 47.
156. (a) Z. Arnold, *Experientia*, 1959, **15**, 415; (b) Z. Arnold, *Collect. Czech. Chem. Commun.*, 1960, **25**, 1308.
157. J. Zemlicka and Z. Arnold, *Collect. Czech. Chem. Commun.*, 1960, **25**, 2838.
158. M. Sreenivasulu and G. S. K. Rao, *Indian J. Chem., Sect. B*, 1989, **28B**, 584.
159. C. Jutz and W. Mueller, *Chem. Ber.*, 1967, **100**, 1536.
160. R. Sciaky, U. Pallini, and A. Consonni, *Gazz. Chim. Ital.*, 1966, **96**, 1284.
161. S. Makoto, T. Kiyobumi, and H. Takatsu, Jpn. Kokai, Tokkyo Koho JP 63253065; *Chem. Abstr.*, 1989, **110**, 145145u.
162. C. Jutz, W. Muller, and E. Muller, *Chem Ber.*, 1966, **99**, 2479.
163. Z. Arnold and A. Holy, *Collect. Czech. Chem. Commun.*, 1963, **28**, 2040.
164. P. C. Traas and H. Boelens, *Recl. Trav. Chim. Pay-Bas*, 1973, **92**, 985.
165. A. L. Cossey, R. L. N. Harris, J. L. Huppatz, and J. N. Phillips, *Angew. Chem., Int. Ed. Engl.*, 1972, **11**, 1110.
166. (a) A. L. Cossey, R. L. N. Harris, J. L. Huppatz, and J. N. Phillips, *Angew. Chem., Int. Ed. Engl.*, 1972, **11**, 1098; (b) A. L. Cossey, R. L. N. Harris, J. L. Huppatz, and J. N. Phillips, *Aust. J. Chem*, 1976, **29**, 1039.
167. M. Mittlebach and H. Junek, *J. Heterocycl. Chem.*, 1982, **19**, 1021.
168. V. N. Proshkina, Z. F. Solomko, and N. Ya. Bozhanova, *Khim. Geterotsikl. Soedin.*, 1988, 1288; *Chem. Abstr.*, 1989, **111**, 39322s.
169. (a) O. Meth-Cohn and B. Narine, Tetrahedron Lett., 1978, 2045; (b) O. Meth-Cohn, B. Narine, and B. Tarnowski, *J. Chem. Soc., Perkin Trans. 1*, 1981, 1531.
170. C. Paulmier and F. Outurquin, *J. Chem. Res. (S)*, 1977, 318.
171. K. Tabakovic, I. Tabakovic, N. Ajdini, and O. Leci, *Synthesis*, 1987, 308.
172. A. L. Cossey, R. L. N. Harris, J. L. Huppatz, and J. N. Phillips, *Angew. Chem., Int. Ed. Engl.*, 1972, **11**, 1099.
173. O. Fischer, A. Mueller, and A. Vilsmeier, *J. Prakt. Chem.*, 1925, **109**, 69.
174. T. L. Wright, U. S. Pat. 4540786 (1985); *Chem. Abstr.*, 1986, **104**, 68762w.
175. R. A. Pawar, R. Bajare, and S. B. Mindade, *J. Ind. Chem. Soc.*, 1990, **67**, 685; *Chem. Abstr.*, 1991, **114**, 143184w.
176. D. R. Adams and C. Adams, *Syn. Commun.*, 1990, 469.
177. O. Meth-Cohn, S. Rhouati, and B. Tarnowski, *Tetrahedron Lett.*, 1979, 3111.

178. O. Meth-Cohn, S. Rhouati, and B. Tarnowski, *Tetrahedron Lett.*, 1979, 4885.
179. O. Meth-Cohn, S. Rhouati, and B. Tarnowski, *J. Chem. Soc., Perkin Trans. 1*, 1981, 1537.
180. F. Korodi and Z. Cziaky, *Org. Prep. Proc. Int.*, 1990, **22**, 579.
181. D. Adams, J. Dominguez, V. Lo Russo, N. Morante de Rekowski, *Gazz. Chim. Ital.*, 1989, 281; *Chem. Abstr.*, 1989, **112**, 55557y.
182. M. J. Grimwade and M. G. Lester, *Tetrahedron*, 1969, **25**, 4535.
183. Z. Arnold and J. Zemlicka, *Collect. Czech. Chem Commun.*, 1959, **24**, 2378.
184. M. Julia, *C. R. Hebd. Seances Acad. Sci., Ser. C,* 1952, **234**, 2615.
185. A. Brack, *LiebigsAnn. Chim.*, 1965, **681**, 105.
186. H. Ahlbrecht and C. Vonderheid, *Chem. Ber.*, 1975, **108**, 2300.
187. O. Meth-Cohn, *Tetrahedron Lett.*, 1985, **26**, 1901.
188. O. Meth-Cohn and D. L. Taylor, *Tetrahedron Lett.*, 1993, **34**, 3629.
189. O. Meth-Cohn, *Heterocycles*, 1993, **35**, 529.
190. W. Wiegrebe, D. Sasse, H. Reinhart, and L. Faber, *Z. Naturforsch.*, 1970, **B25**, 1408.
191. T. Koyama, T. Hirota, Y. Shinohara, S. Matsumoto, M. Yamato, and S. Ohmori, *Chem. Pharm. Bull. Jap.*, 1975, **23**, 2029.
192. T. Koyama, T. Hirota, Y. Shinohara, S. Matsumoto, and M. Yamato, *Chem. Pharm. Bull. Jap.*, 1977, **25**, 2838.
193. T. Koyama, T. Hirota, Y. Shinohara, M. Yamato, and S. Ohmori, *Chem. Pharm. Bull. Jap.*, 1975, **23**, 497.
194. T. Koyama, T. Hirota, I. Ito, and M. Toda, *J. Pharm. Soc. Jap.*, 1969, **89**, 1492.
195. T. Koyama, T. Hirota, I. Ito, M. Toda, and M. Yamato, *Tetrahedron Lett.*, 1968, 4631.
196. T. Hirota, T. Koyama, T. Namba, M. Yamato, and T. Matsumura, *Chem. Pharm. Bull. Jap.*, 1978, **26**, 245.
197. A. J. Liepa, *Austr. J. Chem.*, 1982, **35**, 1391.
198. W. Ziegenbein and W. Franke, *Angew. Chem.*, 1959, **71**, 573.
199. W. Ziegenbein and W. Franke, *Angew. Chem.*, 1959, **71**, 628.
200. W. Ziegenbein and W. Lang, *Chem. Ber.* 1960, **93**, 2743.
202. K. Morita, M. Ochiai, and R. Marumoto, *Chem. Ind. (London)*, 1968, 1117.
203. M. Ochiai, R. Marumoto, S. Kobayashi, H. Shimazu, and K. Morita, *Tetrahedron*, 1968, **24**, 5731.
204. M. Ochiai, S. Kobayashi, H. Shimazu, and K. Morita, *Tetrahedron Lett.*, 1970, 861.
205. S. Kobayashi, *Chem. Pharm. Bull. Jap.*, 1973, **46**, 2385.
206. (a) Z. Csuros, R. Soós, J. Palinkas, and I. Bitter, *Acta Chim. Acad. Sci. Hung.*, 1970, **63**, 215; (b) Z. Csuros, R. Soós, J. Palinkas, and I. Bitter, *Acta Chim. Acad. Sci. Hung.*, 1971, **68**, 397; (c) Z. Csuros, R. Soós, J. Bitter, and I. Palinkas, *Acta Chim. Acad. Sci. Hung.*, 1972, **72**, 59.
207. H. Bredereck, H. Herlinger, and J. Renner, *Chem. Ber.*, 1960, **93**, 230.
208. M. A. Volidina, A. P. Terent'ev, and V. A.Kudrjasov, *Khim. Geterotsikl. Soedin., Sb 1*, 1967, 369; *Chem. Abstr.*, 1969, **70**, 87719h.

209. M. Julia and M. Delepine, *C. R. Hebd. Seances Acad. Sci., Ser. C,* 1952, **235**, 662.
210. K. E. Schulte, R. Reich, and U. Stoess, *Angew. Chem., Int. Ed. Engl.,* 1965, **4**, 1081.
211. R. R. Crenshaw and R. A. Partyka, *J. Heterocycl. Chem.* 1970, **7**, 871.
212. R. L. N. Harris, J. L. Huppatz and T. Teitei, *Aust. J. Chem.,* 1979, **32**, 669.
213. R. L. N. Harris and J. L. Huppatz, *Angew. Chem., Int. Ed. Engl.,* 1977, **16**, 779.
214. S. Kobayashi, *Bull. Chem. Soc. Jap.,* 1975, **48**, 302.
215. A. Monge, I. Aldana, I. Lezamiz, and E. Fernandez-Alvarez, *Synthesis*, 1984, 160.
216. T. L. Wright, U. S. Pat. 4581455 (1986); *Chem. Abstr.*, 1986, **105**, 24271.
217. M. Ahmed, J. Ashby, and O. Meth-Cohn, *J. Chem. Soc., Chem. Commun.,* 1970, 1094.
218. W. Flitsch, J. Lauterwein, R. Temme, and B. Wibbeling, *Tetrahedron Lett.,* 1988, **29**, 3391.
219. R. Brehme, *Chem. Ber.,* 1990, **123**, 2039.
220. H. Harnisch, *Liebigs Ann. Chim.,* 1971, **751**, 155.
221. T. R. Kasturi, H. R. Y. Jois, and L. Mathew, *Synthesis,* 1984, 743.
222. T. R. Kasturi, S. Arumugam, L. Mathew, S. K. Jayaram, P. Dashidar, and T. N. G. Row, *Tetrahedron,* 1992, **48**, 6499.
223. M. Ichiba, K. Senga, and S. Nishigaki, *J. Heterocycl. Chem.,* 1977, **14**, 175.
224. K. Senga, Y. Kanamori, S. Nishigaki, and F. Yoneda, *Chem. Pharm Bull. Jap.,* 1976, **24**, 1917.
225. R. Krecher, G. Vogt, and M.-L. Shultz, *Angew. Chem., Int. Ed. Engl.,* 1975, **14**, 821.
226. M. Weissenfels, *Z. Chem.,* 1964, **4**, 458; *Chem. Abstr.,* 1965, **62**, 9135b.
227. M. Weissenfels and G. Dill, *Z. Chem.,* 1967, **7**, 456; *Chem. Abstr.,* 1968, **68**, 49659k.
228. S. M. Jain, and R. A. Pawar, *Indian. J. Chem.,* 1975, **13**, 304.
229. G. Alverne, B. Langlois, A. Laurent, I. Le Drean, A. Selmi, and M. Weissenfels, *Tetrahedron Lett.,* 1991, **32**, 643.

chapter five

Transformations to give New Ring Systems

An unusual ring-expansion occurs upon reacting 1-methyloxindole with a Vilsmeier reagent (scheme 5.1).[1]

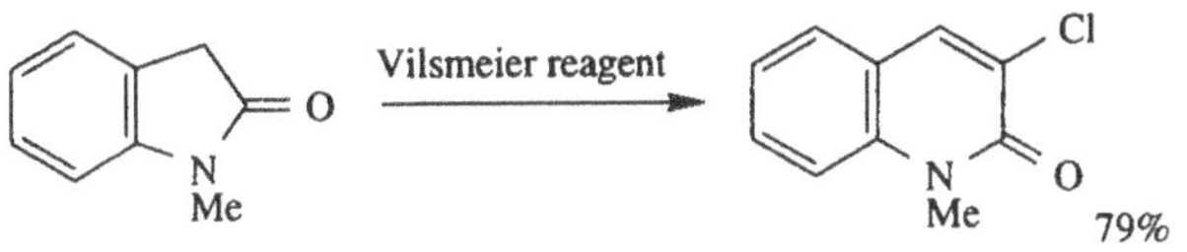

(5.1)

79%

1,3-Oxazin-4-ones **2**, useful as fungicides and analgesics, were prepared by a Vilsmeier reaction of the amide **1**. 1,3-Thiazin-4-ones were similarly prepared from the corresponding 3-hydroxyisothiazoles.[2]

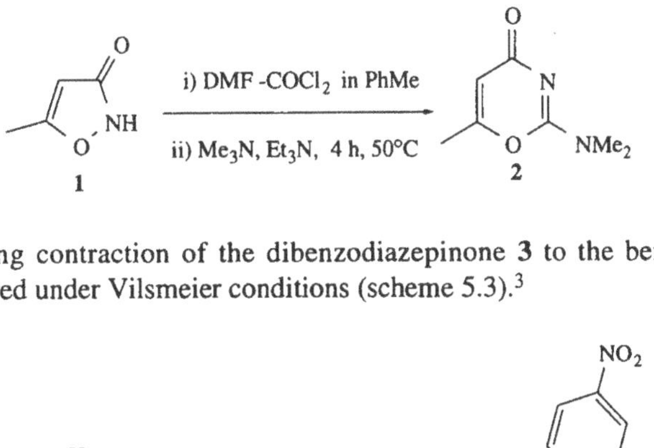

i) DMF -COCl$_2$ in PhMe

ii) Me$_3$N, Et$_3$N, 4 h, 50°C

(5.2)

Ring contraction of the dibenzodiazepinone **3** to the benzimidazole **4** occurred under Vilsmeier conditions (scheme 5.3).[3]

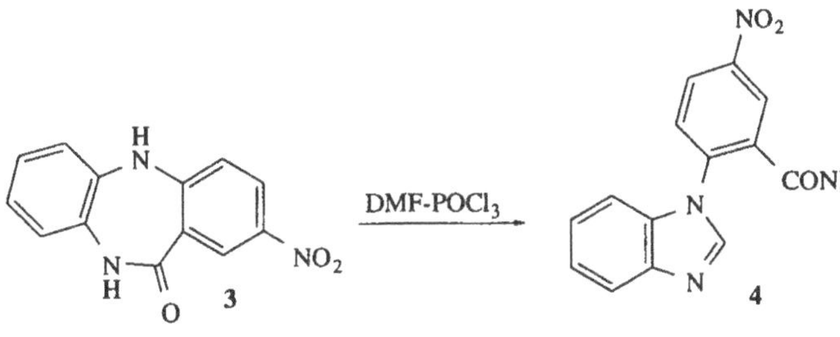

DMF-POCl$_3$

(5.3)

New routes to 1,3-oxazine-6-ones **6**, by the ring-expansion of isoxazolin-5-ones **5** mediated by Vilsmeier reagents, have been developed (scheme 5.4).[4]

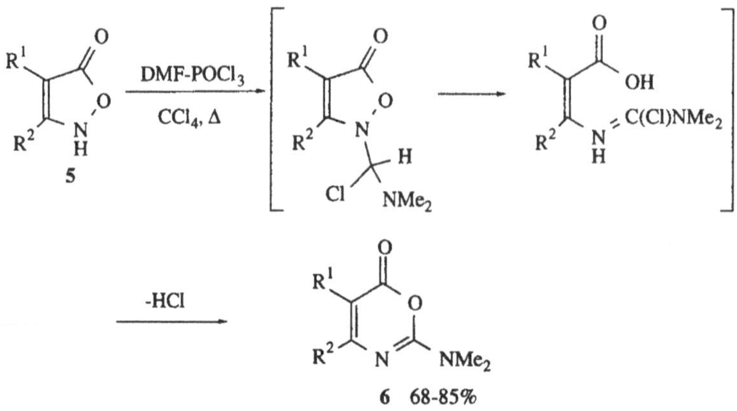

(5.4)

The Vilsmeier-Haack reaction on 4-alkylideneisoxazolin-5-ones **7** gives 1,3-oxazin-6-ones **8** which are useful precursors of α-pyrones, 2-pyridones and pyridines (scheme 5.5).[5]

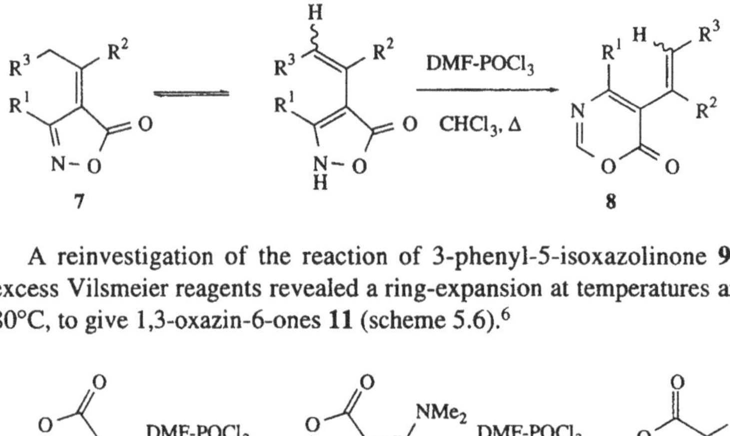

A reinvestigation of the reaction of 3-phenyl-5-isoxazolinone **9** with excess Vilsmeier reagents revealed a ring-expansion at temperatures around 80°C, to give 1,3-oxazin-6-ones **11** (scheme 5.6).[6]

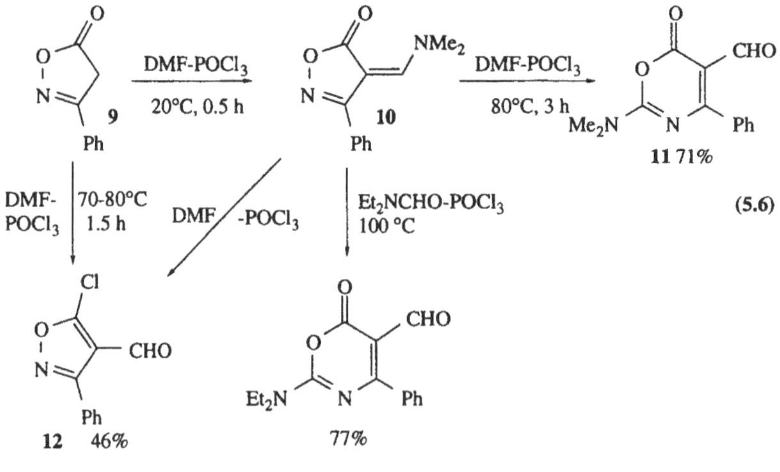

The following rationale has been proposed: the isoxazolone **9** reacts *via* its enol form with the chloromethyleneiminium species to give, after loss of HCl, the dimethylamino-derivative **10**, which under more forcing conditions reacts with excess Vilsmeier reagent at the ring nitrogen atom (scheme 5.7). Ring-opening, subsequent ring-closure, and lastly hydrolysis affords the 1,3-oxazin-6-one **11**. Formation of the 5-chloro-4-formylisoxazole **10** evidently proceeds *via* phosphorylation of the exocyclic oxygen atom of **10**.[6] In certain cases compounds of the form **10** could be isolated.[7]

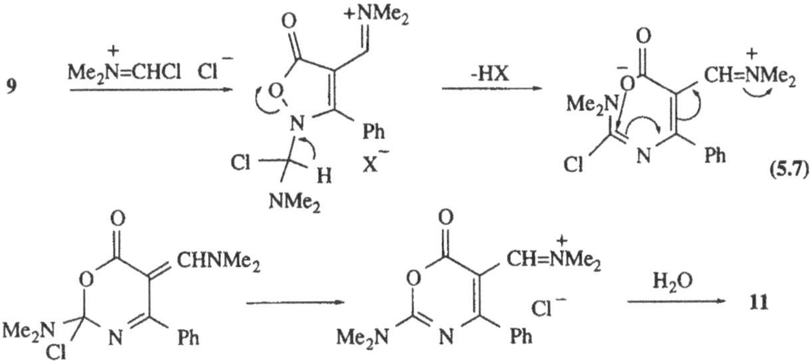

3-Arylisoxazole-5-4*H*-ones **13** have also been reported to rearrange on reaction with DMF-POCl₃ to afford a series of 2,4-dichloroisoquinolines (scheme 5.8).[8] The precise reaction conditions are crucial to the outcome of the reaction; contrast the reaction of **9** in scheme 5.6 under similar conditions.

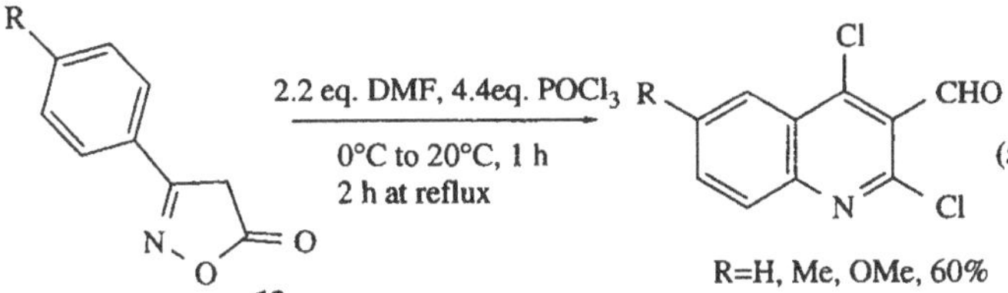

An α-methyl group of pyrylium species can be iminoalkylated; reaction of **14** with DMF-POCl₃ gave enaminic salt **15** (50%). Reaction of **14** with NH₄Cl or *p*-methylaniline gave pyrimidinones **17**.[9] After a ring-opening and ring-closing sequence, fragmentation of an intermediate **16** would afford the pyrimidinone. This elegant application of Vilsmeier chemistry could find use in preparing other heterocyclic rings.

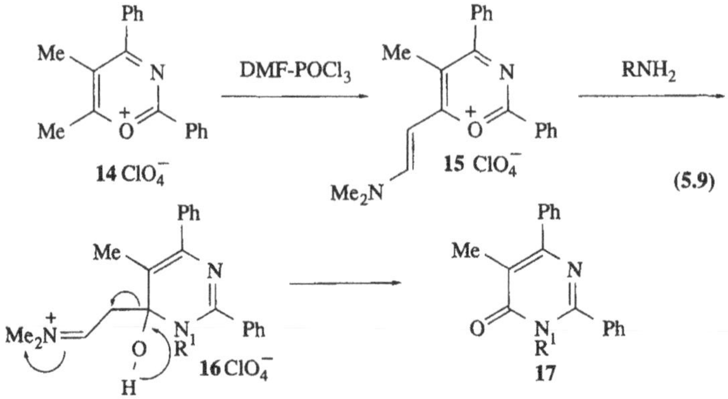

$$(5.9)$$

The dithiadiene **18** is converted into the thiophene-2-aldehyde **19** in a reaction involving extrusion of sulfur.[10]

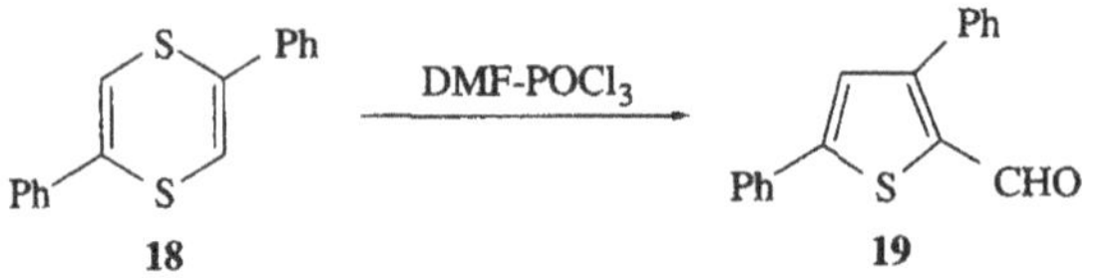

$$(5.10)$$

References

1.   A. Andreani, D. Bonazzi, and M. Rambaldi, *Boll. Chim. Farm.*, 1976, **115**, 732; *Chem. Abstr.*, 1976, **89**, 24107.
2.   K. Tomita and T. Murakami, Jpn. Tokkyo Koho 7920504 (1979); *Chem. Abstr.*, 1979, **91**, 157755b.
3.   K. Nagarajan and R. K. Shah, *Ind. J. Chem., Sect. B*, 1976, **14B**, 1.
4.   E. M. Beccalli, A. Marchesini, and H. Molinari, *Tetrahedron Lett.*, 1986, **27**, 627.
5.   E. M. Beccalli, A. Marchesini, and T. Pilati, *Tetrahedron*, 1989, **45**, 7485.
6.   D. J. Anderson, *J. Org. Chem.*, 1986, **51**, 945.
7.   E. M. Beccalli and A. Marchesini, *J. Org. Chem.*, 1987, **52**, 3426.
8.   K. Ashok, G. Sridevi, and Y. Umadevi, *Synthesis*, 1993, 623.
9.   I. I. Nechayuk, N. V. Shibaeva, S. V. Borodaev, A. I. Pyschev, and S. M. Lykyama, *Khim. Geterosikl. Soedin*, 1990, 134; *Chem. Abstr.*, 1990, **113**, 78318f.
10.   W. E. Parham and W. J. Traynelis, *J. Am. Chem. Soc.*, 1954, **76**, 4960.

Index